GOVERNANCE IN THE INFORMATION ERA

Policy informatics is addressing governance challenges and their consequences, which span the seeming inability of governments to solve complex problems and the disaffection of people from their governments. Policy informatics seeks approaches that enable our governance systems to address increasingly complex challenges and to meet the rising expectations of people to be full participants in their communities. This book approaches these challenges by applying a combination of the latest American and European approaches in applying complex systems modeling, crowdsourcing, participatory platforms, and citizen science to explore complex governance challenges in domains that include education, environment, and health.

The traditional norms of hierarchical governance, expert-driven decision-making, institutional control, and centralization have begun to fade as the costs of community engagement have declined, data have become increasingly available and accessible, and computational and analytical skills have increased dramatically. Further, stakeholders and publics are becoming more diverse, unequal, vocal, and polarized. Against this backdrop, the book explores how to enable public values to flourish by promoting new forms of effective governance and exploring the changes necessary in technology, processes, institutional capacity, and social norms to realize that future.

Erik W. Johnston is the Director of the Center for Policy Informatics at Arizona State University and an Assistant Professor in the School of Public Affairs. He also holds joint or affiliated positions in Applied Mathematics for the Life and Social Sciences, the Center for the Study of Institutional Diversity, the School for Social Dynamics and Complexity, and the Decision Theater.

GOVERNANCE IN THE INFORMATION ERA

Theory and Practice of Policy Informatics

Edited by Erik W. Johnston

Routledge
Taylor & Francis Group

NEW YORK AND LONDON

First published 2015
by Routledge
711 Third Avenue, New York, NY 10017

and by Routledge
2 Park Square, Milton Park, Abingdon, Oxon OX14 4RN

Routledge is an imprint of the Taylor & Francis Group, an informa business

Library of Congress Cataloging-in-Publication Data
Governance in the information era : theory and practice of policy
    informatics / edited by Erik W. Johnston.
        pages cm
    1. Public administration—Technological innovations.   2. Communication
in politics—Technological innovations.   3. Electronic government
information.   I. Johnston, Erik W., 1977– editor of compilation.
    JF1525.A8.G685   2015
    352.3'84—dc23   2014029227

ISBN: 978-1-138-83207-7 (hbk)
ISBN: 978-1-138-83208-4 (pbk)
ISBN: 978-1-315-73621-1 (ebk)

Typeset in Bembo
by Apex CoVantage, LLC

To R. F. "Rick" Shangraw, Jr.—For his vision and support of policy informatics.

CONTENTS

LIST OF FIGURES, TABLES, AND BOXES

Figures

Tables

Boxes

FOREWORD

I started my academic career at a pivotal time in the career of all academics—the introduction of the personal computer. The personal computer transformed the way we analyzed data and conducted research. It also completely modified the process of compiling and writing research, not to mention fundamentally changed the way we conducted our academic administrative processes.

The introduction of the personal computer in the 1980s led to an explosion in research on the interactions between information technology and public management. Ken Kraemer at University of California, Irvine, Wil Gorr at Carnegie Mellon University, Stu Bretschneider at Syracuse University, David Garson at North Carolina State University, and several others led the way in defining this new field. Their work put a public sector twist on the management information systems (MIS) and decision support systems (DSS) research being conducted at business schools. Initially, most of the research focused on the effects of computing on public sector administrative and managerial processes, topics most Public Administration scholars considered mostly descriptive at its core and mundane at best.

Another transition occurred again in the late 1990s with the rise of the Internet. This period fractured the young field of computing and public management. The Internet gave rise to new lines of research, including participatory democracy (or e-democracy), equity and access to Internet-based public services, the effectiveness and efficiency of e-government, and the effects of the Internet on bureaucratic politics. Research on information and communication technologies essentially dissolved as a distinct discipline and integrated into most mainstream public administration disciplines.

When all this was happening, I was a private sector practitioner providing policy and management strategy consulting services to Federal agencies and Fortune

500 companies. It was a time of significant growth in the information technology industry. The academic literature was racing to keep up with dramatic changes occurring in the delivery and management of public services. However, as we moved into the new century, attention increasingly shifted to the role of information and communications technology in the formulation of public policy. Information technologies were being used to design, analyze, evaluate, implement, and even influence the adoption of policies (at the intersection with politics).

When I reentered academia in 2005 and came to Arizona State University (ASU), the field of public management information systems was still splintered. At ASU we addressed this by creating the Center for Policy Informatics, intended to bring definition to the field and distance it from more traditional, descriptive research. We also adopted new techniques such as simulations—particularly agent-based models—to reimagine the way we study complex public policy and management issues. For example, the intersection of complexity sciences with public policy and management is leading to a number of insights into very perplexing public policy issues. Initially at ASU, we took a very informatics-centric approach and defined "policy informatics" as the study of techniques, methods, and theories related to the storage, retrieval, sharing, and optimal use of policy data, information, and knowledge for problem solving and decision-making. However, over the past decade, this definition has continued to evolve and is best articulated in several of the works in this volume.

This edited volume is a tribute to the progress that has been made over the past decade to create and formalize an area of study known as "policy informatics."

R. F. "Rick" Shangraw, Jr.
Founding Director
Center for Policy Informatics
Arizona State University

ACKNOWLEDGMENTS

This book is not the product of individual effort. Over the past several years, I have had the privilege of collaborating with an extraordinary group of scholars and practitioners who all care deeply about various facets of policy informatics. Since it is impossible to thank each of them individually, I will not even attempt to try that here. I do, however, want to express my sincere gratitude to each and every one of them.

All contributors graciously gave their time and energy to this book project. Contributors put up with an aggressive schedule, email reminders, and revision requests. Each contributor went above and beyond the call of duty to prepare chapters that are of the highest quality. I am grateful to all contributors as they worked on their chapters in addition to their regular research and teaching assignments.

Kevin Desouza was a catalyst for the book and was instrumental in the early stages of pushing the project forward. He was valuable in helping to connect a community of scholars and a number of contributions in this book were identified through his early work in the field.

Chase Treisman served as the project manager. I am indebted to her for all that she has done to make this book a reality, including editing chapters, attending to contributor queries, contributing ideas, keeping me honest with deadlines, and stepping in to complete tasks at a moment's notice. Near the end stages of developing the book, Justin Longo stepped up to ensure that the book would be completed professionally.

I am grateful to the administrative support received from my colleagues at Arizona State University (ASU). Michael Crow (President, ASU), Rick Shangraw, Jr. (CEO, ASU Foundation), Sethuraman "Panch" Panchanathan (Senior Vice President, ASU Office of Knowledge and Discovery), and Jonathan Koppell (Dean,

College of Public Programs, ASU) have been terrific supporters and advocates of policy informatics and have generously supported the advancement of the concept both within and beyond ASU.

The job posting for Assistant Professor in Policy Informatics at ASU was the first time I became aware of the term *policy informatics*; the job description was a natural fit for someone searching for an academic home after graduating with a Ph.D. in Information. I thank Yushim Kim, who was hired with me to develop policy informatics. We worked together on creating our original policy informatics paper at the Minnowbrook III conference, hosting the first workshop on policy informatics at ASU in February 2010, and co-editing the special issue of *The Innovation Journal* on policy informatics. Yushim has led the efforts to have APPAM (Association for Public Policy Analysis and Management) conferences include policy informatics in its regular program. At the American Society for Public Administration (ASPA) conferences, I always enjoy developing the ideas behind policy informatics, complexity, and network studies with Chris Koliba, Naim Kapucu, Asim Zia, Jack Meek, and Aaron Wachhuas. I look forward to ongoing collaborations with Darrin Hicks, Ajay Vinze, Ning Nan, Cliff Lampe, Derek Hansen, Hari Sundaram, Spiro Maroulis, and Libby Hemphill. I am grateful to the following individuals and organizations for investing or collaborating in research in this field, including Susan Winter and Jacqueline Meszaros (National Science Foundation), Christopher Barrett (Virginia Tech), Dave White (Decision Center for Desert Cities), Beth Noveck and Stefaan Verhulst at The GovLab at NYU (MacArthur Foundation Research Network on Opening Governance), and Doug Rand (White House Office of Science and Technology Policy). A highlight of the overall process is seeing the development of great students, including my first research assistant, Mitali Ayyangar, the first policy informatics Ph.D. students, Dr. Qian Hu, Dr. Jennifer Auer, and Dr. Jusil Lee, who were vital in the first years of my research and all of our efforts to develop policy informatics, and our current Ph.D. students and Post-Docs, Tanya Kelley, Chase Treisman, Chul Hyun Park, Joshua Uebelherr, Jordan Bates, Dr. Justin Longo, Dr. Dara Wald, and Dr. David Hondula, who represent the future of the field.

Erik W. Johnston
Phoenix, Arizona

PART I
Introduction

1

CONCEPTUALIZING POLICY INFORMATICS

Erik W. Johnston

Policy informatics is the study of how computation and communication technology is leveraged to understand and address complex public policy and administration problems and realize innovations in governance processes and institutions. Policy informatics takes a systemic approach to addressing governance challenges and their consequences, which span the seeming inability of governments to solve complex problems to the disaffection of people from their governments. The objective of the field is to respond to these governance challenges in any domain by applying a combination of computational thinking, complex systems modeling, data analytics, and participatory science. The goal of this book is to present the perspectives of a collection of policy informatics scholars, illustrating various elements of what policy informatics means in theory and in practice in the information era.

A key element of the policy informatics approach is to apply responses in the appropriate context. This must include a robust understanding of how governance is currently manifest and how it is changing. The traditional norms of hierarchical government, expert-driven decision-making, institutional control, and centralization have begun to fade as the capacities of computation and communication technologies have increased, the costs of community engagement have declined, data have become increasingly available and accessible, and digital literacy and analytical skills have expanded. Further, stakeholders and publics are becoming more diverse, unequal, vocal, and polarized. The rapid advance of technology and its widespread diffusion in everyday use have transformed who creates and has access to information and how we leverage information toward constructing knowledge and acting on evidence. Against this backdrop, policy informatics seeks to enable public values to flourish by promoting new forms of effective governance and exploring the changes necessary in technology, processes, institutional capacity, and social norms to realize future innovations.

Consequently, the near-term objective of the policy informatics movement is to leverage technology to engage expertise wherever it exists, taking a broad view of what constitutes expertise. Success will be measured by how effective the platforms, interventions, and decision environments developed are in facilitating broad-based action and responsible distributed governance. The field's approach is applied, system-based, and participatory, moving beyond problem identification to solution generation, implementation, and administration. In the long term, the objective is to create proven mechanisms that incorporate expertise regardless of discipline, institution, or community into governance to provide effective support for social decision-making and collective action, while ensuring the legitimacy of widespread public participation. Presented in this book are a number of proof-of-concept projects that exemplify innovative governance approaches (e.g., participatory platforms, citizen sourcing advice, community engagement, knowledge incubation), explore the drivers of collaborative failure (e.g., power disparities, complexity, uncertainty, conflicting goals), and leverage computational and communication approaches (e.g., agent-based modeling, system dynamics, interactive simulations, participatory modeling, personal health sensors, information visualization) to enhance collaborative, evidence-based decision-making and facilitate good governance.

This introductory chapter has a few simple aims. First, to outline what general societal trends are driving the need for policy informatics. Second, to differentiate clusters of research that have already begun to emerge from various disciplines. And finally, to preview how the contributions of the chapters in this book reveal the different dimensions of this growing community of practice.

Trends Driving the Emergence of Policy Informatics

The ability to interact meaningfully with each other on the Internet is rapidly being diffused among all members of society, driven by the decreased cost of communication technologies and their increased power. According to Pew Research, in 2014 cell phone ownership exceeded 90 percent in such diverse countries as China, Jordan, Russia, Chile, and South Africa (Pew Research, 2014). A report from the United States Census Bureau (File, 2013) shows that in 2011 over 75 percent of American households had a computer and almost 72 percent had Internet access from home, and a recent Pew Research report on the 25th anniversary of the World Wide Web estimated that 81 percent of Americans use Internet-connected laptop and desktop computers daily (Fox & Rainie, 2014). The rapid expansion of Internet connectivity using mobile devices such as smart phones continues to increase basic access, although a small but persistent 'digital divide' remains (United States, 2013). Dispersed access to communication technology enables people to connect and coordinate their personal and professional lives in ways that collapse the traditional geographic boundaries imposed by time and distance and gives rise to a globalized network society (Castells, 2011; Malone,

2004). As we become more interconnected, the challenges that face governmental institutions and public decision-makers are increasingly information-intensive, diverse, complex, interdependent, and resource-constrained.

In response, over the last few years, we have seen positive developments in the commitment of public agencies to leverage the power of technology and information to address social issues. On his first full day in office, President Barack Obama signed and issued a Transparency and Open Government memorandum to the heads of the executive offices (White House, 2009). This memo directed that government should operate in a transparent, participatory, and collaborative fashion by liberating data and information and engaging the public in addressing shared challenges.

When data is considered a public good it affords an opportunity to develop technologies that leverage information to address problems through processes that are more effective, legitimate, and efficient. Governance institutions, public networks, government agency actors, private institutions, and nonprofit and social organizations are available components of a complex system that can be mobilized to collectively address public issues in a manner that is more organic, innovative, precise, and democratic. With the erosion of fixed traditional roles and the increase of permeable organizational boundaries, public and non-governmental organizations are no longer simply passive recipients of technology, data, public problems, and policy solutions distributed top-down by experts and elites. They are increasingly co-creators of knowledge, co-makers of decisions, and co-deliverers of solutions.

Designing advanced communication and computation technology to create a communicative nexus between government and citizens enables the collaborative blending of data, expertise, and experience concerning public problems. This can manifest in the form of more innovative public institutions, policy solutions, and public actions that more legitimately, effectively, and efficiently govern society. Today, we are witnessing the emergence of platforms in the public sector that seek to take advantage of crowdsourcing through mobile apps and platforms. However, the traditional structures, processes, and tools of government characterized by bureaucracy, centralization, hierarchy, formalization, and specializations were not designed with the flexibility required to meet the changing reality of how society interacts. Consequently, they are inadequate in the face of increasingly complex and dynamic public problems, and inflexible in accommodating innovative solutions to long-standing public problems. Government can no longer effectively and legitimately serve the public as the sole proprietor of public information, expertise, and decision-making power.

Managing a growing interconnected population in a diverse society is significantly more challenging than if we had to contend with a smaller, more homogenous population. Iceland's experiment with crowdsourcing its new constitution was ultimately unsuccessful, although it was at least possible because of its specific context; adopting such a model in a setting with a more diverse population might prove quite challenging although not impossible (Landemore, 2014). The same type of social media that was leveraged in the Arab Spring movement to organize political protests was used in the United States to engage youth and others in connecting with and electing President Barack Obama in 2008. However, adaptable and scalable governance institutions and processes can be designed to better address public issues. Thus we are presented with the unique opportunity to overcome these hurdles and advance society by designing institutions and processes that leverage communication and computation technology. Doing so will engage, inform, connect, and empower the diversity of public institutions, non-governmental organizations, and citizens with one another in dynamic, strategic, and innovative ways. This can promote a multi-directional process of sharing, capturing, aggregating, analyzing, and understanding complex problems and dynamic information in new ways, on different platforms, and in immersive environments to develop viable solutions and collaborative actions that address shared public challenges.

Essential Elements of Policy Informatics

Policy informatics is built on the fundamental premise that information can be efficiently and effectively mobilized to enable evidence-driven policy design, implementation, and analysis in a legitimate governance environment. Although a relatively new concept, three distinct research clusters are emerging: analysis, administration, and governance infrastructures. *Analysis* focuses on (1) harvesting information reservoirs to generate evidence and insights, (2) visualizing information and relationships between heterogeneous information sets to make sense of problem spaces, and (3) simulating and modeling complex environments to understand the efficacy of policy interventions under various scenarios and their associated outcomes. *Administration* focuses on (1) understanding how the infusion of technology changes policy processes at the individual and group levels, (2) providing information in and around participatory administrative processes, and (3) leveraging the power of networks through technologies that support collaborative governance. *Governance infrastructures* is a concept that focuses on (1) building the next generation of public organizations, (2) designing open, collaborative, distributed, and public governance frameworks and platforms, and (3) advancing the innovative capacity of public institutions using technology.

Each of the foregoing research strands has at its core an appreciation for the fact that we must navigate complex systems and systems of systems (Miller & Page, 2007). This perspective is central to how a policy informatics approach is distinct from more linear approaches to policy analysis. Furnas (2000) asserts that we must think in terms of adaptive systems, including individuals, organizations, communities, markets, and cultures as a whole. He articulates a Mosaic of Responsive Adaptive Systems (MoRAS) that consist of systems within systems, coupled together in an interdependent network where they can interact and adapt to changes in any one system. In other words, systems are embedded within a larger collection of systems; each system is complex in and of itself. As such, it is no longer sufficient to think of problems and opportunities from a linear or singular perspective of an individual agency, as an isolated problem, a one-dimensional challenge, or even a singular system. We must now think in terms of *complex adaptive systems of systems*. Those who work with public institutional design, management processes, and analytical tools must be equipped to understand and address multi-dimensional public challenges through associated networks that are dynamic and constantly evolving.

In the past, these types of problems might have been categorized as wicked and simply dismissed as too difficult to approach. However, by embracing the perspective of complex systems and harnessing technological and methodological advances, we can: (1) model systems to study both the intended and unintended consequences of policy interventions, (2) redesign governance processes and institutions to take advantage of the affordances of new technologies and information environments, and (3) take a more comprehensive approach when including stakeholders in the design process, providing them with tools to contribute actively in the shaping of our public institutions and policies.

Analysis

Solving complex public policy problems in a networked society requires deliberate and sophisticated information analysis. Faced with conflicting proposed solutions to complex problems, policy makers must test their assumptions, interventions, and resolutions. As such, it is critical to have an information-rich interactive environment that most accurately represents the situational-contingent reality of public problems.

Advances in computation and communication technology; the abundance, availability, and distributed nature of information, knowledge, and expertise in a network society; and the new tools and approaches that have emerged to analyze data have enabled policy makers to answer old questions in new ways as well as asking new types of questions. Consequently, these approaches are opening up new lines of science and inquiry, enabling policy makers to examine problems and their underlying governing dynamics in a more rigorous manner than previously possible by simulating, modeling, and visualizing complex systems. We can now examine phenomena at a much finer level of granularity, with greater precision, and with more accurate and relevant data. To address the changing dynamics in the policy-making process, we also need additional perspectives to ameliorate the potential for structural groupthink created by dependence on any particular research methods to the exclusion of others.

Our institutions and the environments in which they operate are also increasing in complexity. Thankfully, over the last few years, we have witnessed the emergence of a science charged with examining complex adaptive systems. Although the field goes by many names (complex systems, emergence, system dynamics), its central premise is shared: by examining the behavior of individual components, we can uncover the emergent dynamics of the aggregation of their interactions in complex environmental systems. Today, the term *generative science* represents the collection of modeling methods that focus on heterogeneous individual actors, their decision-making capacities, decision outcomes under conditions of incomplete information and uncertainty, and the resulting interactions that emerge between sets of actors and their environment that give rise to emergent stable and predictable patterns within systems (Epstein, 2011). The objective is to arrive at explanations for the emergent phenomenon in complex environmental systems, rather than simply producing a descriptive or predictive explanation of its state (Axtell, Axelrod, Epstein, & Cohen, 1996). Put another way, the computational approaches commonly used in generative science, like system dynamics simulations or agent-based modeling, are about illuminating trends, probabilities, and trajectories that are emergent from social phenomena within a particular system, rather than on set point predictions.

Generative science has an important role to play in the context of public administration and policy informatics. First, use of its methods allows us to arrive at counterintuitive hypotheses and to ask new questions about systems (Axelrod, 1997; Forrester, 1971). Second, we can engage in theory stressing (Davis, Eisenhardt, & Bingham, 2007) by taking an existing theory and pushing its boundaries and underlying assumptions. By altering assumptions on individual variables (e.g., individual decision and behavioral probabilities, temporal horizons, number of participants, and system conditions) and interacting with them through simulation models, we can test the robustness of hypotheses that validate existing theories. Third, simulation models allow us to uncover unintended effects of policy intervention and why systems are policy-resistant (Ghaffarzadegan, Lyneis, & Richardson, 2011; Sterman, 2001).

Whereas this section has so far emphasized improving the analytic capabilities of policy makers, public administrators, and professional stakeholders, there is growing evidence that increasing the analytic capacity of the public—also known as digital literacy—is a source of untapped potential. For instance, in 2008 Chicago had a significant digital divide, where broadband use in low-income communities was at 38 percent compared to 61 percent across the city (Mossberger, Tolbert, & Anderson, 2014a). However, a Smarter Communities program was targeted for increasing digital literacy and the result of a neighborhood level analysis found that communities affected by the program significantly increased their online activities, including accessing information about jobs, health, and mass transit (Mossberger, Tolbert, & Anderson, 2014b). The public has also shown that they will develop tools and resources to analyze, make sense of, and act in response to government data about public concerns. Some of the most innovative uses of technology to increase transparency and civic engagement through public data occur at the local level. Integration of data, social media, and visualizations enable novel approaches to neighborhood watches, tracking the spread of illnesses, city maintenance, and policing. As the next two sections emphasize, the public is going to become increasingly important and to realize its potential it will be important to value, on par with traditional forms of literacy, the analytic skills that are associated with digital literacy.

With a rise of digital literacy, the possibility of crowdsourcing and citizen science has implications for how we think about citizen engagement. Public participation has been singled out as a unique approach to arriving at innovative collective decisions for resolving social problems (Chambers, 2003). In this age of globalization, it has become important for public agencies to combine the capabilities of their resources with the knowledge and abilities of the communities they serve (Glaser, Yeager, & Parker, 2006). Public agencies are realizing the importance of developing community-based solutions for social problems (Blair, 2004; Glaser et al., 2006). Platforms that help foster citizen collaboration and facilitate communication between public agencies and citizens are slowly becoming a best practice adopted by public agencies for a variety of purposes.

Administration

Whereas the analysis cluster focused on *what* could and should be done, the administration cluster emphasizes the *how*. A common thread in traditional democratic theory (Dahl, 2000), participatory democracy (Pateman, 1970), deliberative democracy (Bohman, 1998), and cosmopolitan democracy (Held, 1995) is the desire for increased inclusion of diverse *team and citizen participation* in the process of administering solutions to public issues. As the challenges facing socionatural systems within the network society become more complex, it is imperative that the participation of stakeholders with relevant expertise and experience is increased and deepened. When designed well, advances in technological platforms can lower the barriers to meaningful participation and extend the range and diversity of stakeholder voices that participate in governance. This can occur through a commitment that ensures the governance system is inclusive, manages minority perspectives and power differentials, delivers easily digestible information concerning public issues and instructions for participation, and encourages richer dialogue, more sustained and collaborative interactions, and more useful contributions.

Many contemporary problems, such as climate change, disease eradication, sustainable water management, pollution, and land management, require coordination and collaboration among different systems that span local and national boundaries. A collaborative approach will increase avenues for policy deliberation involving diverse stakeholders at different levels with legitimate interests, diverse powers, knowledge, and an ability to share information and communicate openly in the development of a common understanding of the issues at stake.

Collaboration between diverse stakeholders is difficult and requires that individual stakeholders analyze massive amounts of information in the process of decision-making. Thus, it is useful to design governance deliberation spaces with appropriate incentives and rules that will enable stakeholders to effectively manage and coordinate information, build shared understanding, and collaborate. The design should be mindful of fostering open dialogue, trust, a sense of shared identity among group members, and accountability (Johnston, Hicks, Nan, & Auer, 2011).

Toward those ends, *participatory modeling* becomes a viable administrative tool of governance. Participatory modeling switches the role of policy stakeholders from being the consumers of generative science models to the co-producers. The construction of models enables us to engage holistically with the stakeholders and the theories that are relevant to the challenge at hand. When constructing models, one has the opportunity to contribute expertise and local context in the construction of both the problem and the potential solution. To develop a participatory model, stakeholders help build the model, validate it, and even analyze the results. By involving actor groups in this manner, not only do researchers engage in richer problem-solving exercises, but also a diversity of participants have the opportunity to interactively explore, legitimate, and understand others' perspectives and claims concerning shared public issues. Participation in simulation models affords stakeholders a common language and framework for approaching a situation and with the potential to engender empathy and thereby produce more effective collaborative decisions concerning a shared public good or resource (Hu, Johnston, & Hemphill, 2013). By strategically considering the affordances of administrative advances we have yet another category of governance innovations that become possible.

Governance Infrastructures

Governance infrastructures in the network society consist of a collection of communication and computation technologies, systems, people, policies, practices, and relationships that interact to support governing activities (Johnston, 2010; Johnston & Hansen, 2011). The integrated nature of these systems means that changes in one dimension will have consequences on each of the others. The mass diffusion of and continuous intense reliance on advanced technologies to instantaneously carry out our everyday personal and work life intimately and dynamically interconnect us all as we deal with the realities of life and the challenges we face as individuals, organizations, sectors, and societies. Policy informatics affords an opportunity to design novel governance infrastructures within existing institutions and beyond. As such, future governance infrastructures should be smart, open, collaborative, and participatory in nature.

Toward this end, policy informatics is fortunate to play a role in the MacArthur Foundation–supported *Research Network on Opening Governance*,[1] which seeks to advance fundamental understanding of how to make governance more effective and legitimate in the context of the information technology revolution and the rise of the network society. This open governance agenda has two key dimensions: opening governance (as processes of social decision-making) to more diverse sources of knowledge, more avenues of interaction, enhanced social understanding and distributed responsibility; and opening government (indeed, any formal institution) as knowledge organizations, promoting the conditions where knowledge is shared and used, collaboration encouraged, expertise sought from diverse sources, and capacity throughout the policy cycle enhanced.

For instance, the next big initiative for our work on *distributed governance infrastructures* research is to shift the focus in environmental hazards research from a framework centered on populations and places to one that focuses on individuals, personalization, and public empowerment. Sensor and communication technologies have advanced to the point where micro-customization of interventions and policies centered on hazards are becoming possible, potentially even at the individual level. These advances could lead to more effective use of public funds and ultimately lead to fewer injuries and deaths when hazards occur. We are addressing this need by developing and executing a range of statistical and real-life experiments aimed at creating a new pool of information regarding individual exposure and risk and the personalization of warning messages and intervention efforts. When the necessary data are collected, analyzed, and returned to individual users and governance entities, we believe that informed local control and self-organization will lead to reductions in adverse health outcomes associated with hazards. Executing this work will require expertise in the platforms and technology that make personal monitoring and personalized communication possible, working relationships with regional- and state-level hazards, health, and preparedness stakeholders, and multidisciplinary expertise that includes the physical sciences, social sciences, and health dimensions of the threats posed by natural and anthropogenic hazards. Our immediate focus is on the regionally relevant challenge of extreme heat, but the research framework is applicable to a much broader range of other environmental and health issues. Successful implementation of this initiative requires investments in sensor and information technologies that could be components of smarter environmental hazard governance strategies in the future, as well as personnel who can design and conduct experiments that determine the utility of such devices and how they would be implemented in operational settings.

Utilizing Internet-based communication technologies can work to engage a diverse group of expert stakeholders to share and learn about information concerning public problems and opportunities and to leverage the resulting collective intelligence to identify innovative solutions and take action. The design of participatory platforms and "government-as-a-platform" (O'Reilly, 2011) is integral to successfully engender civic engagement that legitimately and effectively addresses public issues and fulfills the act of public governance. The ability to coordinate efforts in a manner similar to plug-n-play, where diverse approaches seamlessly work together and are interoperable across heterogeneous environments, will be a crucial design feature of next-generation governance infrastructures.

The rise of the Web has given citizens the opportunity to spend their spare time, mental and physical energy, and enthusiasm more constructively by providing participation opportunities that are proactive, educational, creative, and generative rather than passive, consumptive, and depleting. The cognitive surplus that citizens spend on consumptive activities, such as watching television, can now be used in collaborative efforts for creativity and problem-solving (Shirky, 2010). The public is assuming an active role in examining complex problems, collecting data that can be analyzed, critiquing tentative findings, and promoting the dissemination of knowledge. People choose to participate in platforms as citizen scientists because they gain enjoyment from the experience (Kim, Lee, Thomas, & Dombrowski, 2009) as well as for intrinsic, norm-oriented, and reputation-seeking motivations (Nov, Arazy, & Anderson, 2011). Crowdsourcing leverages these motives by engaging the surplus interests, activities, and productivity of the public to actively address public issues and challenges through participatory platforms.

Virtual platforms that constructively crowdsource the cognitive surplus of the public and exemplify citizen science abound. For example, Wikipedia[2] is a compelling virtual artifact representing millions of hours of cumulative human thought devoted to organizing information and solving problems (Oomen & Aroyo, 2011). Galaxy Zoo[3] is an online platform that contains images of almost a quarter of a million galaxies and asks the public to visually assess and classify images of galaxies according to a typology of characteristics. It has been clearly demonstrated that the distributed and collective knowledge of citizen volunteers can help make sense of large amounts of data: within a month of its launch in July 2007, 80,000 Galaxy Zoo volunteers had classified more than 10 million images, and in its first year the public produced 50 million classifications of over one million galaxies that were as accurate as those completed by astronomers (Fortson et al., 2011). Today there are over one million registered users on Zooniverse.org (the successor to Galaxy Zoo), with volunteers able to contribute to more than 30 different projects, from classifying galaxies to transcribing First World War soldiers' diaries (Zooniverse, 2014).

The motivation to participate in these platforms can range from a sense of duty to curiosity, and even to enjoyment. Platforms designed to engage the public through game structures and play have been used by organizations to complete specific objectives or tasks. Platforms with game structures afford participants the incentive- and reward-based opportunities to compete or collaborate to solve a challenge in areas of personal interest (Kim et al., 2009). Gamification of citizen participation in the public policy and administration process has become a mainstay in our society (Kelley & Johnston, 2012). Today, we have serious games that are employed for training and development, are played on various social networks, and leverage the collective intelligence of players to solve complex problems.

Games engage users in a deep manner and keep them connected through the provision of incentives and rewards and the evolution of levels from simple to complex that serve as challenges to test their skills.

A game framework can provide us with an opportunity to engage citizens to not only learn more about public institutions but also immerse them in complex policy deliberations. Some games are based on real information and choices made by public administrators. For example, the Maryland Budget and Tax Policy Institute developed the Maryland Budget Game (O'Malley, 2010) and invites citizens to act as the state governor to evaluate budget proposals—including those currently being proposed at the state level—and make choices about funding options for the next fiscal year. This game not only informs citizens of actual proposed budget options, but also exposes citizens to the plethora of considerations and tough choices inherent in the public decision-making process, and provides a space that can collectively influence real budget allocation decisions, illustrating that games can provide an enticing platform for individuals to play and contribute to the public good.

Although features of transparency and collaboration are important for realizing the goal of participatory platforms, they are not sufficient. Care must be taken when designing participatory platforms that incentives are offered that will attract a diverse group of individuals and entities. In addition to incentives, consideration must be given to the modes of engagement that will benefit each party's interest while simultaneously advancing the common good. Finally, it is imperative that we lower both the real and opportunity costs of participation. Ultimately, the network society requires participatory platforms that facilitate information sharing, learning, inclusion, collaboration, and innovation to design the next-generation of institutions, products, and services that generate public value.

Roadmap for the Book

This book is a collection of creative essays that highlight various essential elements of policy informatics. By no means do we claim that this book is representative of all the elements of policy informatics. Doing so would require several volumes. We have specifically asked our authors to be illustrative rather than comprehensive, highlighting opportunities and challenges for policy informatics through examples. The chapters are divided into four sections that cover the three dimensions of policy informatics described earlier—analysis, administration, and governance infrastructures—preceded by a 'basics' section.

All chapters share the common threads of policy informatics. Specifically, they all take a system perspective of socio-natural reality and discuss how advances in computation and communication technology can contribute to improving the structure, process, and tools of public governance, policy analysis, and administration to more legitimately, effectively, and efficiently address complex public problems. Ultimately, this involves liberation of data and engagement of diverse stakeholders and perspectives.

Part II: The Basics

The basics section introduces the challenges and opportunities presented by information technology in the public policy analysis process, and how a system perspective can exploit the opportunities to improve our understanding, decisions, and evaluations of public issues. It begins with Sharon Dawes and Natalie Helbig in chapter 2 examining the current *value and limits of information* in the public policy and administration process. They assert that information is often treated as a black box in the traditional public decision-making process—trusted as given without critical examination of its validity within context. This problem is magnified for the less structured but richly diverse information sources that are emerging from the open data movement and the advent of social media applications. After describing several classic information problems (i.e., conventional wisdom and assumptions; provenance, metadata, and out-of-context information; data collection, management, access, and dissemination practices), Dawes and Helbig outline a framework for ensuring information quality and fitness for public decision-making. Government must carefully consider the role and implications of information resources, policies, and management practices that produce, disseminate, and use data as evidence for public decisions.

Although it seems self-evident that good policy decisions are informed by the best available knowledge and sound and authoritative evidence, research indicates this is infrequently true. In chapter 3, Anand Desai and Kristin Harlow argue that because policy problems are complex and the definition of actionable evidence is ambiguous, the link between evidence and policy action is often unclear and the products of policy inquiry by themselves are not sufficient to determine the course of policy action. Instead, evidence can inform and influence decisions and actions only when accompanied by an argument that interprets the evidence for the given context and it is given import by arguments that confirm the evidence is legitimate support for decisions and actions.

Concluding this section, in chapter 4 Evert Lindquist explores how the visual representation of information, data, and evidence can enhance the process of deliberating and evaluating public policies and actions. This chapter provides background, perspectives, and suggestions for public sector executives and analysts seeking to either invest in or explore the potential and limitations of visualization techniques for various types of policy work. Lindquist begins with a survey of visualization techniques and then outlines a framework that connects visualization to various types of policy work (undertaking analysis, providing advice to elected leaders, and engaging citizens on policy matters). His concluding section identifies challenges inherent in the use of visualization in these environments and suggests guidelines on how best to proceed with strategically investing in expanding visualization capabilities in government.

Part III: Analysis

The analysis section includes a collection of chapters that examine computational modeling—an analytical technique that has gained prominence in public policy circles. To begin this section, in chapter 5 George Richardson discusses the origin of system dynamics, starting with Forrester's original statement of the foundations of system dynamics, which emphasized four threads: computing technology, computer simulation, strategic decision-making, and the role of feedback in complex systems. Subsequent work has expanded on these to expose the significance in the system dynamics approach of concepts such as dynamic thinking, stock-and-flow thinking, and operational thinking. But the critical foundation of systems thinking and system dynamics—the *endogenous point of view*—lies deeper than these and is often implicit or even ignored.

Simulation model-based analysis that relies exclusively on parameter sensitivity testing may ignore how parameter changes in a computer model can be implemented by public organizations in the real world. In chapter 6, David Wheat explores how model-based implementation planning and analysis during the policy design stage of system dynamics simulation modeling can lead to the discovery of *emergent patterns* when exploring policy problems, more operational thinking during model-based policy design, and the development of models that are less reliant on wishful thinking and are ultimately more useful to policy makers. The entire process is illustrated with models of two public health issues: a flu epidemic and automobile pollution. The final section highlights the application of useful modeling insights gleaned from the implementation literature, forewarning that the successful implementation of policies is not guaranteed and requires a separate analytical process.

David N. Ford, Ivan Damnjanovic, and Scott T. Johnson investigate in chapter 7 how a system dynamics modeling approach that considers the *tipping point of risk allocation* among diverse stakeholders can be used to better understand how a public-private partnership between a government agency and a private developer can best deliver and manage a large toll road transportation infrastructure project. A toll road case study is presented, and simulations of three extreme risk allocation policies and four traffic growth patterns are used to describe how risk allocation policies impact the performance from the perspective of the private developer and the public. Model results describe wide performance ranges for an unstressed project. The model was expanded to reflect financial stress and showed reduced returns to both developers and the public when compared to returns from an unstressed project.

In chapter 8, Navid Ghaffarzadegan, John Lyneis, and George Richardson emphasize the characteristics of certain policy challenges that make identifying a resolution difficult when using traditional approaches. Such challenges include policy resistance, the importance and cost of experimentation, the need to achieve consensus between diverse stakeholders, overconfidence, and the need for an endogenous perspective. They argue that a policy challenge that exhibits these characteristics, such as urban dynamics and welfare policy, can benefit from system dynamics modeling using *small models* that can be conceptually understood because they focus on a specific part of the overall system. The authors conclude by developing a set of arguments about how and why small system dynamics models can uniquely address characteristics of public policy problems and complement other policy informatics approaches, shedding light on the factors that modelers should consider when developing effective models for policy makers.

Where politically oriented group violence is concerned, accurate assessment of and response to a violent event require the identification of responsible parties. Often this information is not readily available due to conflicting or missing knowledge about these groups. In chapter 9, Christopher Bronk and Derek Ruths describe how a Bayesian approach to modeling can be used to understand system dynamics and analyze seemingly disparate data collected on political violence behavior from local *open-source* events to connect groups with documented behavioral patterns with an event where participants (or culpable parties) are unknown.

Modelers often say that the process of working on a model to support policy is at least as important as the final product. Despite this realization, analysts rarely capture the experiential learning that led to the end results in final publication or other products. Kimberly Thompson argues in chapter 10 that the process of *developing and using a model* among diverse stakeholders through a deliberative process to support policies is as important as any final product that emerges from the model. With over a decade of experience using modeling approaches to manage the global risks of the polioviruses, she articulates the importance of adequately understanding the system, framing the analysis, engaging a diversity of stakeholders and managing expectations, seeing the pathways forward, creating and communicating shared insights, and iterating and learning throughout the process.

Part IV: Administration

The administration section includes chapters that discuss the role and value of utilizing computation and communication technology to engage a diversity of stakeholders and situational perspectives in the collective and collaborative process of administering and evaluating public policy solutions. The section begins with chapter 11, in which Christopher Koliba and Asim Zia call for a growing appreciation of the roles that complex governance networks play in both the design and execution of public policies. They highlight that the entire process of *governance informatics* projects can help practitioners work to collectively address wicked problems by developing a conscious *situational awareness* of network structures, policy tools, accountability ties, and performance measures to create conditions uniquely situated for conversation, collaboration, and discovery.

Analysis of current education outcomes suggests that the STEM (science, technology, engineering, and mathematics) skills of the future workforce will be insufficient, and the gap between what society needs from its citizens and what is produced by the system will continue to increase. In chapter 12, Nora Sabelli, William Penuel, and Britte Cheng explore why many well-intentioned education policy interventions have had limited success due to policy resistance. They argue that there is a need for a broader systemic perspective and the inclusion of a greater diversity of stakeholders for more innovation as well as better implementation of future changes. The use of a policy informatics approach can create a context that empowers stakeholders to understand the dimensions of policies within their own local contexts and make informed, data-driven decisions on whether and how to implement them. In addition to these factors, one crucial aspect in building collaboration is providing individual stakeholders with the experience of perspective taking, which enables them to understand the claims of other stakeholders as legitimate, rather than a threat to their identity.

In chapter 13, Petra Ahrweiler, Andreas Pyka, and Nigel Gilbert present an agent-based *innovation policy model* (SKIN) developed by a team of scientists that has made a number of unique contributions simulating the effects and impacts of policy-making on the structure, composition, and outputs of research and innovation networks. The SKIN model allows policy makers to examine the contributions of specific research and innovation policies to societal goals, and it represents one of the most successful policy models used in Europe over the last decade. The authors show how five distinct policy-relevant research threads have been developed and experimented with, and what outcomes arose from the efforts. Using real-world datasets and carefully analyzing questions put forward by stakeholders, SKIN can provide precise, detailed information on the effects of specific policy instruments, on how research and innovation networks operate, and how to understand and manage the relationship between research funding and policy goals. The chapter concludes with the reflections of modelers who have decades of experiences as to why they think policy informatics should be more regularly included as a bridge between research and policy.

The diffusion of the Internet and Web has expanded the potential for crowdsourcing and mass collaboration. However, whether these approaches are viable in the public sector remains an open question. Based on a variety of case studies and participant-observation of public initiatives, William Dutton argues in chapter 14 that crowdsourcing can be well suited to capturing the value of distributed intelligence in support of public policy and regulation. To accomplish this, networks of individuals must be cultivated and managed, not as crowds, but as *collaborative network organizations*. Three types are distinguished: CNO 1.0 (sharing information), CNO 2.0 (capturing and contributing information), and CNO 3.0 (co-creating knowledge). The benefits of crowdsourcing information are presented using illustrative examples: (1) directly connecting public officials with a greater diversity of independent experts, (2) exploiting the convening power of governments, (3) creating compatibility with popular open government initiatives, (4) tapping distributed expertise, (5) serving as a complementary mechanism for outsourcing, (6) moving with speed and urgency, and (7) co-creating policy.

Critical lessons to manage crowdsourcing initiatives are outlined including using existing e-infrastructures created by open source software to support collaboration, emphasizing the criticality of top management to support CNOs, describing the importance of managing collaboration, and modularizing tasks. This chapter seeks to clarify a workable vision and strategies for tapping distributed public intelligence—a concept that presents a significant prospect for the public sector to address critical problems—and outlines lessons learned from early cases, including key opportunities and risks.

Part V: Governance Infrastructure

The governance section includes a collection of chapters that discuss how computation and communication technologies are transforming the design of governance institutions and processes. This section begins with chapter 15, in which Christopher Barrett and colleagues discuss how *synthetic information environments* can be designed to enable a *distributed cognition* of complex systems, specifically in public health and pandemic preparedness planning. The complexity and coupling between behavior and infrastructure during a public health epidemic (the 2009 H1N1 outbreak) are described, and the authors argue that policy-making in this context can be abstracted as a cognitive problem. Socially coupled policy domains, such as public health, are influenced by the collective behavior of millions of individuals who participate in them and respond to policy plans and interventions. When there is understanding about the diversity of perspectives, agendas, motivations, and capabilities of stakeholders, synthetic information environments can combine information from mutable unstructured data sources to create models for forecasting and for *in-silico* experimentation of different intervention policies. Participating in the simulations creates a shared perspective of the system, thereby enabling better understanding, interactions, and planning of approaches to deal with real problems. The authors describe synthetic information environments as a systematic solution to this problem by presenting a distributed cognition perspective. They describe two different case studies (pandemic response training, and the H1N1 outbreak of 2009) in the context of pandemic preparedness planning, which illustrate the use of the system and describe its distributed nature and the constraints induced by this perspective.

In chapter 16, Gerard Learmonth, Sr., and Jeffrey Plank review the evolution of modeling and simulation in support of policy and decision-making, especially in modeling the relationship between human behavior and natural processes, and present their experience in the development of the *participatory simulation* known as the *UVa Bay Game*. Their simulation included stakeholders in the design of the model to generate hypotheses about system drivers and to test innovative solutions to the policy challenges in the Chesapeake Bay area. To develop the science underlying the game, they drew from a team of scientists that represented academic disciplines not conventionally associated with modeling and simulation. Hundreds of participants played different roles in the management of the watershed, and the outcomes are based on their interactive contributions. Participants found the simulation to be a reasonable interpretation of the dynamics of the Bay and soon appreciated the complexity of the issues, realizing that others are not acting as adversaries and empathizing with the legitimate motivations for their behaviors, which is a useful precondition for finding common ground to shared challenges. The authors argue that the inclusion of live agents in participatory simulations changes the process of modeling and the model application to one of social integration with fundamental cultural significance.

One major challenge of enabling civic participation is pairing those who are willing to take action with those who have needs. Jes Koepfler and colleagues present a successful case study in chapter 17 of an innovative *participatory platform*, ACTion Alexandria, where local communities motivate and coordinate local volunteerism and problem-solving through *action brokering*. Action brokering is both an informal and formal process of civic participation where intermediaries connect and negotiate the exchange between individuals or organizations that can take action and those that are in need. Findings from a year-long study of ACTion Alexandria demonstrated that its success hinged on many system factors, including a competent community manager, institutional support from an existing agency, effective use of social media, partnerships with other nonprofits to collaboratively grow networks, and an emphasis on promoting immediate actions and soliciting ideas. The authors identify the social practices and technical features that can be used to implement action brokering and assess their impact, highlighting key factors that contribute to successful collective action, the challenges, and policy implications of platforms that promote action brokering, concluding with a recap of best practices learned.

Chapter 18, by Ines Mergel, reviews the challenges of incorporating innovative knowledge into government, and the prospects for using social media technologies to access knowledge both from inside government and from the public sphere, especially as it pertains to the mode of interaction between citizens and public agencies. Mergel argues that although we have new technologies, public agencies use them for traditional activities (e.g., broadcasting information or educating audiences), thereby under-exploiting their full potential. Insights are provided into emergent forms of knowledge incubation from selected pockets within government and beyond. She concludes with insight into how public agencies can embrace *bi-directional knowledge flows* that take full advantage of emerging technologies to help realize active citizen engagement.

In the concluding chapter, Justin Longo, Dara Wald, and David Hondula assess where the field appears to be heading and what the prospects are for its future. They start with an evaluation of the place of policy informatics in the spectrum of disciplines, interdisciplinary fields of study, and research areas and look at what fields help to frame the current state of policy informatics. They then look forward to anticipate how policy informatics might continue to develop in response to a number of modern policy challenges and what the field can offer society.

Concluding Thoughts

This book seeks to catalyze a field of scholarship, both research and training, around the concept of policy informatics that integrates public administration, policy analysis, information technology, behavioral economics, law, complex systems, decision sciences, and political science. The chapters have both methodological and domain-specific strengths that enable advancement of governance itself and its application and manifestations across multiple contexts. As an early entrant in the space of policy informatics with content expertise in computational thinking, open governance, participatory modeling, and smarter governance infrastructures, this volume is uniquely positioned to connect seemingly disparate researchers and perspectives for solving important societal problems. This publication is the start of what we hope is a long conversation with an expanding community of scholars, governance leaders, practitioners, and citizens, about what policy informatics is and what it can offer society. We look forward to building that future together, exploring how computation and communication technology can be leveraged to understand and address complex public policy and administration problems and realize innovations in governance processes and institutions.

Acknowledgments

This material is based upon work supported by the National Science Foundation under grants 0838206, 1143761, 1241782, 1243968, 1322296, and SES-0951366, Decision Center for a Desert City II: Urban Climate Adaptation. The John D. and Catherine T. MacArthur Foundation also supports the research through its Research Network on Opening Governance. The Virginia G. Piper Charitable Trust also supports the research. Any opinions, findings, and conclusions or recommendations expressed in this material are those of the author and do not necessarily reflect the views of the National Science Foundation, the John D. and Catherine T. MacArthur Foundation, or the Virginia G. Piper Charitable Trust.

Notes

1   http://www.opening-governance.org.
2   http://www.wikipedia.org.
3   http://www.galaxyzoo.org.

References

Axelrod, R. (1997). Advancing the art of simulation in the social sciences. *Complexity, 3*(2), 16–22.

Axtell, R., Axelrod, R., Epstein, J.M., & Cohen, M.D. (1996). Aligning simulation models: A case study and results. *Computational and Mathematical Organization Theory, 1*(2), 123–141.

Blair, R. (2004). Public participation and community development: The role of strategic planning. *Public Administration Quarterly, 28*(1), 102–147.

Bohman, J. (1998). Survey article: The coming of age of deliberative democracy. *Journal of Political Philosophy, 6*(4), 400–425.

Castells, M. (2011). *The rise of the network society: The information age: Economy, society, and culture* (2nd ed., Vol. 1). Oxford, UK: Wiley-Blackwell.

Chambers, R. (2003). Participation and numbers. *PLA Notes, 47*, 6–12.

Dahl, R.A. (2000). *On democracy.* New Haven, CT: Yale University Press.

Davis, J.P., Eisenhardt, K.M., & Bingham, C.B. (2007). Developing theory through simulation methods. *Academy of Management Review, 32*(2), 480–499.

Epstein, J.M. (2011). *Generative social science: Studies in agent-based computational modeling.* Princeton, NJ: Princeton University Press.

File, T. (2013). *Computer and Internet use in the United States.* Washington, DC: United States Census Bureau. Retrieved from http://www.census.gov/prod/2013pubs/p20-569.pdf

Forrester, J.W. (1971). Counterintuitive behavior of social systems. *Technology Review, 73*(3), 52–68.

Fortson, L., Masters, K., Nichol, R., Borne, K., Edmondson, E., Lintott, C., Raddick, J., Schawinski, K., & Wallin, J. (2011). Galaxy Zoo: Morphological classification and citizen science. *arXiv preprint.* arXiv:1104.5513.

Fox, S., & Rainie, L. (2014). The Web at 25 in the US. *Pew Research Center's Internet & American Life Project,* 5. Retrieved from http://www.pewinternet.org/2014/02/27/the-web-at-25-in-the-u-s/

Furnas, G.W. (2000). Future design mindful of the MoRAS. *Human–Computer Interaction, 15*(2–3), 205–261.

Ghaffarzadegan, N., Lyneis, J., & Richardson, G.P. (2011). How small system dynamics models can help the public policy process. *System Dynamics Review, 27*(1), 22–44.

Glaser, M.A., Yeager, S.J., & Parker, L.E. (2006). Involving citizens in the decisions of government and community: Neighborhood-based vs. government-based citizen engagement. *Public Administration Quarterly, 30*(1/2), 177–217.

Held, D. (1995). *Democracy and the global order: From the modern state to cosmopolitan governance.* Stanford, CA: Stanford University Press.

Hu, Q., Johnston, E., & Hemphill, L. (2013). Fostering cooperative community behavior with IT tools: The influence of a designed deliberative space on efforts to address collective challenges. *Journal of Community Informatics, 9*(1). Retrieved from http://ci-journal.net/index.php/ciej/article/view/699/977

Johnston, E.W. (2010). Governance infrastructures in 2020. *Public Administration Review, 70*(1), s122–s128.

Johnston, E.W., & Hansen, D.L. (2011). Design lessons for smart governance infrastructures. In A. Balutis, T.F. Buss, & D. Ink (Eds.), *Transforming American governance: Rebooting the public square?* (pp. 197–212). Armonk, NY: M.E. Sharpe.

Johnston, E.W., Hicks, D., Nan, N., & Auer, J.C. (2011). Managing the inclusion process in collaborative governance. *Journal of Public Administration Research and Theory, 21*(4), 699–721.

Kelley, T., & Johnston, E. (2012). Discovering the appropriate role of serious games in the design of open governance platforms. *Public Administration Quarterly, 36*(4), 504–554.

Kim, J., Lee, E., Thomas, T., & Dombrowski, C. (2009). Storytelling in new media: The case of alternate reality games, 2001–2009. *First Monday, 14*(6). Retrieved from http://journals.uic.edu/ojs/index.php/fm/article/view/2484/2199

Landemore, H. (2014). Inclusive constitution-making: The Icelandic experiment. *Journal of Political Philosophy.* Early view (published online February 25, 2014). doi: 10.1111/jopp.12032

Malone, T.W. (2004). *The future of work: How the new order of business will shape your organization, your management style, and your life.* Boston, MA: Harvard Business Press.

Miller, J.H., & Page, S.E. (2007). *Complex adaptive systems: An introduction to computational models of social life.* Princeton, NJ: Princeton University Press.

Mossberger, K., Tolbert, C. & Anderson, C. (2014a, October 27). Digital literacy is a game changer: A neighborhood-level analysis of Chicago's smart communities. Brookings Institution blog, *Tech Tank.* Retrieved from http://www.brookings.edu/blogs/techtank/posts/2014/10/27-chicago-smart-neighborhoods

Mossberger, K., Tolbert, C. & Anderson, C. (2014b). *Measuring change in Internet use and broadband adoption: Comparing BTOP smart communities and other Chicago neighborhoods.* Chicago, IL: Smart Communities, Chicago Digital Excellence Initiative. Retrieved from https://copp-community.asu.edu/sites/default/files/REVChicagoSmartCommunitiesCHANGE042514-final%20%282%29.pdf

Nov, O., Arazy, O., & Anderson, D. (2011, July). *Technology-mediated citizen science participation: A motivational model.* Paper presented at the AAAI International Conference on Weblogs and Social Media, Barcelona, Spain.

O'Malley, M. (2010, Mar 19). *How would you balance the state's budget? Maryland budget and tax policy institutes updates "The Maryland budget game."* Press release. Office of Governor Martin O'Malley. Baltimore, MD. Retrieved from http://www.governor.maryland.gov/pressreleases/100319.asp

Oomen, J., & Aroyo, L. (2011). *Crowdsourcing in the cultural heritage domain: Opportunities and challenges.* Paper presented at the Proceedings of the 5th International Conference on Communities and Technologies, Brisbane, Australia.

O'Reilly, T. (2011). Government as a platform. *innovations, 6*(1), 13–40.

Pateman, C. (1970). *Participation and democratic theory.* Cambridge, MA: Cambridge University Press.

Pew Research. (2014). *Emerging nations embrace Internet, mobile technology.* Washington, DC: Pew Research, Global Attitudes Project. Available at http://www.pewglobal.org/2014/02/13/emerging-nations-embrace-internet-mobile-technology/

Shirky, C. (2010). *Cognitive surplus: How technology makes consumers into collaborators.* New York: Penguin Group.

Sterman, J.D. (2001). System dynamics modeling: Tools for learning in a complex world. *California Management Review, 43*(4), 8–25.

United States, Department of Commerce. (2013). *Exploring the Digital Nation: America's Emerging Online Experience.* Washington, DC: U.S. Department of Commerce. Retrieved from http://www.ntia.doc.gov/files/ntia/publications/exploring_the_digital_nation_-_americas_emerging_online_experience.pdf

White House. (2009). President's memorandum on transparency and open government—Interagency collaboration. February 24, 2009. Retrieved from http://www.whitehouse.gov/sites/default/files/omb/assets/memoranda_fy2009/m09-12.pdf. (Archived by WebCite at http://www.webcitation.org/6TUzwzARy)

Zooniverse. (2014, February 14). One million volunteers. *Zooniverse Blog.* Retrieved from http://blog.zooniverse.org/2014/02/14/one-million-volunteers/

PART II

The Basics

2

THE VALUE AND LIMITS OF GOVERNMENT INFORMATION RESOURCES FOR POLICY INFORMATICS

Sharon S. Dawes and Natalie Helbig

Introduction: The Information Dimension of Policy Informatics

Policy problems, alternatives, and decisions encompass myriad topics and issues. They can focus on education, public health, transportation, environmental stewardship, economic development, and many other areas. Different substantive considerations, forms of expertise, and interests must be considered for each policy. Regardless of these differences, however, all instances of policy analysis have one common element—they need, use, and generate information. Ironically, information is often treated as a black box in policy analysis. Stakeholders, analytical techniques, and technology tools all receive considerable attention, but analysts and decision-makers seldom critically assess the inherent suitability, strengths, and weaknesses of the information sources they use. Instead, information is often seen as a given, used uncritically, and trusted without examination. This can be a serious problem even when that information is well defined and carefully managed throughout its lifecycle. The problem is greatly magnified for the less structured, but richly diverse information sources that are emerging from the open data movement and the advent of social media. These newly available government data sources, if managed and used with care, can contribute new insights into the complexities of policy problems and ways to address them. In this chapter, we discuss important issues associated with the *information* component of policy informatics.

A basic assumption in policy informatics is that more intensive and creative use of information and technology can improve policy making processes and lead to better policy choices, especially under conditions of complexity. An interdisciplinary research community is coalescing to look beyond traditional econometric-based approaches to different computational tools, such as agent-based modeling, simulation, system dynamics, data visualization, network analysis, and data mining, which are all better suited for displaying and understanding complexity (Johnston & Kim, 2011).

Increasing attention is being paid to these tools because they offer the ability to analyze large heterogeneous datasets, the very kind of data that is now being released from administrative systems and amassed from social media interactions. However, this kind of data, which is usually collected or created for other purposes, presents substantial risks concerning validity, relevance, and trustworthiness of analytical results. We explore the nature, limitations, and role of information in policy informatics and illustrate the issues, risks, and potential benefits with selected cases in human services, land records, and financial accountability.

Government Information Resources for Policy Analysis and Evaluation

The government is the major data resource for policy making and policy evaluation. Government information resources are defined as the data, information content, systems, and information services that emanate from the day-to-day administration of government programs (Dawes & Helbig, 2010). Each information resource is created to serve the needs of a specific law, regulatory scheme, or service program and thus reflects the requirements and intentions of those purposes and the people responsible for them. These sources reflect specialized definitions, limitations on sources and frequency of data collection, pre-conceived notions about the players and dynamics of a policy arena, and beliefs about how policies work and what constitutes a good or bad result.

The use of these information resources can and does extend well beyond the government itself to a very diverse multi-stakeholder society. While the day-to-day value of this information comes from its use in specific government programs and services, the societal value of these information resources is derived primarily from unpredicted and flexible uses of the data content by all stakeholders (Bellamy & Taylor, 1998). These information resources offer value for policy debates, choices, and products and services, both planned and unexpected.

Certain sources of government information are commonly used in traditional policy analysis. These information sources include the United States Census Bureau, Bureau of Labor Statistics, and National Center for Health Statistics, as well as similar organizations with the formal responsibility and professional skill to collect, manage, maintain, and disseminate data for public use. These sources are well understood and readily usable because they apply social science research standards in data collection and management. They collect well-defined data on specific topics using well-documented methodologies that follow a logical design. The data files are managed, maintained, and preserved according to explicit plans that include formal rules for access, security, and confidentiality. These data resources date back many decades and reflect a long-standing government commitment to collect and provide certain kinds of social, economic, and demographic information to the public.

Two new trends are attracting great attention for their potential analytical value, both inside and outside government: open data programs embodied in the free release of administrative data, and the emergence of entirely new information sources related to government use of sensors and social media.

Administrative data reflect government operations and its many programs. This type of data is increasing in volume (George & Lee, 2002). The automation of government activities and the advent of electronic government services brought with it not only online convenience, but also the ability to capture enormous amounts of digital information about services, individuals, organizations, and transactions (Snellen, 2005; Bekkers, 1998; Frissen, 1992). Emails in government record systems, for example, number in the billions. Transactional data reveals the workflow activities of case management systems or customer service exchanges, much of it collected in real-time. The government deploys sensors on highways, in toll booths, at border crossings, and in ports and airports that collect data about transport and travelers. Air quality is monitored by a nationwide network of sensors that record levels of pollution every hour.

Such data is created or collected by organizational activities at the environmental, strategic, tactical, operational, and transactional levels at every level of government across different agencies and jurisdictions. Snellen (2005) reports that governments take advantage of several different types of information systems, including *database technologies* as data repositories or for file sharing; *tracing and tracking technologies* for workflow management and monitoring purposes; *desktop technologies* like text processors, digital personal assistants, and email; *decision support technologies* such as spreadsheets, various task-directed software, and expert systems; and *network technologies* that include websites, homepages, call-centers, and email. Constantly improving computational tools allow these data types to be easily integrated, shared, and manipulated (Dawes & Pardo, 2006).

Meanwhile, the recent open government movement is making tens of thousands of these administrative datasets available to the public through programs like Data.gov, which makes data from federal government agencies readily accessible for external use. Its central Web portal provides electronic access to machine-readable information about government finances, program performance, trends, transactions, and decisions. The goal is to allow people and organizations outside government to find, download, analyze, compare, integrate, and combine these datasets with other information in ways that provide value to the public. This phenomenon is not limited to the federal level. States and municipalities are experiencing similar growth in data holdings and take advantage of new technologies to gather and analyze data from routine operations. Citizen hotlines (i.e., 311-systems) record millions of complaints and requests for information at the county and state levels. Vital records, land records, voting systems, and licensing programs all generate or collect enormous amounts of detailed data that could be used for policy analysis.

Government use of social media applications is emerging as another type of information resource, adding to the volume and variety of data available for policy-oriented users. Interaction between government agencies and citizens through two-way exchanges of text, video, or pictures, and related commenting, voting, or tagging provide new streams of *government-related* information. Government agencies use such applications as Facebook, Twitter, IdeaScale, YouTube, and Google+. They also build their own platforms (We the People, Regulation Room, and Peer-to-Patent, for example) to elicit greater public participation. Agencies use these sites to solicit feedback on services or policy issues by encouraging people to comment, "like," tweet, or blog about various issues and questions. These interactions create real-time streams of unstructured data that, alone or combined with other sources, have the potential to reveal connections among political, personal, and professional preferences and attitudes.

As is the case with administrative data, social media data is very diverse and potentially very valuable, both on its own and when compared or combined with data from other internal or external sources. However, non-governmental data resources are unpredictable, casual, and diffuse in their application. Government may be the steward of some of these data sources, but it is not the primary creator. As a consequence, government cannot vouch for or assure the quality of such information compared to the precise, structured data products expected of the Census Bureau or Bureau of Labor Statistics. The data may not be well-defined and may be difficult to aggregate. The topics are diverse and emergent, and collection or production processes may not conform to certain standards. Moreover, rarely are these sources managed, maintained, or preserved according to formal rules for records retention, security, privacy, or confidentiality. These data offer new windows into topics of policy concern, but are more difficult to use and interpret, and, therefore, subject to misunderstanding and misuse (Dawes, 1996; Ballou & Giri Kumer, 1999).

Sources of Information Problems

Information problems stem from conventional beliefs and assumptions about information in general and from the provenance and practices that surround government information in particular.

Conventional Wisdom

A set of common beliefs and unstated assumptions are often substituted for critical consideration of information. In the positivist tradition of social sciences, quantitative data is automatically preferred or assumed to have better quality than qualitative data (Heinrich, 2012), although this view can be challenged as overly narrow and inappropriate for understanding complex social phenomena. Radin (2006) identifies several common beliefs about information that lead to unrealistic and even false expectations and results. These include assumptions that needed information is available and sufficient, objectively neutral, understandable, and relevant to the task of evaluation. Left unchallenged, these beliefs compromise all forms of program assessment and policy analysis.

Recent open government initiatives like Data.gov present similar problematic beliefs. They convey an unstated assumption that large, structured, machine-readable datasets are intrinsically better than processed data, and data in electronic form suitable for delivery on the Internet is superior to other formats. Thus the "low-hanging fruit" of available machine-readable raw datasets receives more attention than better-defined and potentially more suitable traditional datasets that reflect some interim processing or that cannot be easily posted onto the Web. Meijer (2009) illustrates the impact of these assumptions and argues that instead of providing more transparency, the consumption of this computerized, raw data can actually threaten public trust because data is removed from shared social experiences, takes an overly structured and predominately numerical form, or directs attention to narrow and sometimes irrelevant (but quantifiable) concerns.

Provenance

Much of the information becoming available for policy analysis and evaluation is emerging from activities and contexts that are far different in purpose, context, and time from the policy analysis and evaluation processes. Taken out of context, data loses meaning, relevance, and usability. Although the public may be offered thousands of datasets from one convenient Web address, these information resources derive from different government organizations, locations, systems, and custodians and thus are defined and collected in different ways. The datasets represent different time frames, geographic units, and other essential characteristics. Most come from existing information systems that were designed for specific operational purposes. Few were created with public use in mind.

One way to address this problem is through the creation and maintenance of high quality, detailed metadata—a description of data collection practices and rules, formats, data definitions, historical changes, gaps, overlaps, weaknesses, and other characteristics—that will make it possible for users to better understand the nature of the data and what it can, cannot, or should not be used for. Unfortunately, metadata receives little attention in most organizations. An administrative or operational dataset is usually defined at the point of creation in just enough detail to support those who operate the system or use the data directly. As the underlying dataset or system changes over time, corresponding maintenance of metadata is low-priority. The idea of fully describing data for the benefit of some unknown future user is rarely considered.

Even when metadata exists in reasonably complete form, it often fails to capture contextual knowledge that can have a powerful effect on its quality and usability. Consider, for example, how the intake process for a person entering a homeless services shelter can affect the data collected. That person is likely to be under stress, have no records or a hazy recollection of past welfare benefits, service programs, or previous jobs, and is unlikely to have a social security card or birth certificate to verify facts usually required for a case record. Consequently, the case file is replete with guesses, gaps, and errors that may not be rectified during their stay in the shelter, but will remain in the dataset.

Practices

Research shows that in order to understand data, one must understand the processes that produce the data (Dawes, Pardo, & Cresswell, 2004). Data collection, management, access, and dissemination practices all have strong effects on the extent to which datasets are valid, sufficient, or appropriate for policy analysis or any other use (Dawes & Pardo, 2006).

Data collection schemes may generate weekly, monthly, annual, or sporadic updates. Data definitions and content could change from one data collection cycle to the next. Some datasets may go through a routine quality assurance (QA) process, others do not. Some QA processes are rigorous, others superficial. Some datasets are created from scratch, others are byproducts of administrative processes, or may be composites of multiple data sources, each with their own data management practices.

Datasets may be readily accessible to internal and/or external users, require an access application or authorization process, or be disseminated with or without cost. Access may be limited to certain data subsets or for limited time periods. In addition, data formats are likely suitable for the organization that creates and manages the data and may not be flexible enough to suit other users with different capabilities and other interests.

Understanding Information in Context

One way to understand these challenges is to consider how data sources are intertwined with policies, management practices, and technologies, and how this ensemble of factors is embedded in social, organizational, and institutional contexts that influence data quality, availability, and usability (see Figure 2.1).

The challenges in using government information can be understood as policy problems that examine the balance and priority of internal government needs against the needs of secondary users, the resources allocated to serve both kinds of uses, and traditional concerns such as confidentiality, security, and authenticity. Policy considerations drive whether and how government agencies can collect personal information. Policies guide how often and in what situations data is collected, summarized, and published. Policies determine whether data is available to the public or protected.

Organizational and management activities influence the design of technology systems, security protocols, and access points, as well as the meaning and context of the programs and services in which the data is created. Data is affected by management choices about the rationales and internal processes of data collection, analysis, management, preservation, and access.

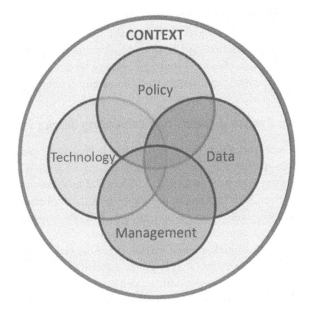

FIGURE 2.1 Conceptual framework

Technology also plays an important role in the way agencies handle information gathering, storage, access, inquiry, and display. Technology tools are generally selected to fulfill a business need of the government agency collecting the data. Technology often limits the ability to integrate data across sources, organizations, and jurisdictions. For example, in complex data integration projects, legacy systems often do not easily connect to new technologies. Consequently, more agile and functional new systems may run side-by-side with old systems rather than immediately replace them.

The institutional, organizational, and social contexts of information collection, management, and use all affect the value of government information resources for policy analysis. Policies, governance mechanisms, data management protocols, data and technology standards, and a variety of skills and capabilities both inside and outside government are needed if policy informatics is to contribute to a better understanding of critical social and economic issues and ultimately better policies. Policy informatics must see itself as more than big data and powerful forms of computation. As a field of study and application it must take into account information policies and management practices that produce data.

The following three cases show how examining information through the framework above brings into focus different policy, management, and technological influences that affect data quality and fitness for use in policy analysis. We also show how the contexts of data collection make a difference. Even when information sources are well-managed, they bear the limitations and characteristics of their respective institutional and organizational contexts. We explore some of these factors in the cases below regarding homeless services (CTG, 2000), land records (Cook et al., 2005), and financial accountability (Helbig et al., 2010).

Homelessness: Assessing Trends and Evaluating Service Programs

Each night in New York State, more than 40,000 homeless people, both families and single adults, receive emergency shelter and support services. They require assistance in dealing with their immediate incidence of homelessness as well as for a variety of other problems including domestic violence, alcoholism or substance abuse, poor parenting skills, mental illness, and a lack of education or employment skills.

Professionals in the homeless services field believe the various service programs they provide to homeless people reduce public assistance costs by helping people achieve independence. But there is little evidence to either support or challenge this belief. Program managers do have quarterly aggregated statistics for payment purposes. However, information about service effectiveness is mostly anecdotal. Detailed data about individuals, programs, and services resides in the shelters in various separate electronic and paper record systems. These organizations vary greatly in size, specialization, and scope of service. Each provider has its own naming conventions for specific data elements and each has individualized business rules that dictate how work is done and what types of data are collected. As a result, it is unclear whether self-sufficiency, reduced recidivism, reduced dependence on public assistance, and improved overall life skills are being systematically achieved. Other outcomes that were rated as highly desirable include increasing permanent housing placements, reducing lengths of stay, and increasing sobriety.

Clearly, there are many issues with the data and lack thereof for this programming. Consequently, in the late 1990s, the state human services agency and service providers considered the feasibility of a new information resource to combine and compare information on services to information about client outcomes. A fundamental issue was whether the state agency and the homeless shelter providers could agree on a service evaluation model that would satisfy their various assessment needs. They needed to agree on standard service definitions and evaluation measures and to define, for example, what kind of behavior or outcome constituted "success" for a deeply troubled individual compared to a relatively stable family. They first focused on simply identifying 66 distinct services and then began to work toward standard definitions and attributes of those services so they could be compared.

Some services were straightforward with few attributes that varied from one place or client type to another; others had many variables. An attribute such as service location could be applied to all services and was considered useful for comparing the outcomes of similar service programs offered in different locations. But temporary housing was identified in 26 different ways (types of beds, family or single units, or special population characteristics).

Data quality, context, definitions, and usability all emerged as problems. One common source of data errors was the stressful situation of the client at the point of admission to a shelter. Going to a shelter is frequently a last resort for a client. In some cases, clients may deliberately provide false information to protect their identity. More commonly, the stress associated with the situation can cause clients to forget or have no record of dates, social security numbers, or histories. Thus, the information provided to a case manager at intake can be fraught with gaps and errors. In many cases, data for a client remains incomplete.

Another challenge was finding commonality among the data elements used by the different organizations. The design team needed to understand how data was collected, what similarities existed among data sources, and how data would be aggregated in the new system in order to document all this with reliable metadata. Finally, each data code for the integrated system had to be reviewed in the context of its related programmatic issues. Some data was collected based on unique policies or business rules specific to a provider. For example, each system contained information regarding a client's ethnicity, which usually consisted of five categories, but one provider used 12 due to federal regulations and the funding requirements of its programs.

Over the course of several years, these and other information, policy, management, and quality issues were resolved and prototyped. However, due to a lack of funding, the integrated system was never built and the problems of understanding program effectiveness remain.

Real Property Tax Administration: Seeking Multiple Forms of Public Value from Land Records

Parcel information pertains to the smallest unit or lowest common denominator of land ownership. Parcel identification, description, and ownership information is collected as a function of real property laws and tax administration. Typically, deeds are recorded by county clerks to create an official record of land ownership and ownership transfer. Assessors employed by cities or towns conduct real property value appraisals of parcels in their jurisdictions to establish their value for tax purposes.

The collected information that originates with these real property recording and tax functions has great utility for many other practical and policy uses, including transportation planning, emergency response, economic development, and protection of green space. However, parcel data is subject to errors and inconsistencies in different contexts, thereby reducing its quality and usability for uses other than tax valuation. All users want "high quality" data, but this universal desire masks a great deal of variation. For example, accuracy is an important data quality characteristic. "Accurate" data for an engineering firm planning a residential subdivision means very detailed survey-grade information. In contrast, for a town attempting to designate a rough boundary for a new municipal park, "accurate" information may be no more than the names of the adjacent streets. Tax collection organizations require assessment data only from last year's final annual tax roll, but a realtor selling a home needs to know the current tax status of a property, current annual taxes, and who pays them.

From the very beginning of parcel data collection, inconsistent numbering and indexing systems, inconsistent terminology, and factual errors all affect the value of the data (i.e., its quality and usability). For example, different municipalities adopt their own numbering or identification system for tax parcels (e.g., a key term like "location" could be defined as "street address," "tax map number," or "geographic coordinates"). The quality of property sales information is frequently criticized by assessors who contend that buyers, sellers, and attorneys pay little attention to the accuracy of required documents. Assessors also report difficulty obtaining complete and accurate information because property owners can deny access to buildings with the result that an assessor's best judgment is used for an appraisal rather than direct and detailed observation of the property.

Parcel data users include state agencies, county and municipal governments, nonprofit organizations, and private sector companies. Users can spend considerable resources obtaining, improving, and standardizing parcel data before it is usable for their needs. Even when original data is high quality, it may not have sufficient detail or be readily comparable to data from other sources, or it may be derived from systems that are technically incompatible.

One of the most troublesome issues for data users is the inverted relationship between geographic coverage and timeliness, detail, and completeness of parcel data. At the original point of collection at the municipal level, parcel data is most current and contains the most detail regarding a variety of attributes. However, assessment data is reported only once a year and some municipalities do not report all of their parcel inventory and improvement data up the hierarchy to the county and state levels. At the county level, an additional kind of information is produced by turning the municipal reports into tax maps, but the county tax maps and assessment rolls are not as detailed as the information in the municipal assessor's office. When the state tax agency receives data files from the counties and shares them with the statewide GIS Clearinghouse, the files generally contain only 25 percent of the originally recorded attributes with the parcel centroid (the coordinates of the approximate geographic center of the parcel), but none of the mapping information.

Thus, a user seeking statewide information from a single source (the state) has access to only a small amount of the information originally collected, while a user whose purpose is limited to the small geographic area of a single town can access the greatest amount of information. As a consequence, any use that requires regional or statewide information will require a user to make numerous separate requests from different data suppliers. Usually requests go to counties where tax maps and associated attribute data offer relatively good coverage for most applications. However, this process is time-consuming, costly, and unpredictable because counties do not follow uniform procedures or policies for dealing with data requesters. Some of the costs are associated with fees charged by various localities or private data suppliers for the data itself, although there are no standard policies about fees and wide variations exist. Some providers require a formal data sharing agreement or subscription while others provide data on request. Some require formal Freedom of Information Act (FOIA) requests; others treat these requests as routine.

The absence of update and feedback mechanisms prevents data improvements and is a serious data quality issue. With the exception of the statewide GIS Data Sharing Cooperative, users are neither expected nor allowed to return data corrections, enhancements, or other improvements to the government data sources. When users obtain updates from their data sources, data that users may have improved for their own purposes can be replaced by old errors that still exist in the source files. The difficult choice is to forgo updates in order to maintain corrections, lose some of their corrections in order to obtain updated files, or engage in very costly and time consuming matching and integration activities. These problems of quality and consistency are shared by all secondary users, but are minor issues in the context of tax administration and costly for small municipalities to address, and therefore receive little or no attention.

American Recovery and Reinvestment Act: Tracking the Results of Stimulus Spending

The American Recovery and Reinvestment Act (ARRA) became law in 2009. Its purpose is to distribute $800 billion to stimulate the United States economy in the wake of the worldwide financial crisis. ARRA also institutionalized financial reporting requirements that promised strict and detailed accounting to the public of all funds spent, right down to the level of local program operators and contractors. Novel reporting requirements were implemented to collect relatively real-time data on all tiers of awards and post it on Recovery.gov, a consolidated national website. The detailed, frequent reporting produces an extraordinary amount of policy-relevant data about stimulus spending, recipients of funds, the flow of funds, and economic effects at the local, state, and national levels.

ARRA disburses funds through different mechanisms (e.g., grants, contracts, and reimbursements); not all funds are accounted for or reported on in the same way. Prime-recipients must report quarterly data on funds used for projects, whereas federal agencies are required to report data on compliance, communications, formula block grants, and other information in weekly updates. The reporting required by prime-recipients is detailed in Section 1512 of ARRA ("1512 reporting").

Typically, federal reporting is an annual affair where data is gathered and summarized all year long and reported after the fiscal year ends. Section 1512 stipulates that prime-recipients must report on more than 90 data elements every quarter. The data requirements fall into two main categories: expenditure data on contracts, grants, and loans, and job creation data. Contract, grant, and loan data includes agency name, agency code, sub-recipient name, financing used, address information, status, purpose, planned activities, date of completion, dollars spent, and the five most highly compensated employees for each project. Job creation data includes the number of newly created and saved jobs and the description of the jobs. All data elements are defined in a uniform data dictionary that applies to all programs and recipients nationwide.

We traced the implementation of ARRA reporting through the experiences of the New York State Department of Transportation (NYS DOT), which was the prime-recipient of $1.12 billion for highways and bridges. As a prime-recipient, NYS DOT must meet Section 1512 reporting requirements across the entire chain of spending. Approximately 50 percent of the $1.12 billion was disbursed to sub-recipients, primarily local governments and contractors. When ARRA was enacted, NYS DOT began using spreadsheet templates recommended and provided by FederalReporting.gov (the centralized back-end system that collected all reporting data) to report on its 470 federally funded projects.

Transportation projects are complex undertakings and require multiple parties to execute the work. NYS DOT is responsible for only part of the lifecycle of a project (primarily the contracting part), while most of the actual construction work is carried out by municipal and county transportation departments and private contractors. Over 400 *projects* were identified. The reports required for these projects demanded data be further broken down to report on the multiple sub-recipient awards, almost doubling the number of *reporting entities*. NYS DOT soon realized that providing one spreadsheet for every ARRA award was not feasible.

NYS DOT first called upon its own IT department to identify and aggregate data from 13 existing legacy systems and migrate it into a new data warehouse. To do this, they mapped critical reporting requirements to existing system components and business processes.

The possibility of enormous data quality issues loomed, if only for the volume of data being collected. NYS DOT had to map the reporting data standards to its own core data elements. Managers had to call on the deep institutional knowledge and experience of its staff throughout the organization to fully understand the different data elements and their semantic and contextual meaning. The data elements required for stimulus reporting did not match other uses or system architectures and the process of determining the correct source from which to feed the federal reporting needs was time consuming and challenging. NYS DOT also had to coordinate data gathering from more than 700 sub-recipients who were unfamiliar with federal reporting processes or did not have access to enter data directly into the agency's legacy systems.

To ensure the submission of usable data, the United States Office of Management and Budget (OMB, 2009) instituted a series of ARRA feedback mechanisms at the federal and prime-recipient level, including rolling reporting periods to correct errors and provide an opportunity for data quality improvement. Technical mechanisms to check for data quality were employed at both federal and state levels, as were mechanisms for responding to calls from citizens. When issues of accuracy or discrepancies were identified by OMB in NYS DOT's reports, someone from NYS DOT needed to first diagnose the issue, determine the source of the discrepancy, and whether it was internal or external to the agency. If it was external, they had to reach out to sub-recipients to resolve the problem.

Data standards for performance metrics presented another challenge. Recipients often expressed frustration about the definition of impact metrics (such as jobs created or saved or the number of shovel-ready projects). Various interpretations of the OMB standards were re-released after several rounds of reporting. While OMB was the primary issuer of this guidance, other federal agencies issued additional guidance. Collecting data on the number of jobs created or saved was especially difficult. There were inconsistent definitions of *saved jobs* and guidance changes meant review of hundreds of pages to determine what had changed and how the changes affected NYS DOT reports. Moreover, educating sub-recipients about different data elements and metrics revealed confusion about data quality and caused delays in reporting.

The federal effort to create Recovery.gov was important for several reasons, including economies of scale, coordination, and control. However, many of the entities with reporting responsibility also wanted to provide direct access to the data in support of their own transparency agendas. As a consequence, data about ARRA projects can be found across the Web. The impact of these multiple "authoritative" sources is still unknown, but the potential for confusion is gaining attention from government watchers, and often makes its way into the news.

NYS DOT wanted to create one data source at the state level and opted for a technical solution that would synchronize its data warehouse with the reporting data on its website. This allowed it to point constituents to the public website with confidence that the information would be as current as their internal systems. However, this process was complicated by the need to synchronize certain data elements with the Office of the State Comptroller, which must approve all state contracts and has its own authoritative source for contract information, Open-Book New York. The two state agencies must ensure that the two systems are synchronized at all times so that the same information on funding and contracts is displayed on multiple New York State websites.

Other state agencies invented their own ways to meet ARRA reporting requirements. The national data source, Recovery.gov, implies a straightforward collection of specified data. In fact, it masks indescribable diversity and complexity in data gathering and processing throughout the nation. ARRA data thus presents many pitfalls for analysts and policy makers trying to understand the effects of stimulus spending on jobs and economic recovery, the advisability of economic stimulus as policy choice, and the various effects of ARRA spending on diverse industries or regions.

Ensuring Information Quality and Fitness for Use

Given the above realities, even when government information resources are well defined and managed, substantial problems for use in the entire policy making lifecycle cannot be avoided. As the cases illustrate, significant challenges remain before most government data is usable for policy analysis. The parcel data case reveals a comprehensive view of a single data resource and all of its primary and secondary providers, uses, and users. The homelessness case illustrates the extensive diversity in settings, capabilities, definitions, and understanding that can hide behind seemingly standardized state-level data. It shows that even when the path to good quality data is known, we may not get there. The ARRA reporting case extends from the local to the national level and documents how dynamic information needs demand ongoing retrofitting of data sources and data management systems. If we think of the situations in these cases as representative of the thousands of datasets created in similar ways, the magnitude of the challenge grows, but its general characteristics become more apparent and therefore more amenable to understanding and improvement.

The quality of government information resources is a fundamental consideration and is clearly a factor affecting use or misuse of information. The term *quality* generally means accuracy, but research studies identify multiple aspects of information quality that go well beyond simple data accuracy. A data quality problem can be defined as any difficulty encountered along one or more quality dimensions that renders data completely or largely unfit for use (Strong, Lee, & Wang, 1997).

Most practical research on data quality originates in information resource management (IRM) and management information systems (MIS) and takes an organizational or managerial perspective (Redman, 1998). Traditionally, the focus in IRM is to assure or improve quality of information systems and information management practices (Horton, 1979; Caudle, 1990). In most MIS research, information is considered in the context of private sector firms and strategic goals. However, Wang and Strong (1996) offer a useful framework for assessing data quality that encompasses both organizational and user perspectives. They adopt the concept of *fitness for use*, assessing fitness from the point of view of data users, while acknowledging how data use has implications for the way organizations manage their data resources.

The model comprises four categories of data quality, each with specific attributes: intrinsic quality, contextual quality, representational quality, and accessibility. Together they denote the importance of data quality in its own right, but they also address quality within the context of use and the quality of data systems and management practices.

- *Intrinsic quality* most closely matches traditional notions of information qual- ity. It includes accuracy and objectivity, but also involves believability and the reputation of the data source.
- *Contextual quality* refers to the context of the task for which data will be used. It considers timeliness, relevancy, completeness, sufficiency, and value-added to the user. Often there are trade-offs among these characteristics, for exam- ple, between timeliness and completeness (Ballou & Pazer, 1995).
- *Representational quality* relates to meaning and format. It requires that data be not only concise and consistent in format, but also interpretable and easy to understand.
- *Accessibility* comprises ease, means, and security of access.

These categories comprise both subjective perceptions and objective assess- ments, which all bear on the extent users are willing and able to use information. Table 2.1 (Pipino, Lee, & Wang, 2002, p. 212) defines the individual dimensions of data quality that comprise these categories.

Intrinsic information quality involves not only factual accuracy, but also believ- ability, objectivity, and the reputation of the sources. But even information with high intrinsic quality must be appropriate in the context of use and therefore be relevant, timely, concise, and complete enough for the work at hand. Information must be task-appropriate in terms of interpretability, accessibility, and security (Wang & Strong, 1996). Users may need to make choices or trade-offs among these characteristics (Ballou & Pazer, 1995), but they need good data descriptions to help them decide (e.g., they may choose to use information that is timelier but less complete for some purposes and vice versa for others).

An additional set of issues is associated with summarized information that is intended to stand for a collection of other measures. For example, *report cards* and *benchmarks* reduce complex phenomena to simple numbers or letter grades that necessarily ignore scale, scope, and context, and can mask other data qual- ity problems (Bannister, 2007; Meijer, 2009). In order for users to comprehen- sively assess data quality, they must be able to ascertain the nature of the data. Because data producers cannot anticipate all users and uses, the existence of good quality metadata is as important as the quality of the data itself (Dawes et al., 2004). All of these points must be considered when making a judgment about the extent to which information is fit for use (Wang & Strong, 1996; Pipino et al., 2002).

Legislation guiding the development of government information resources for public consumption makes explicit the need to consider external data users. The Data Quality Act of 2001 (Section 515 of Public Law 106–554) and subsequent OMB rules (2002) established a set of principles that agencies are required to implement in their own mission-appropriate ways. The Data Quality Act and OMB rules do not mandate specific standards or practices, but "provide pol- icy and procedural guidance to Federal agencies for ensuring and maximizing the quality, objectivity, utility, and integrity of information (including statistical

TABLE 2.1 Data quality dimensions

Dimensions	Definitions
Accessibility	Extent to which data is available, or easily and quickly retrievable
Appropriate Amount of Data	Extent to which the volume of data is appropriate for the task at hand
Believability	Extent to which data is regarded as true and credible
Completeness	Extent to which data is not missing and is of sufficient breadth and depth for the task at hand
Concise Representation	Extent to which data is compactly represented
Consistent Representation	Extent to which data is presented in the same format
Ease of Manipulation	Extent to which data is easy to manipulate and apply to different tasks
Free-of-Error	Extent to which data is correct and reliable
Interpretability	Extent to which data is in appropriate languages, symbols, and units, and the definitions are clear
Objectivity	Extent to which data is unbiased, unprejudiced, and impartial
Relevancy	Extent to which data is applicable and helpful for the task at hand
Reputation	Extent to which data is highly regarded in terms of its source or content
Security	Extent to which access to data is restricted appropriately to maintain its security
Timeliness	Extent to which the data is sufficiently up-to-date for the task at hand
Understandability	Extent to which data is easily comprehended
Value-Added	Extent to which data is beneficial and provides advantages from its use

Source: Pipino, Lee, and Wang (2002, p. 212). Reprinted with permission.

information) disseminated by Federal agencies." OMB (2002) defined the key information quality principles as follows:

- *Objectivity* addresses whether disseminated information is presented in an accurate, clear, complete, and unbiased manner. This includes whether information is presented within a proper context and, in the case of scientific, financial, or statistical information, developed using sound statistical and research methods.
- *Utility* refers to the usefulness of information to its intended users, including the public. In assessing the usefulness of information, an agency must consider uses from not only the perspective of the agency, but also that of the public.

- *Integrity* refers to the security of information—protection of the information from unauthorized access or revision to ensure that the information is not compromised through corruption or falsification.

The guidelines address the need to maintain underlying documentation that provides data transparency and develop administrative processes for users seeking corrections to data. Agencies are instructed to implement these principles in ways that support their particular missions and operations and to periodically report progress to OMB and the public. While these guidelines are flexible and non-prescriptive, they do represent a formal acknowledgment and first step toward quality improvements that are relevant to external data users. However, these requirements must be met using existing funding and no formal assessment has yet been conducted to determine how or how well they work.

Discussion and Conclusion

The current emphasis on opening access to more government information and the evolving capability of technological analysis tools offer many opportunities to apply the tools of policy informatics to big data and complex public problems. However, as shown, the path to well-grounded, circumspect policy decisions necessarily includes the need to confront and account for information problems and stakeholder views, technology choices, and other aspects of problem formulation and analysis. Our conceptual framework is a holistic socio-technical view of government information. By emphasizing how policy, management, data, and technology interact in a social context, it can help in two ways. It can guide information users to ask critical questions about the data resources they use and adjust their analyses and expectations accordingly. It can also guide information producers and stewards in government to adopt information policies and practices that make the strengths and weaknesses of information resources more apparent and help identify ways to improve data quality for both governmental and external use.

For most government agencies, providing information for public use is an extra responsibility that competes for resources with the demands of mission-focused operations. As our cases illustrated, vast amounts of useful information are contained in government data systems, but the systems themselves are seldom designed for use beyond the primary agency's own needs. With few exceptions, making data holdings available to others in a meaningful and usable way will demand thoughtful investments in skills, tools, and policies and changes in processes and practices. New roles may be required to facilitate the coordination of agency-level and government-wide programs of information dissemination and user-support services. Two practice improvements that are especially critical pertain to creating and maintaining good quality metadata and building and using formal feedback mechanisms that better connect data users to data sources. The detail and accuracy of metadata must be commensurate with the value of the data for multiple uses and the likelihood that some of those uses cannot be predicted in advance. Providing a formal mechanism for data users to report errors and enhancements to data sources will improve overall quality and integrity of the data and benefit all future users, including the government itself.

Likewise, expert policy analysts must take responsibility for "looking under the hood" of data sources and adjust their expectations and assumptions to more closely match the realities of data quality and fitness for use. Government data must be approached with the same skepticism and evaluative skills that social scientists apply to other aspects of the research process. The strengths, weaknesses, limitations, and peculiarities of these new datasets must be considered in the design and execution of policy analyses so that data value is maximized, but not overstated.

Finally, a rich research agenda is also apparent. Some examples:

- While we know good quality metadata enhances usefulness and validity of data for policy analysis, we do not know how to translate "good quality" into practice across data sources that vary dramatically in content, volume, specificity, etc. What should be the essential core elements of metadata for all government data sources? Under what circumstances should additions or elaborations from the core be included? And what should those enhancements contain?
- If a feedback mechanism is adopted for government datasets, what should it look like? How much would it cost to implement? How can the benefits be calculated, both for government and external stakeholders? What effects would feedback have on the programs or processes that produce the data?
- Data used for policy analysis will inevitably be flawed. How can flaws be recognized, categorized, and evaluated? How can this evaluation be meaningfully reflected in the presentation and consideration of analytical results?

The information dimension of policy informatics presents significant challenges for information providers, analysts, and consumers, but it also offers an under-appreciated opportunity to explore and understand both the context and possible effects of policy choices. The many new sources of government data offer potential value for society by contributing to better policies, but their value will be realized only if government information policies and practices are better aligned with the needs of policy-oriented data users. In turn, analysts must treat information critically in the policy analysis process so that we are more likely to gain the benefits of new approaches without naively committing design and interpretation errors that threaten both validity and usefulness.

References

Ballou, D.P., & Giri Kumer, T. (1999). Enhancing data quality in data warehouse environments. *Communications of the ACM, 42*(1), 73–78.

Ballou, D. P., & Pazer, H. L. (1995). Designing information systems to optimize the accuracy-timeliness tradeoff. *Information Systems Research, 6*(1), 51–72.

Bannister, F. (2007). The curse of the benchmark: An assessment of the validity and value of e-government comparisons. *International Review of Administrative Sciences, 73*(2), 171–188.

Bellamy, C., & Taylor, J. A. (1998). *Governing in the information age.* Philadelphia: Open University Press.

Bekkers, V.J.J.M. (1998). New forms of steering and the ambivalency of transparency. In I.T.M. Snellen & W.B.H.J. Van der Donk (Eds.). *Public administration in an information age: A handbook* (pp. 341–357). Amsterdam: IOS Press.

Caudle, S. L. (1990). Managing information resources in state government. *Public Administration Review,* September-October, 515–524.

Center for Technology in Government (CTG). (2000). Building trust before building a system: The making of the homeless information management system. Retrieved from http://www.ctg.albany.edu/static/usinginfo/Cases/bss_case.htm

Cook, M. E., Dawes, S. S., Helbig, N. C., & Lishnoff, R. J. (2005). Use of parcel data in New York State: A reconnaissance study. Retrieved from http://www.ctg.albany.edu/publications/reports/use_of_parcel_data/use_of_parcel_data.pdf

Data Quality Act. (2001). *Public Law No. 106–554 § 515.*

Data.gov Website. (2012). Retrieved from http://www.data.gov

Dawes, S. S. (1996). Interagency information sharing: Expected benefits, manageable risks. *Journal of Policy Analysis and Management, 15*(3), 377–394.

Dawes, S. S., & Helbig, N. (2010). Information strategies for open government: Challenges and prospects for deriving public value from government transparency. In *Proceedings from 2010 IFIP: eGovernment Conference.* Springer: Lecture Notes in Computer Science.

Dawes, S.S., & Pardo, T. A. (2006). Maximizing knowledge for program evaluation: Critical issues and practical challenges of ICT strategies. Paper presented at the *5th International Conference, EGOV.* Krakow, Poland.

Dawes, S.S., Pardo, T.A., & Cresswell, A.M. (2004). Designing electronic government information access programs: A holistic approach. *Government Information Quarterly, 21*(1), 3–23.

Frissen, P.H.A. (1992). Informatization in public administration: Introduction. *International Review of Administrative Sciences, 58,* 307–310.

George, R.M., & Lee, B.J. (2002). Managing and cleaning administrative data. In M. ver Ploeg, R. A. Moffitt, & C.F. Citro (Eds.), *Studies of welfare populations: Data collection and research issues.* Committee on National Statistics, Division of Behavioral and Social Sciences and Education, National Research Council. Retrieved from http://aspe.hhs.gov/hsp/welf-res-data-issues02/index.htm

Heinrich, C.J. (2012). How credible is the evidence, and does it matter? An analysis of the Program Assessment Rating Tool. *Public Administration Review, 72*(1), 123–134.

Helbig, N., Stryin, E., Canestraro, D., & Pardo T. A. (2010). Information and transparency: Learning from recovery act reporting experiences. In *Proceedings of the Eleventh Annual International Conference on Digital Government Research.* May 2010, pp. 1–10.

Horton, F. W. (1979). *Information resources management: Concept and cases.* Cleveland, OH: Association for Systems Management.

Johnston, E., & Kim, Y. (2011). Introduction to the special issue on policy informatics. *The Innovation Journal: The Public Sector Innovation Journal, 16*(1), 1–4.

Meijer, A. (2009). Understanding modern transparency. *International Review of Administrative Sciences, 75*(2), 255–269.

Office of Management and Budget (OMB). (2002, January 3). *Guidelines for ensuring and maximizing the quality, objectivity, utility, and integrity of information disseminated by federal agencies.* Retrieved from http://www.whitehouse.gov/omb/fedreg_reproducible

Office of Management and Budget (OMB). (2009, April 3). *Updated implementing guidance for the American Recovery and Reinvestment Act of 2009.* Washington, DC: Executive Office of the President of the United States. Retrieved from http://www.whitehouse.gov/sites/default/files/omb/assets/memoranda_fy2009/m09-15.pdf

Pipino, L.L., Lee, Y. Y., & Wang, R.Y. (2002). Data quality assessment. *Communications of the ACM, 45*(4), 211–218.

Radin, B. A. (2006). *Challenging the performance movement: Accountability complexity and democratic value.* Washington, DC: Georgetown University Press.

Recovery.gov Website. (n.d.). Retrieved from http://www.recovery.gov/

Redman, T.C. (1998). The impact of poor data quality on the typical enterprise. *Communication of the ACM, 41*(2), 79–82.

Snellen, I. (2005). E-government: A challenge for public management. In E. Ferlie, L.E. Lynn Jr., & C. Pollitt (Eds.), *The Oxford handbook of public management* (pp. 389–421). Oxford: Oxford University Press.

Strong, D. M., Lee, Y. W., & Wang, R.Y. (1997). Data quality in context. *Communications of the ACM, 40*(5), 103–110.

Wang, R.Y., & Strong, D.M. (1996). Beyond accuracy: What data quality means to data consumers. *Journal of Management Information Systems, 12*(4), 5–34.

3

EVIDENCE FOR POLICY INQUIRY

Anand Desai and Kristin Harlow

Introduction

This chapter focuses on the use of evidence in policy inquiry. There are two beginning premises here. First, the purpose of policy inquiry is to inform decisions and recommend actions for practitioners, decision-makers, or fellow researchers. Second, unlike usual scientific research where problems are well defined, the focus of policy inquiry is on problems that are unstructured, intractable, and messy. That is, policy problems are difficult to define and do not lend themselves to neat solutions.

It seems self-evident that good policy decisions are informed by the best available knowledge and sound and authoritative evidence. However, research indicates that policy decisions are infrequently based on the best available evidence (Bogenschneider & Corbett, 2010; Owens, Petts, & Bulkeley, 2006; Sutherland et al., 2012). The dearth of evidence-based decision-making results in part because empirical policy research does not provide the broad range of information that practitioners require to make policy decisions and defend their actions (Bridges & Watts, 2009).

Further, there are unresolved theoretical and practical issues underlying the notion of evidence and its use (Achinstein, 2010; Majone, 1989). Although research provides evidence of a correlation between policy actions and outcomes, claims of causation in the policy realm are at best limited and at worst incoherent and inapplicable (Kincaid, 2009). To bridge the gap between evidence and action, policy researchers should provide argument at various stages of the policy inquiry process to buttress the evidence for decision-making. Argumentation can effectively identify assumptions and their alignment with the data and methods that constitute evidence in support of research claims (Kezar, 2004).

In years past, public officials were trusted to use their best judgment in making policy decisions. However, in more recent years, we as a society expect decision-makers to support their choices with *objective* evidence. In this chapter, we argue that because policy problems are complex and the definition of actionable evidence is ambiguous, the link between evidence and policy action is often unclear. Although the essential research practices of policy inquiry are similar to those in the natural, physical, or social sciences, the products of policy inquiry by themselves are not sufficient to determine the course of policy action. If the purpose of policy inquiry is indeed different from traditional forms of inquiry, and our understanding of the nature of the resulting evidence is unclear, then we need to clarify the role of evidence. We shall argue that evidence, as a product of policy inquiry, is not sufficient for informing policy decisions and action. Instead, evidence can inform and influence decisions and actions only when accompanied by an argument that interprets the evidence for the given context and is given import by arguments that contend the evidence is legitimate support for decisions and actions (Majone, 1989).

Along the way, we shall argue that although evidence-based medicine is an effective tool for health care providers, an uncritical transfer of that concept to our context might be ill-advised. In effect, our understanding of "what works" and evidence-based practice as sources of practical advice in the policy context is problematic. Policy recommendations based on best practices that were developed in different communities or contexts are, at best, premature. Evidence-based policy making can potentially lead to a sense of complacency or hubris regarding our understanding of the problems we attempt to address and the solutions we seek.

Policy Decisions

The main purpose of policy inquiry is to affect and foster change. Policy problems are identified when an individual or group decides that the current state of affairs is unsatisfactory and a path must be found to reach a preferred state of affairs. Thus, this unsatisfactory state of affairs is the policy issue, concern, or problem that a decision-maker seeks to address. In order to most effectively address the policy problem, the decision-maker enlists the assistance of a researcher to provide information so that an informed decision about how to proceed can be made.

It is the role of the policy researcher to give form to the policy problem so that it lends itself to systematic inquiry. The researcher then conducts that inquiry to identify the path toward achieving a more desired state of affairs and to propose criteria for determining whether and when the original problem has been satisfactorily addressed. This description is a simplification of the researcher's task, but it illustrates that policy inquiry is research with practical ends and that the goal of policy inquiry is implementable solutions.

Policy inquiry, although often framed as objective research, is in fact a value-laden process (Le Grand, 1990; Okun, 1975). When a decision-maker declares a state of affairs to be unsatisfactory, the decision-maker has made a value judgment. Subsequently, the decision-maker effects additional value judgments when establishing what a preferred state of affairs might be and when determining *the* best path to achieving that state. Further, there is an assumption that policy inquiry is a value-free objective process informing the decision-maker. However, deciding how to structure the policy problem into a researchable question requires the policy researcher to make value-based decisions. These decisions are often untested or unquestioned because the research process has the aura of objectivity.

Policy inquiry is conducted within a neutral framework for combining means, ends, and goals, as if the underlying values are uncontested or stipulated from outside the inquiry process. Acceptable results for use in the decision-making process will reflect the prevailing values of the decision-makers and stakeholders. The task of values clarification is difficult enough for an individual. However, most policy decisions and actions are implemented in contexts that require collective action, which is rife with the potential for value conflicts.

Policy makers must contend with significant uncertainty when making policy decisions. There is ambiguity inherent in policy contexts due to the lack of a clear link between policy actions and their consequences. In addition, there exist multiple sources of uncertainty about the future: stochastic uncertainty due to chance and scientific uncertainty due to lack of knowledge. These variations of uncertainty result in imperfect knowledge about the future consequences of our actions. During the last century, a number of endeavors to manage nature—aggressively fighting forest fires in the American northwest, changing the flow of rivers in Egypt and China, draining swamplands in the southwestern United States, New Zealand, and Europe, and deforestation in Africa and South America—have provided benefits while also resulting in unforeseen consequences that have led many to question the relative value of those benefits (Balint et al., 2011; Nixon, 2004).

Policy inquiry, as a decision support tool, helps decision-makers—be they researchers or practitioners—choose from a set of potential actions not only when there is uncertainty, but also when there is value conflict (Raiffa, 1968). A decision is generally made where a decision-maker chooses an action in a given context to produce an outcome that maximizes utility or return. For the decision-maker, utility is a complex construct that must embody the values of all who have a stake in the decision. Complications arise when: (1) the relationship between actions and outcomes is ambiguous, (2) there is uncertainty about stakeholder values, thus uncertainty about return or utility for the decision-maker or stakeholders, or (3) there is uncertainty regarding outcomes. In other words, one or more of the potential actions, contexts, outcomes, or the utility of the outcomes is unknown or not known with a high degree of confidence (Mason & Mitroff, 1973). Researchers can assist with these complications that are due to value incongruence and uncertainty, but cannot solve them. For policy inquiry to support such a decision, the policy researcher must produce evidence that would help resolve any potential complications and identify the actions deemed best or more likely, deemed satisfactory (Simon, 1997).

Consider the decision to order an evacuation to prevent the loss of life in the event of a forest fire. Compared to most nontrivial policy decisions, how to react to a forest fire is a fairly well-defined problem. In areas prone to forest fires, the set of options available to the local emergency management officials is similarly well defined. In most well-organized jurisdictions, not only are the options known, but the managers also have experience from previous fires that has perhaps been supplemented with simulation exercises and practice drills. The science of forest fires is well understood and the set of potential states of nature or contexts is usually known in advance.

However, establishing in advance which of these possible states will materialize by predicting the path, speed, spread, and intensity of fires is not easy or reliable due to uncertainty in future events. Despite the depth of knowledge regarding forest fires generally, it is extremely difficult to predict which of these states of nature will materialize and further, when it will materialize. Even when a decision is made regarding how to react to a fire, no one knows at the time of the decision what the outcomes will be, not only because there is uncertainty about the severity of the fires, but because there is ambiguity about how people will react to the decision to evacuate. Over time, conditions change and so the fire fighters must adapt to the changing conditions. Public managers' decisions regarding evacuation must take into consideration possible behaviors of residents in the path of the fires. Recent history and experience, personal resources, the reaction of one's neighbors, and a host of other unpredictable factors will influence a resident's behavior. Consequently, outcomes will depend not only on which state of nature materializes, but also on individual reactions to an evacuation order. The extent of uncertainty makes it extremely difficult to determine the set of possible utilities that might result and almost impossible to assign an unambiguous utility to a policy decision.

It is not whether officials have made a good or bad decision. The point is that it is extremely difficult to make a decision even when a problem has fairly well-defined parameters because policy decisions are rife with uncertainty. The complexity of decisions is further compounded by the fact that the notion of evidence is itself nebulous.

Evidenced-Based Practice

Evidence-based practice is the most recent manifestation of empirical investigation, as it introduced the term evidence-based medicine (EBM) into the literature. EBM is use of scientific evidence, in contrast to personal opinion or experience, to inform medical practice and teaching decisions. It has taken hold not only in medicine, but in a variety of other contexts, including public policy (Heinrich, 2007; Meier & O'Toole, 2009; Oliver et al., 2005; Pawson, 2002).

Users of EBM have developed a hierarchy by which to judge the quality of research that yields the evidence upon which medical decisions are made (Guyatt et al., 1995). Doctors are encouraged to assess the quality of the research and the resulting evidence according to this hierarchy. At the top of the methodologically based hierarchy are randomized clinical trials and systematic reviews or meta-analyses of such trials, while expert opinion is at the bottom. In between are trials without proper randomization, natural experiments, quasi-experimental designs, time-series analyses, with or without intervention, and regression discontinuity designs that rely upon methodological rigor without a randomized experiment. Of lower status than quantitatively rigorous studies, but more authoritative than expert opinions, are well-conducted case studies. Scholars and practitioners in the social and policy sciences have applied this hierarchy in evidence-based practice and evidence-based policy (Boruch & Rui, 2008; Leigh, 2009).

EBM, however, is not without its critics (Clinicians for the Restoration of Autonomous Practice Writing Group, 2002). Obvious criticisms of EBM include practical considerations such as cost, timeliness, feasibility, and distribution of resources, as well as intangible considerations such as tradition and experience. Reliance on randomized clinical trials as the gold standard, and research-based decision-making generally, might be feasible in medicine, but it is not practical in most other contexts, especially in policy and management. Even specialized areas of medicine may not be well served by EBM. For instance, in mental health care, relationship-focused therapy, such as psychotherapy, does not lend itself to randomized clinical trials in the same way as technique-focused therapy, such as cognitive behavioral therapy (Tanenbaum, 2005). As a result, when using EBM as guidance for clinical decision-making, process-based therapies are necessarily marginalized. Tonelli (2006) has suggested broadening the notion of EBM to include evidence not restricted to research findings and clinical trials.

Proponents of good government have argued that common sense, good judgment, good intentions, intuition, and experience are not sufficient for defending policy claims that promote one policy option over another. Evaluation researchers have celebrated the introduction of evidence-based practice and evidence-based policy (Pawson, 2002). However, identifying best practices is more straightforward in medicine than in policy contexts. The literature includes substantial evidence on "what works" from various state level "experiments" in welfare reform (Blank, 2008), the provision of Medicaid services, programs to reduce recidivism in the criminal justice system, education (Ladd, 2012), and a variety of other public programs (Bogenschneider & Corbett, 2010). Unfortunately, success in a given context or identification of a best practice under a specific formulation of a problem does not guarantee success in a different context or with a slightly different problem formulation.

In policy contexts, no two situations are alike, particularly across time and space. The complexity of policy issues is such that the issues lend themselves to multiple problem formulations, where each formulation determines what might be considered relevant evidence. Unlike medicine, where there is considerable uniformity in the nature of the problems and approaches for addressing them, policy contexts tend to be unique. Even in instances where there is debate regarding the efficacy of one medical procedure over another, well-established protocols exist to help resolve the conflict. Unfortunately, empirical support is not effectively transferable from one policy context to another because evidence is so often context-dependent for policy inquiry. Each policy problem is unique and the nature of the need and available supply of resources is different in every community. Context matters, meaning general principles or laws, even if they are determined through rigorous research, do not apply in every situation. As a result, evidence requires a thoughtful argument to describe how the evidence will be applicable and effective in a specific context.

Beneath the "genteel veneer" (Lindblom, 1980, p. 17) of policy analysis is a decision-maker with certain perspectives who sets agendas (Kingdon, 1995) and frames problems (Haas, 2004), thereby determining the types of analyses and solutions that are considered acceptable. Further, the evidence derived from policy inquiry is imbued with assumptions that are value-driven rather than evidence-driven. Decisions about what questions to ask and which observations to take into account when answering a policy question allow for variation in outcomes not attributable to objective data. Although there is some measure of disagreement in health care about the definition of health, the discussion is narrower than the disagreements among decision-makers and policy researchers regarding the definition of the health of society. As a result, evidence gleaned through policy inquiry will necessarily be less focused and less easily transferable than evidence derived from medical research.

So, What Next?

Although it may not be possible to *a priori* claim something as evidence to conclusively justify a belief or to seamlessly transfer what has been found to work in one context to a different context, all is not lost. We would instead argue that it is possible for policy inquiry to produce evidence to inform and influence, *not determine*, policy decisions and actions. Effective policy decision-making is based on evidence embedded in an argument that connects the evidence to decision and action claims necessary to address the policy issue.

The first step is to recast the role of the policy researcher. The policy researcher is not a mere technician, producing objective knowledge in a value-free environment. According to Majone (1989), policy inquiry is conducted by someone who is a:

> producer of policy arguments, more similar to a lawyer—a specialist in legal arguments—than to an engineer or a scientist. His basic skills are not algorithmical but argumentative: the ability to probe assumptions critically, to produce and evaluate evidence, to keep many threads in hand, to draw for an argument from many disparate sources, to communicate effectively.

He recognizes that to say anything of importance in public policy requires value judgments, which must be explained and justified and is willing to apply his skills to any topic relevant to public discussion.

(21–22)

Policy arguments are how policy debates are conducted (Dunn, 1981). Arguments and argumentation are based on an informal logic of reasoning, which provides a framework for the critical evaluation of individual worldviews as well as claims made by those making the arguments (Toulmin, 1958).

Toulmin's (1958) theory of *argumentation* is an alternative to formal syllogistic logic. A syllogism is simply a logical set of premises that lead to a logical conclusion, but it never asks *how* or *why* and, therefore, it does not advance knowledge. In Toulmin's view, people do not communicate in syllogisms. He developed the theory of argumentation in order to explain how argument occurs in the natural process of everyday life and as a method for gaining knowledge.

Policy arguments advance reasoned claims using the following six elements (adapted from Dunn, 1981, pp. 41–42):

1. ***Policy Relevant Information***: Policy-relevant information serves as the basis for policy arguments. It comes in many forms. For instance, outcomes of forest management policy analysis might be expressed in the form of a statistical generalization: *Results of firefighting experiments show that creating gaps in the forest is more effective than attempting to fight fires without such breaks*; as the conclusion of experts: *The panel of experts reports aggressive firefighting resulted in fires that burned with greater intensity and caused more damage than if small fires had been allowed to burn through*; or an expressed value or need: *Logging, if properly done, contributes to both the economic vitality of a community and controlling forest fires.* Policy-relevant information, when used in an argument, becomes the basis of the evidentiary claims in the context of the policy issues.

2. ***Claim***: A claim is the final product of an argument. Various types of claims such as claims about values or about which actions to take or policies to pursue can result from arguments. Claims are typically subject to disagreement or conflict among different segments of the community. For example, the advocative claim that the Forest Service should invest in protecting endangered species may not be a value shared by all stakeholders. The argument provides a logical link, usually in the form of a "therefore," between relevant information and the ensuing claim. So, if we value biological diversity or information demonstrating that the spotted owl is in danger of extinction, then it follows that (therefore) the government must protect the owl in some fashion. Hence, policy claims are the logical consequence of policy-relevant information.

3. *Warrant*: A warrant is a conceptual link between policy-relevant evidence and a policy claim. It is a reason for accepting the claim. A warrant may be an assumption regarding causes, values, or simply pragmatism. For example, a warrant for the claim that the Forest Service must fight forest fires might be expressed as simply as "life and property must be protected."
4. *Backing*: If the warrant is not accepted at face value, then backing provides support through additional assumptions or arguments. The backing for warrants could be in the form of information, principles, expert knowledge, or appeals to ethical or moral consideration.
5. *Rebuttal*: A rebuttal is a second conclusion, assumption, or argument that states the conditions under which an original claim is unacceptable or under which it may be accepted only with qualifications. Taken together, policy claims and rebuttals form the substance of policy disagreements among different segments of the community about alternative courses of government action. The consideration of rebuttals helps the researcher anticipate objections and serves as a systematic means for criticizing one's own claims, assumptions, and arguments.
6. *Qualifier*: Often expressed in probabilistic terms, a qualifier expresses the degree to which the researcher is certain about a policy claim. Qualifiers are only necessary when there is uncertainty about a claim, in which case the level of certainty is expressed as a probability, such as "95 percent level of confidence."

Effective policy decision-making uses evidence embedded in an argument that connects the evidence to the decision and action claims necessary to address the policy issue. In order to do so, both a conceptual and an empirical case must be made. Hence, the case we propose to make rests on the following claim: *Evidence-based arguments require both a conceptual model and a data model.*

This claim is neither novel nor new. All empirical research rests upon some conceptual understanding of the issues and data on which to base the claims. What is perhaps different about this claim is that it suggests the need for the careful development of arguments supported by evidence that *link* various aspects of the conceptual and data models.

Conceptual and Data Models

These conceptual and data models include a series of arguments and supporting information. Because we just argued that the notion of evidence is not well defined, we follow Weiss (1977), for whom evidence serves as an enlightenment function, not as definitive proof of the veracity of a claim.

Figures 3.1 and 3.2 illustrate that both unstructured problems and data exist independent of human interpretation or decision-making. Societal conditions underlying the unstructured problem have an independent existence in reality and the researcher or decision-maker determines that a policy issue or concern is a "problem." A forest fire, health care, or the economic welfare of the elderly is not a public policy concern until someone decides to make it so. Similarly, data has an independent existence and it is up to the researcher to identify which data can help provide the evidence that will help address the problem. Hence, both data and unstructured problems are found, but evidence is constructed. The purpose of policy inquiry is: (1) to construct evidence that will help address the unstructured problem and (2) to find data that can be converted into the necessary evidence.

Conceptual Model

When a researcher constructs evidence using the conceptual model (Figure 3.1), the purpose is to determine whether the unstructured problem has been addressed. The conceptual model illustrates the process of developing such evidence by constructing a series of arguments supported by information. This allows the researcher to defend the choices being made when developing an operational abstraction of the problem. The researcher builds an argument to claim that certain facts, values, behaviors, interpretations, relationships, and interdependencies constitute evidence when viewed within the context of the unstructured problem. Constructing that evidence entails the following five steps.

Step 1: Convert an Unstructured Problem into a Researchable Problem

The policy researcher uses extant theories, frameworks, experience, analogies, and other forms of abstraction to convert the unstructured policy problem into a

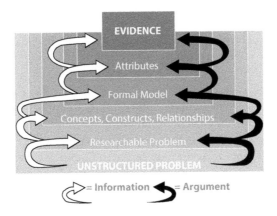

FIGURE 3.1 Conceptual model

researchable problem. Addressing a researchable problem can lead to its resolution, dissolution, or solution.

Structure can be imposed upon an unstructured problem in a number of ways. Hence, conversion of an unstructured problem does not result in a unique researchable problem. The researchable problem is the product of a series of arguments about where to place boundaries and constraints, about drawing analogies between this problem and other problems for which solutions may be known, and other ways of framing the question. Policy researchers make these arguments, supported by information, to inform decision-makers about more or less promising paths.

Argumentation may be used to convince others that the researchable problem is a reasonable abstraction of the unstructured problem and that it provides an adequate representation of the current (undesirable) state of affairs. By effectively modeling the unstructured problem, researchers and decision-makers can vary the inputs to the model or test various decisions to gain an understanding of the potential consequences of those decisions and ultimately improve the situation with respect to the unstructured problem.

For the Forest Service, managing the national forests is an unstructured problem in which a variety of interests—recreation, economic development, ecological sustainability, aesthetics, preservation of flora and fauna, safety, etc.—must be evaluated and balanced. If the Forest Service formulates the problem in terms of economic development or ecological sustainability, it is arguing for one set of values over another and taking into account information to support one formulation over another. Therefore, to effectively frame an unstructured problem as a particular researchable problem, the researcher should explicitly outline the values and assumptions underlying their choices in defining the researchable problem.

Step 2: Identify Concepts, Constructs, and Relationships Underlying the Researchable Problem

In the process of building a formal model, a researcher should define the concepts and constructs used in the model and clarify the underlying relationships among these entities. Building a model is a multistage process in which the researcher decides which constructs, concepts, and relationships will be operationalized. Policy management, including forest management, draws upon multiple disciplines, each with their own approaches to modeling.

In the presence of scientific uncertainty and the absence of a "best" problem formulation, researchers would debate and arrive at a consensus or compromise regarding which concepts and constructs to include in the model and how to represent the relationships among them. For example, if the Forest Service wants to measure success in preventing forest fires, it may choose to measure year-over-year change in the number of acres of forest burned, the monetary value of property destroyed, and/or the impact of the fires on particular animal species. As in defining the researchable problem, this process of operationalizing the problem requires making choices based on the values of the decision-maker, researchers, and stakeholders.

Step 3: Construct the Formal Model

Constructing the formal model is separated from operationalizing the variables because constructing the model entails the choice of the analytical engine. The formal model is the "working" representation of the problem. Thus, if the formal model is a simulation, the researcher specifies the various interactions to be simulated. If it is a statistical model, the researcher identifies the functional form of the relationships among the variables or the statistical procedures to be used to identify patterns in the data. If the formal model is a narrative, the researcher establishes story lines and the main players and decides which narrative threads, relationships, interactions, behaviors, and interdependencies are to be described.

At each step, the policy researcher makes an argument to defend his choices, and buttresses those arguments with information. Fighting fires, for instance, can be modeled as a problem in resource allocation for which a policy researcher would likely use a mathematical programming model. In contrast, fighting fires might be modeled using a simulation that allows the exploration of different scenarios and associated strategies, which would produce different types of information for addressing the problem of forest fires. The researcher may argue in favor of one model over another based on a variety of factors, such as the skill sets of the researchers, the availability of resources, time frames, and the nature of the answers being sought.

Step 4: Identify Attributes to Be Measured

In steps 2 and 3, the researcher identifies the units of analysis and measurement together with how to integrate them into an operational representation of the problem. The attributes are the specific measurements of quality or quantity representing the constructs, concepts, and relationships. At this stage, the researcher provides arguments regarding which attributes of the underlying concepts should be measured. These arguments could be informed by experience from similar modeling exercises in which the researcher has identified the suitability of one measure over another based on select criteria, such as ease of measurement, availability, and validity.

Arguments regarding how to measure a variable may be less about values, as was relevant in earlier steps, and more about feasibility. For example, when assessing whether damage from forest fires has decreased, access to information about the monetary value of damages from insurance claims or information about approximate square footage of forest may be accessible to the researcher. In contrast, even if the policy problem included the impact on specific animal species, collecting data about the number of animals killed or displaced may not be feasible.

Step 5: Define Evidence from the Attributes

Evidence is created through the culmination of the previous steps. The researcher uses the formal model to process the qualities and quantities of the attributes, which then produces evidence. Numbers are a source of information, but the policy researcher also provides context and a story that gives meaning to the information gleaned from the models. For purposes of policy inquiry, the researcher creates the most effective evidence when argumentation and values are integrated with specific measurements that they determine best represent the policy issue.

Data Model

The data model (Figure 3.2) illustrates the methodological steps required to produce the data that make up the evidence. The conceptual model establishes what evidence is necessary to determine whether change toward the desired state of affairs has been effected. The progression from data to evidence can be compared to organizing a disordered pile of bricks into a stack of bricks that are reorganized according to a conceptual scheme to form walls and the floors of a brick house. A home, which implies the practical use of a house, is akin to evidence.

Practical knowledge that can be used to inform and influence decisions and action is evidence. Knowledge is obtained by interpreting information in a given context. Information is obtained by giving form to data by organizing them

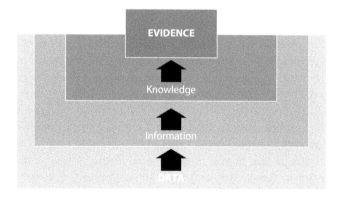

FIGURE 3.2 Data model

according to a system. This model is not new; there are many manifestations of the data, information, knowledge, wisdom (DIKW) hierarchy in the literature (Ackoff, 1989; Cleveland, 1982; Rowley, 2007; Weinberger, 2012; Zeleny, 1987), including perhaps what T.S. Eliot wrote in his 1934 poem, *The Rock*:

> Where is the wisdom we have lost in knowledge?
> Where is the knowledge we have lost in information?

Developing evidence within the data model requires the following three steps.

Step 1: Organizing Data into Information

The researcher uses the conceptual model to determine what kinds of evidence can be best used to address a policy problem. The researcher then collects the data, but data themselves are not useful on their own. To obtain information from the data, they must be organized according to established practice in a way that can answer specific questions dictated by the formal model. For example, if the Fire Service created a simulation of a forest fire in a specific area, the researcher would enter the parameters into the model and the model would provide information about the impact of that scenario.

Step 2: Contextualizing Information into Knowledge

A researcher may convert information into knowledge by providing a context in which to make sense of the information. For example, regression results are information created by a statistical model, but the information only has meaning if it is presented with the definitions of the variables, describes how and why the data were collected, and concerns the relevant policy environment.

Step 3: Imbuing Knowledge with Values to Create Evidence

Policy researchers create new knowledge to inform decisions and actions, keeping in mind the purpose or goal that is to be achieved. As discussed in the conceptual model, defining a purpose or goal involves negotiating values to distinguish between desirable and undesirable states. Policy researchers and decision-makers find evidence by imposing a value structure on knowledge. Arguments regarding which values to use, based on evidence regarding individual and collective preference structures, yield evidence from the knowledge.

To summarize, the data model describes how researchers organize, contextualize, and place data within a value structure to produce evidence. The conceptual model is used to make sense of the evidence to determine whether the problem has been addressed.

Regardless of the ontological or epistemological assumptions researchers make, the connections between the evidence and how people make sense of the unstructured problem should be explicitly established. Similarly, policy researchers create evidence through the organization of data as an element of the research process. In an ethnographic or interpretive study, the order in which the steps are taken might change, but the essential nature of the role of evidence and its links to the unstructured problem and raw data would not. Hence, regardless of the methodological propensities of researchers, empirical studies require clarification of how raw data will produce evidence that will be used to determine whether the policy intervention did or did not result in change from an undesirable to a desirable state of affairs.

Conclusion

Researchers using policy inquiry are often able to determine what did and did not work and perhaps even shed light on why. In some instances, where problems are well defined and the contexts are consistent, policy inquiry might even indicate what works and explain why, but it is a long way away from offering general evidence-based recommendations that may be applicable without adaptation across time and space.

There are many influences on policy decisions that go beyond evidence and argument. Even within a well-crafted argument, scientific knowledge is inconclusive because of the under-determination of theory. Rationality unaccompanied by argument is unlikely to sway minds (Fischer & Forrester, 1993). The value-laden nature of policy decisions, the lack of structure and uncertainty in the definition of policy problems, the role of power and influence in policy decisions, and the lack of clarity regarding what constitutes evidence for justifying belief lead to the pessimistic conclusion that there is no theoretical rationale for believing that policy inquiry yields definitive answers.

However, the influence of lobbying, media, advocacy coalitions, and tradition cannot be underestimated. For Flyvbjerg (1998), change is brought about by the effective marshaling and use of power.

Further, as discussions on determinants of climate change, causes of crime, origins of poverty, sources of global economic crises, and other policy discussions illustrate, arguments backed by evidence rarely change people's minds. Research on assimilation bias has demonstrated that people interpret new information in terms of their existing mental models (Munro & Ditto, 1997). They have systematic biases that influence the way in which they assimilate information such that when they encounter new information that supports their beliefs, they give it additional weight and discount or even ignore information that does not comport with their beliefs. People are likely to reinterpret the information to fit their mental model rather than modify the mental model in light of the new information. In some instances, it is not the evidence or the argument that they believe, but the person who is making the argument.

As researchers, we claim to offer scientific evidence obtained through conceptual rigor, the use of appropriate tools, and the application of systematic practice. However, in public policy contexts, power and values have a legitimate role in determining which policies are implemented. Arguments can accommodate values, but we remain uncertain about how to judge the multiple sources of power and how various stakeholders use power. By systematically building conceptual frameworks and associated data models, researchers can craft evidence-based arguments that are mindful of prevailing values. Whether such arguments prevail will depend upon power and its use. We need to better understand how power, values, arguments, and evidence interact in the process of making of policy decisions and influencing change.

Acknowledgments

We are grateful to Hank Wilson for producing the figures and thank Jos Raadschelders and Stephen Roll for their comments on an earlier draft. They, of course, do not share in the blame for the many shortcomings of this paper.

References

Achinstein, P. (2010). *Evidence, explanation and realism.* New York: Oxford University Press.

Ackoff, R. (1989). Presidential address: From data to wisdom. *Journal of Applied Systems Analysis, 16*(1), 3–9.

Balint, M., Domisch, S., Engelhardt, C.H.M., Haase, P., Lehrian, S., Sauer, J., Theissinger, K., Pauls, S. U., & Nowak, C. (2011). Cryptic biodiversity loss linked to global climate change. *Nature Climate Change, 1*(6), 313–318.

Blank, R.M. (2008). Presidential address: How to improve poverty measurement in the United States. *Journal of Policy Analysis and Management, 27,* 233–254.

Bogenschneider, K., & Corbett, T. J. (2010). *Evidence-based policymaking.* New York: Routledge.

Boruch, R., & Rui, N. (2008). From randomized controlled trials to evidence grading schemes: Current state of evidence-based practice in social sciences. *Journal of Evidence-Based Medicine, 1*(1), 41–49.

Bridges, D., & Watts, M. (2009). Educational research and policy: Epistemological considerations. In D. Bridges, P. Smeyers, & R. Smith (Eds.), *Evidence based education policy.* Chichester, UK: Wiley-Blackwell.

Cleveland, H. (1982). Information as resource. *The Futurist,* December, 34–39.

Clinicians for the Restoration of Autonomous Practice (CRAP) Writing Group. (2002). EBM: Unmasking the ugly truth. *British Medical Journal, 325,* 1496–1498.

Dunn, W.N. (1981). *Public policy analysis.* Englewood Cliffs, NJ: Prentice Hall.

Fischer, F., & Forester, J. (Eds.). (1993). *The argumentative turn in policy analysis and planning.* Durham, NC: Duke University Press.

Flyvbjerg, B. (1998). *Rationality and power: Democracy in practice*. Chicago, IL: University of Chicago Press.

Guyatt, G.H., Sackett, D.L., Sinclair, J.C., Hayward, R., Cook, D.J., & Cook, R.J. (1995). Users' guides to the medical literature IX: A method for grading health care recommendations. *Journal of the American Medical Association, 274*(22), 1800–1804.

Haas, P. (2004). When does power listen to truth? A constructivist approach to the policy process. *Journal of European Public Policy, 11*(4), 569–592.

Heinrich, C.J. (2007). Evidence-based policy and performance management: Challenges and prospects in two parallel movements. *The American Review of Public Administration, 37*(3), 255–277.

Kezar, A. (2004). Wrestling with philosophy: Improving scholarship in higher education. *The Journal of Higher Education, 75*(1), 42–55.

Kincaid, H. (2009). Causation in the social sciences. In H. Beebee, C. Hitchcock, & P. Menzies (Eds.), *The Oxford handbook of causation* (pp. 726–743). Oxford: Oxford University Press.

Kingdon, J.W. (1995). *Agendas, alternatives, and public policies* (2nd ed.). New York: Longman.

Ladd, H.F. (2012). Presidential address: Education and poverty: Confronting the evidence. *Journal of Policy Analysis and Management, 31*(2), 203–227.

Le Grand, J. (1990). Equity versus efficiency: The elusive trade-off. *Ethics, 100*(3), 554–568.

Leigh, A. (2009). What evidence should social policymakers use? *Economic Roundup, 2009*(1), 27–43.

Lindblom, C. (1980). *The policy making process*. Englewood Cliffs, New Jersey: Prentice Hall.

Majone, G. (1989). *Evidence, argument, and persuasion in the policy process*. New Haven, CT: Yale University Press.

Mason, R.O., & Mitroff, I.I. (1973). A program for research on management information systems. *Management Science, 19*(5), 475–487.

Meier, K.J., & O'Toole, L.J. (2009). The proverbs of new public management: Lessons from an evidence-based research agenda. *The American Review of Public Administration, 39*(1), 4–22.

Munro, G.D., & Ditto, P.H. (1997). Biased assimilation, attitude polarization, and affect in reactions to stereotype-relevant scientific information. *Personality and Social Psychology Bulletin, 23*(6), 636–653.

Nixon, S.W. (2004). The artificial Nile. *American Scientist, 92*(2), 158–165.

Okun, A.M. (1975). *Equality and efficiency: The big tradeoff*. Washington, DC: The Brookings Institution.

Oliver, S., Harden, A., Rees, R., Shepherd, J., Brunton, G., Garcia, J., & Oakley, A. (2005). An emerging framework for including different types of evidence in systematic reviews for public policy. *Evaluation, 11*(4), 428–446.

Owens, S., Petts, J., & Bulkeley, H. (2006). Boundary work: Knowledge power and the urban environment. *Environment and Planning C, 24*, 633–643.

Pawson, R. (2002). Evidence-based policy: In search of a method. *Evaluation, 8*(2), 157–181.

Raiffa, H. (1968). *Decision analysis*. Reading, MA: Addison-Wesley.

Rowley, J. (2007). The wisdom hierarchy: Representations of the DIKW hierarchy. *Journal of Information Science, 33*(2), 163–180.

Simon, H. (1997). *Models of bounded rationality: Empirically grounded economic reason* (Vol. 3). Cambridge, MA: MIT Press.

Sutherland, W.J., Bellingan, L., Bellingham, J.R., Blackstock, J.J., Bloomfield, R.M., Bravo, M., . . . Zimmern, R.L. (2012). A collaboratively-derived science-policy research agenda. *PLOS ONE, 7*(3), e31824.

Tanenbaum, S.J. (2005). Evidence-based practice as mental health policy: Three controversies and a caveat. *Health Affairs, 24*(1), 163–173.

Tonelli, M.R. (2006). Integrating evidence into clinical practice: An alternative to evidence-based approaches. *Journal of Evaluation in Clinical Practice, 12*(3), 248–256.

Toulmin, S. (1958). *The uses of argument.* Cambridge, UK: Cambridge University Press.

Weinberger, D. (2012). *Too big to know: Rethinking knowledge now that the facts aren't the facts, experts are everywhere, and the smartest person in the room is the room.* New York: Basic Books.

Weiss, C. (1977). Research for policy's sake: The enlightenment function of social research. *Policy Analysis, 3*(4), 531–545.

Zeleny, M. (1987). Management support systems: Towards integrated knowledge management. *Human Systems Management, 7*(1), 59–70.

4

VISUALIZATION MEETS POLICY MAKING

Visual Traditions, Policy Complexity, Strategic Investments

Evert Lindquist

Introduction[1]

Complexity is an inherent challenge of policy development. No matter the policy domain, there is greater appreciation of the manifold linkages of actors affecting problems and those affected by them, the myriad intertwining issues and the importance of context, the time lags of causal factors and interventions, and the surprises and uncertainties associated with problems and policy interventions (Geyer & Rihani, 2010; Rittel & Webber, 1973). Capturing, acknowledging, and addressing such complexity is a challenge for governments and those seeking to inform and advise policy makers.

One promising and increasingly widely acknowledged avenue for grappling with complexity is to tap into visualization techniques. The field of visualization is diverse, rapidly expanding, and its practitioners are moving forward with great enthusiasm. Visualization encompasses efforts to distill and project findings from large datasets, find creative ways to display information, engage staff and communities in recognizing and mapping complexity, and work with key stakeholders inside and outside an organization to identify strategic directions. There has been an increasing use and celebration of visualization and social media tools in the private sector and considerable interest from public service and political leaders about exploiting the potential of these tools. There is a growing expectation that government leaders and the public service institutions that support them should tap into visualization technologies to better analyze complex issues, advise political leaders, and engage citizens and stakeholders.

A premise of this chapter is that public service institutions—with the exception of security and intelligence domains (Thomas & Cook, 2005)—have until recently under-invested in visualization techniques for different kinds of policy work. Notwithstanding declarations about open government, governments do not appear to have aggressively invested in uniform visualization, with uneven take-up across departments and agencies. But many bureaus and agencies do use visualization techniques associated with particular knowledge domains for specific programs.

Visualization can mean very different things to different people. Diverse agendas for visualization are often at play (promoting data, producing beautiful images, securing good advice, showing complexity, engaging citizens and partners in policy and strategy development, etc.), and visual techniques can be misused. As visualization techniques are more widely adopted in a variety of private sector, university, scientific, and public settings (such as newspapers, websites, and specialized magazines), there will be growing expectations that government tap into these alternative ways of sifting through and presenting information. Visualization capability is emerging bottom-up in government organizations, sometimes the byproduct of hiring and interaction for other purposes. Government executives must develop fluency with visualization techniques and have realistic expectations about utilizing them and understanding the impact they will likely have. This will lead to a more strategic perspective when making investments in these methods of sharing, interpreting, and using visual data.

The purpose of this chapter is to provide background, perspectives, and suggestions for public sector executives and analysts seeking to either invest in or explore the potential and limitations of visualization techniques for various types of policy work. It begins with a brief survey of visualization techniques and then outlines a framework that connects visualization to various types of policy work (undertaking analysis, providing advice to elected leaders, and engaging citizens on policy matters) in the context of broader policy making dynamics associated with government and public service systems. The last section identifies challenges inherent in the use of visualization in these environments and suggests guidelines on how best to proceed with strategically investing in expanding visualization capabilities in government.

What Is Visualization?

Anyone invoking the term *visualization* quickly learns that it is adopted by practitioners and scholars working in very different fields. Moreover, reactions to potentially using visualization to assist with policy analysis, advising, and engagement will depend on the techniques users are familiar with. It is therefore important to delineate the different areas of practice and scholarship associated with visualization.

This section summarizes the findings of a survey of three main domains of visualization practice—information visualization and data analytics, graphics and information display, and visual facilitation for thinking and planning strategy—and identifies similarities and overlaps across the domains (see Figure 4.1).[2]

Information Visualization and Data Analytics

The field of information visualization is relatively new and rapidly growing, driven by the latest developments in information and communications technologies. But its origins can be traced to early mapping and graphing techniques (Friendly, 2008; Tufte, 1990). It has developed at the intersection of the fields of computing, engineering, graph analysis, data management, cognitive psychology, software development, human–computer interface, and others. Contributors come from a host of scientific, social science, and humanities disciplines. The field has grown and been institutionalized through conferences, journals, university research centers, courses and programs, and textbooks (e.g., Ware, 2004; Spence, 2007; Ward, Grinstein, & Keim, 2010). Many consulting firms—boutique and large-scale—now count information visualization and data analytics as a core business.

The field of information visualization has been driven by the need to visually represent increasingly large datasets to enhance how humans can analyze and learn from the information. The field is breathtaking and its diversity of techniques and applications can be used in diverse *practice domains* (e.g., geosciences, biology, medical imaging, physics, health, monitoring social media). Visualizations are collectively driven by a fundamental, shared premise: access to data, when found to be accurately transformed, well represented, and properly matched with other streams of data in the form of visualizations, can help inform and improve awareness of issues, analysis, and decision-making. Although information and scientific

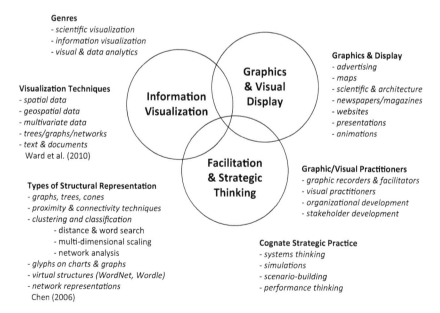

FIGURE 4.1 Three visualization domains

visualization outputs can be stimulating and aesthetically pleasing, at the base they are still representations of data: various scientific measurements, information packets, abstract numbers representing variables, images, text, and documents.[3] Bederson and Shneiderman (2003) identified several tasks that information visualization specialists typically undertake: overview, zoom, filter, details-on-demand, relate, history, and extract.

Chen (2006) identified different forms of structural representation: graphs, trees, and cones; proximity and connectivity techniques (e.g., semantic distance and word search, multi-dimensional scaling, and network analysis); clustering and classification (e.g., dividing data into sub-sets and taxonomies, cluster-seeds); use of glyphs (e.g., using symbols on charts to convey additional information); creating virtual structures (e.g., WordNet, Wordle); and creating networks (e.g., scale, small or large, topological, nodes). However, Chen (2006) also observed that most visualization techniques focus on ascertaining structure from data. He correctly predicted an increased focus on finding and displaying the dynamic and evolutionary properties of data. The distinct sub-field of visual and data analytics has rapidly emerged, driven by the availability of increasingly large datasets and the real-time needs of governments, corporations, and scientific disciplines. This has led to great interest in *data-mining* and the challenges of assembling, representing, linking, and analyzing diverse data in real-time contexts.

The lines of contemporary research are diverse: finding the best ways to represent data; understanding how individuals and groups can better problem-solve with visual displays; how displays and systems (hardware, software, and physical space) can be improved to assist in manipulating and interpreting data; and how techniques developed for one substantive domain can be adapted to others, to name only a few. There have been increased calls for more education and training in information visualization for practitioners, particularly for novices and generalists (Chen, 2005; Grammel, Tory, & Storey, 2010).

Graphics and Information Display

The field of information visualization overlaps with the writing and practice of information and graphics design in two directions: (1) the broad field of graphics, which has long explored and celebrated innovative ways to convey information for scientific, professional, and advertising purposes, and (2) the increasing number of magazines and newspapers investing in visual renderings of issues and stories, now referred to as *infographics*. The number of websites and books has multiplied on this subject, as have gurus, like McCandless (2009) and Baer (2010), who variously generate and convey the best and most intriguing of these efforts. However, the fields of graphics and information display should not be confused with information visualization described above; the latter is wholly data-driven, whereas the former places a greater premium on aesthetics, beauty, and impact as points of departure.

This field, like information visualization, is broad and diverse: exploration of new programs and algorithms for producing visualizations, showcasing the remarkable and beautiful examples of visualization, exploring applications in an ever-increasing array of fields, developing theoretical constructs, and exploring the cognitive dimensions of processing and interpreting visualizations. Baer defines the field as "the translating [of] complex, unorganized, or unstructured data into valuable, meaningful information" (2010, p. 12). Information-design practitioners can include graphic designers, information architects, interaction designers, user experience designers, usability and human-factors specialists, human-computer interaction specialists, and plain language experts (Baer, 2010, pp. 14–15). Practitioners work with diverse media: printed matter (e.g., signs, guides, marketing), information graphics produced for magazines and newspapers, interactive websites and screen-based projects, various types of animation, and advertising. Baer (2010) and Steele and Iliinsky (2010) show this work includes: social and market network analysis, voting patterns in legislatures, aviation flight patterns and subway maps, text-related applications such as Wordle, searching *New York Times* databases, monitoring the editing of entries in Wikipedia, and even autopsies!

Another focus concerns designing visual displays of information to engage audiences with presentations and animations (e.g., Atkinson, 2008; Heer & Robertson, 2007), but Duarte (2010) takes this to new levels by using visuals to analyze how speakers can create emotional and intellectual impact through good visuals, adroit timing and scripting of presentations, oral and visual balance of information flows, and linking data and presentations to good stories and overriding messages that broaden horizons, encourage commitment, and stimulate change. Such assessment and instruction is focused on persuasion. Segel and Heer's (2010) research on how graphics are juxtaposed with newspaper articles leads one to consider the appropriate balance between narratives projected by the author versus exploration of the reader. Fisher (2010) makes a distinction between using animation for presenting versus exploring (and learning), reporting that users take longer to explore and play with animations, but were less accurate when responding to questions compared to when static diagrams were viewed.

To what extent are such compelling and intriguing visualizations relevant, useful, and economical? Beauty is not inconsistent with utility, often arising from its correspondence and assistance to the tasks at hand. Even when beauty predominates as a goal and effect of visualization, such "play" can lead to greater interest in visualization, increased facility with associated technologies, and discovery of more practical applications.

Visual Facilitation for Thinking and Strategy

When the term *visualization* is uttered, another set of practitioners might step forward: a growing community of visual and graphic artists who assist clients grapple with complexity by means of sketching (Walny et al., 2011), often through elaborate renderings of challenges and strategies. Known as graphic recorders, graphic facilitators, and visual practitioners, they sketch the evolution and key conclusions of meetings and conferences that can be over a day in length (Sibbet, 2012). The sketches are engaging and are often substantial dynamic diagrams that attempt to capture the movement, enthusiasm, and vision of participants.

Also included in this broad domain is a large circle of approaches for strategy and organizational development—systems thinking, simulation, scenarios, and performance thinking—each relies heavily on visual techniques. Of three visualization domains, this domain most directly grapples with the challenges confronting policy makers and advisors, although it involves information and perspectives supplied by other visualization domains.

The International Forum of Visual Practitioners (IVFP), founded in 1995 to support the visualization community,[4] hosts an annual conference. Another supportive organization is VizThink,[5] which has a broader mandate, providing advice on compelling presentations with different visual technologies and monitoring different techniques for telling stories. Many visual practitioners have similar styles, but some specialize as recorders, others facilitate, and still others take on broader organizational development and stakeholder engagement projects. Despite considerable overlap in approach, gurus, and literature, there is convergence in approaches and techniques (Blackwell et al., 2008; Hyerle, 2009; Margulies & Valenza, 2005; Sibbet, 2010). These include Venn diagrams, concept mapping, bubble maps, mind maps, thinking maps, systems feedback loops, mind-scaping, thinking hats, visual journeys, assumption trees, icebergs, influence circles, etc. Horn (1998) used sketches to assist policy makers and citizens comprehend and think about how to address complex policy challenges and wicked problems (Horn, 2001).

Visualization practice and writing often taps into *systems thinking*, which seeks to bring more holistic analysis to organizations and communities to address complexity and wicked problems. We can also include well-known techniques that inform the emerging field of policy informatics: simulations, scenario-building, and performance modeling.

- *Systems thinking.* Practitioners seek to work with decision-makers and stakeholders to better understand (in the context of problems and interventions) the issues surrounding complexity, diverse interests and perspectives, and the task and institutional factors at play. Through dialogue, pragmatic ways for improving the situation are identified. A key feature of systems thinking involves encouraging participants to commit perspectives, perceptions, and even emotions to paper. The resulting *rich pictures* and other kinds of sketches can be shared and debated (Chapman, 2004; Chapman, Edwards, & Hampson, 2009; Checkland, 1999; Checkland & Poulter, 2010; Senge, 1990).
- *Simulations.* These include models of how market, social, organizational, and natural systems work and evolve over time, with the ability to alter input and

external variables to understand the properties of complex systems. This can allow users to consider the trajectories of other variables and decision-making quandaries, constraints, and trade-offs. The altered intersections and trajectories of key variables are often conveyed visually (think of how economists display different *runs* of a model) to engage analysts and audiences. Other examples include airplane cockpit training devices; climate-change models; or multi-actor game simulations (see chapters 5, 8, 11, and 16 in this volume).

- *Scenario-building*. Some practitioners have long worked with clients to imagine distinct futures comprised of diverse variables defined by key contingencies and to consider how these futures might be connected to the present. Scenarios seek to develop *shared mental maps*, assist users in thinking broadly and creatively about future possibilities, and better appreciate dynamic, complex environments. Scenario-building is typically very visual, even if more elaborate exercises include speakers, streams of data, and background documents. Participants are encouraged to share ideas on walls and whiteboards, explore connections among variables, and develop coherent narratives and images of future states (e.g., Ringland, 1998; Rosell, 1995).

- *Performance modeling*. Most public executives are familiar with developing *logic models* linking inputs and activities of programs to outputs and desired outcomes as a basis for developing performance measurement and management systems (McDavid & Hawthorn, 2006). Although the final diagrams are linear, delineating logic models is a highly visual and iterative process—often balancing the needs of parsimony and detail to develop a *model* representing a more complex reality, usually leaving out details on the state of organizational capabilities and culture, political dynamics and commitment, resource allocation, client perspectives, and environmental change.

Practitioners in these traditions would not consider themselves *visualists*, but each approach in varying degrees uses group processes to encourage sense-making and strategic dialogue about complex challenges, and relies on different sub-sets of visual techniques to capture complexity at different stages and levels of analysis. Often these practitioners, after having moved through processes and analysis, use visual techniques to convey insights to clients.

Of the three broad visualization traditions, strategic visualization practitioners are typically the closest to addressing the specific challenges of decision-makers: they seek to assist clients in capturing the nature of problems and developing strategies for addressing them, as opposed to only supplying them with data or perspectives driven by data, and strive to assist them in discovering what they know and don't know. Conversely, these approaches do *not* rely on computer-mediated visualizations (simple or complex) of findings from datasets (larger or small), instead relying on hand-sketched renderings to move conversations along. However, strategic visualization practitioners and their clients can be informed by data and rendering from the other visualization practitioners, and some graphics recorders and facilitators have found ways to use digital technologies to draw, store, and transmit information to clients and others.

Looking across Visualization Domains: Some Insights to Consider

There is no overarching theory of visualization, reflecting its status as a user-oriented practice and craft field, much like policy analysis. However, several broad messages arise from all three areas of practice and scholarship which are important for government leaders to consider as they consider how to incorporate and invest in visualization capacities. A key reason for adopting visualization techniques is to see the *whole* in order to analyze the parts when considering complex challenges, giving users the ability to zoom in and out, see and make connections, and explore and reintegrate. Visualization techniques can be used to engage and improve the quality and usability of representations.

There are several challenges, though, particularly with respect to informing different phases of policy development, deliberation, decision-making, and implementation. One challenge is that visualization techniques of all kinds, regardless of their ability to show the whole and complexity, often requires simplifications, distillations of information, choices about look and feel, and how to show underlying detail. These choices and trade-offs might reflect different tastes, techniques, or professional orientations, which could influence interpretations and strategic conversations. Indeed, one person's idea of a good visualization might not square with someone else's representation (Bresciani & Eppler, 2010) or some people like to receive information in text-based, verbal, or more tactile ways. Relatedly, visualizations may simply display the results of analysis and thinking for consumption by others or serve as platforms for engagement and further exploration by other groups—the question here is whether the ultimate audience can manipulate and modify visualizations (Eppler & Pfister, 2010; Segel & Heer, 2010). Although much progress has been made on tackling and representing complexity, visualizations of all kinds often rely on data and representations at particular points in time—the amount of effort required to show trends and evolving relationships *over time* requires different thresholds of effort and trends are highly desirable for better understanding phenomena and arriving at strategic interventions.

Even the most enthusiastic proponents of visualization, regardless of the traditions and techniques they work with, recognize that visualizations can rarely stand on their own. Even the best visualizations are enhanced by story-telling to draw out interesting facts and interesting issues (Atkinson, 2008; Duarte, 2010). Audiences need context, narrative, and often a guide to parse information. Visualizations, unless they combine multiple streams of information, usually only represent patterns or conclusions from limited data and information sources. Visualization can come from multiple lines of data (text, images, and documents), involve diverse perspectives on interpretation, and the appraisal of the final results can be from different vantage points. These caveats are particularly important as we consider how visualization might impact policy making.

Visualization and Policy Work: A Preliminary Framework

Visualization technologies can intersect with policy making in many ways. The three broad and overlapping domains of visualization with three similarly broad but distinct areas of policy work (analysis, advising, and engagement) can be juxtaposed, recognizing the possibility of delineating finer typologies of policy tasks and circumstances (Althaus, Bridgman, & Davis, 2007; Arnstein, 1969; Bardach, 2005; Bishop & Davis, 2002; Lindquist, 2012; Mayer, Bots, & Daalen, 2004; OECD, 2009; Scott & Baehler, 2010).

Organizations and programs adopt visualization in varying degrees and tap into different techniques depending on the challenge at hand (i.e., critical tasks, mission and culture of departments and agencies, and patterns for recruiting staff and engaging networks). In these more limited contexts, adoption can be driven through strategic direction in a top-down way, or bottom-up due to curiosity, talent, work demands, and staff discretion. But it is important to identify public-service-wide strategies to improve capability, utilization, and strategic investments, especially where capacities are distributed and focused on very different challenges. Finally, there are many organizational actors in public sector systems, each with their own focus, needs, and constituencies—although they are part of one system, they often compete for resources and the attention of policy makers.

Figure 4.2 takes each of the factors and identifies trends driving interest in visualization. It emphasizes that all visualization outputs compete with well-known streams of policy-relevant information (data, research, analysis, and reporting from the great variety of government and non-governmental institutions shared in published formats and at events), continuous media reporting and political intelligence, and more generally, the ever-changing swirls of contending beliefs, values, and narratives (Hullman & Diakopoulos, 2011; Roe, 1994; Sabatier, 1987; Segel & Heer, 2010).

Figure 4.2 implies that there are different levels of visualization users: proximate, intermediate, and ultimate. Most information visualization literature focuses on how well visualizations satisfy *proximate* users but, in the policy world, the ultimate test is how well users—decision-makers, citizens, and stakeholders—are served by alternative modes of information presentation. Unless policy analysis units have cultivated specialized capabilities to create visualizations, they can be thought of as *intermediate* users that commission and interpret visualizations for other audiences. That analysts may rely on visual experts creates a different set of issues.

Figure 4.2 suggests that it will be difficult to ascertain the extent of visualization use across policy domains and to assess the impact and effectiveness of visualizations. Factors for this include: (1) the cognitive style of leaders, (2) the

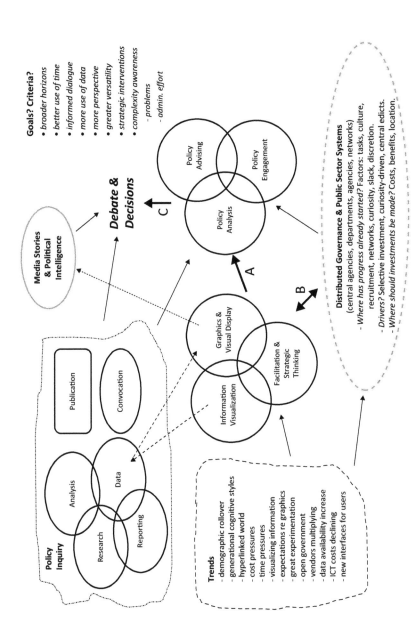

FIGURE 4.2 Visualization and public sector governance

preferences and expectations of citizens and stakeholders, (3) the time available for absorbing and probing information conveyed through visualizations, (4) the availability and quality of other streams of information, and (5) whether visualizations project policy narratives or seek to provide others with the issues to explore and construct or project their own narratives. Visualizations will variously compete with and complement each other and other streams of information, differentially meshing with the cognitive styles and dispositions of users inside and outside of government, and possibly making more effective use of available cognitive and deliberative bandwidth for decision-making and engagement.

It is prudent to have tempered and nuanced expectations about how visualizations might add value and influence policy making. Influence on policy actors may be more subtle and indirect and prove difficult to definitively measure. Visualizations can broaden horizons and the understanding of context, improve appreciation of the complexity of problems and pertinent delivery systems, augment more efficient and productive use of scarce time, encourage greater use of data, provide alternative channels for advising, and encourage more informed and strategic deliberations. Visualization should not be understood as a substitute for policy analysis, advising, and engagement—rather, the domains of visualization offer a menu of tools that can be applied differently by analysts, advisors, and facilitators who undertake policy work in challenging environments.

Taking a Further Step Back: Perspectives on Visualization and Policy Making

There is no shortage of visualization techniques (Lengler & Eppler, 2007) and there is a great deal of research on their effectiveness with different user groups. However, broader perspectives are required to assess whether and how governments should proceed with strategic investments in visualization, including consideration of how visualization techniques intersect with policy work and, more generally, governance.

Overcoming Pre-conceptions: Visualization for What?

There are bound to be very different reactions to the practice and promise of visualization. This may be due to the casual knowledge of specific visualization techniques used for certain purposes (for example, many colleagues and policy makers might understand visualizations simply as marketing or data-mining), by the experience of users and producers in distinct contexts, or that visualizations can be used to promote certain views (Sabatier, 1987).

Visualization techniques and practices are incredibly diverse. When choosing an appropriate visualization tool, proponents must consider what goals they are trying to achieve: supply policy makers with additional cognitive aids for problem definition, horizon-building, illumination, or analysis of complex problems? Synthesize and present results of analysis, deliberation, integration, and options development? Develop vision and strategic direction? Inform real-time decision-makers (e.g., security and emergency services)? Or, succinctly convey administrative complexity (e.g., how government delivers and collaborates, the diversity of needs and points of delivery, and the resources and capabilities that can be mobilized to address challenges)?

Although each goal is legitimate, furthering each may require tapping a different range of techniques and condition how we think about the benefits and effectiveness of investments. For any given policy, these needs may materialize at any point during the policy cycle. We must therefore appraise visualization possibilities for immediate contexts as well as needs across government at different times. A need for cost efficiencies might suggest that horizontal or network strategies could be a better choice than in-house resources.

Visualization and Constraints: Cognitive Styles, Bandwidth, and Channels

The impact of a visualization will depend not only on the data, quality of representation, and how it is delivered (with a narrative and other streams of information), but also on the context and preferences of users. As individuals, we like to absorb information in different ways—text, imagery, and briefings—including receiving, working with, or engaging with the material or the person who delivers them.

Producers of visualizations need to anticipate the communications bandwidth of decision-makers and their advisors, the shifting and contested environments in which they work, and the scale of decisions under consideration, all of which can affect the willingness of individuals to entertain new perspectives and alternative forms of analysis and communication (Lindquist, 1988). In short, those producing and incorporating visualizations not only require deep knowledge of the underlying data, but also the versatility to consider alternative ways of presenting findings and perspectives, or modify and adjust visualizations once policy makers and stakeholders become engaged.

Impact on Decisions: Visualization as Decision-Relevant or Play?

Visualizations can be aligned with policy making needs and bring data-informed insight to the attention of various users and thus are connected to the much broader evidence-based decision-making movement (Nutley, Walter, & Davies, 2007). Visualists may access data streams or zero in on factoids that intrigue them, and whether undertaken for fun or serious purposes, such exploration may lead to interesting visualizations that capture attention and stimulate interest.[6] On the one hand, such 'play' meets the needs of creative practitioners who seek to impress and inform envisioned consumers, over time developing a broader constituency that demands visual techniques. On the other hand, it develops a broader community of individuals who could match these techniques to make headway on problems that, in some instances, they could not imagine. The notion of the importance of *gossip* (non-aligned or non-decision-specific information production and sharing in organizations) in this respect is useful (March & Sevón, 1988). In early seminal work, Feldman and March (1981) noted that information is typically over-produced and under-utilized in organizations, serving many symbolic and organizational functions, including readiness for the future and preparedness for unimagined decisions (Lindquist, 2009).

Some visualizations may appear to be 'hit and miss' or irrelevant. However, we should expect that in the early stages of applying new ways to convey and analyze information, there will be over-promise of their yields and application to all sorts of challenges (recall the metaphor of 'having a hammer to deal with all problems'). Additionally, this dynamic is consistent with the fate and probable impact of much data, research, and analysis produced by government, think tanks, universities, and other sources—the vast majority of this work is not used or read extensively, but is nevertheless functional and helps further scholarly and practice fields (Lindquist, 2009; March and Sevón, 1988; Weiss, 1977, 1980).

Measuring Visualization Impact: Relative Costs and Benefits of Different Techniques

Developing a systematic sense of the costs and benefits of producing credible, high quality visualizations will be difficult. Easy-to-measure costs include the outlays for software, training, staff, and contracting. Studies on the effectiveness of certain visualization are sparse, but it is a growing area of interest. Given the variable cognitive dispositions of users toward visualization in general, and the different techniques in particular, measuring the benefits of visualization as an aid in policy work may be difficult due to the allusive ways in which information feeds into and influences decision-making. Moreover, we should keep an open mind about the value of time-honored methods (traditional maps, stories, personal stories, field visits) to convey information to decision-makers and other audiences, particularly with respect to complex challenges.

Visualization and Open Government: Credibility and Quick Response

Information visualization, particularly the area of data analytics, graphics, and information display, relies heavily on the availability of reliable streams of data and an ability to categorize and manipulate it often to draw informed conclusions. As government and other organizations make more datasets freely available, there will be greater opportunity for analysts to work with, analyze, and present renderings of data.

Whether such visualization adds value from a strategic policy viewpoint depends on insight and finding gold in the dross. Although able to capture and analyze complexity, there is risk of visualization misuse by political and other interests. Government departments and agencies should build sufficient capacity to monitor such use and respond quickly to misleading interpretations. Related is the growing interest in developing *provenance* protocols and reporting (Hullman & Diakopoulos, 2011). An info-graphics community-of-interest from across news and other organizations is working to establish protocols in the United States (Cox, 2011).[7]

As the open-government movement gathers steam, liberating more data for use by outsiders, two related challenges stand: Is there sufficient capacity within government to mine and analyze data of considerable policy relevance and is there sufficient interest or incentive for outsiders to become involved? This suggests that government should find ways to encourage outsiders—perhaps with inducements and competitions—to create visualizations with priority streams of data for certain audiences.

Is Visualization So Different?

The visualization movement presents many exciting possibilities for how governments can approach policy work. But all may not be new; modern visualization techniques expand the range of tools used by policy analysts, researchers, and advisors to reach ministers, stakeholders, and citizens.

The real challenge for public officials will be to increase awareness of the possibilities and become conversant about visual techniques, eventually becoming intelligent and smart consumers of visualization. Likewise, there is a need to develop state-of-the art capabilities in diverse visualization techniques across the public sector system for distinct challenges and users at different stages of policy making. This suggests we should resist the impulse to view visualization as a magic bullet for policy work. Rather, we should depict it more as a situational and rolling craft that can apply diverse tools for informing different strategic conversations.

Investing in and Coordinating Visualization

A premise of this chapter is that public service institutions have under-invested in visualization techniques for policy analysis, advising, and engagement. First, in recent years, governments and observers have focused on performance, achieving identified results by measurable outcomes and relevant outputs. This has shifted attention away from inputs and activities crucial to producing the outputs that make desired outcomes possible and reducing the need to convey internal and process complexity (although some program areas, such online service-delivery operations, have very sophisticated capabilities for tracking and ensuring quality control). Second, many governments may *think* that they have made progress by making reports and websites more attractive from a visual perspective, but the information remains linear and the enhancements are largely adornments, failing to provide more complex and efficient renderings of issues and work. Finally, open-government advocates would likely argue that government conventions and repertoires not only continue to constrain the availability of data, but have also curtailed concomitant and concerted exploration of new ways to analyze, depict, and convey data, thus inhibiting new insight and perspectives on policy challenges.

There are many pockets of innovation that use visualization techniques in different parts of government. Indeed, we anticipate selective investments by departments and agencies where data collection is critical to their missions (Wilson, 1989). For example, national security agencies invested heavily in data-mining and visual analytics as tools (Thomas & Cook, 2005), many science-based organizations (e.g., medical, health, biology, informatics) have made investments to remain on the leading edge of research and analysis, and specialized agencies are increasingly asked to report on data trends to both expert and non-expert audiences. To varying degrees, government leaders inclined toward the possibilities for visualization encourage departments and agencies to incorporate these techniques in briefing material. Two interesting questions to ask are: Will explicit investments in building visualization capacity produce better results[8] than levering latent capabilities in visualization? Is it better to invest in creating forums and networks to facilitate mutual learning, perhaps leading to better coordination of investments and more fulsome leveraging of existing capabilities?

Visualization capacity for analysis, advising, and engagement has been pulled into the executive ranks public service systems, either by external individuals and units who are aware and want to tap into or develop such capabilities, or by purpose-oriented initiatives to address proximate, specific challenges. But many visualization capabilities may have long existed in professional or disciplinary departments and agencies (e.g., geographic information systems, modeling, simulations). Software for undertaking visualization is readily available on the Web; only basic programming and data manipulation skills are required to use them. Then, there are curious individuals, unafraid to learn about and experiment with visualization, even if they were not hired for that purpose. For these reasons, we should anticipate that the likelihood of bottom-up experimentation and innovation will increase in more data-rich and science- or engineering-based areas. More public policy, public administration, and business schools at universities are introducing program content pertaining to information visualization and data analytics (Dawes, Helbig, & Nampoothiri, 2014).

In short, there are diverse and manifold needs for policy development and a great variety of visualization techniques that can be accessed, depending on the tasks, circumstances, personalities, expertise, and objectives at hand. What is missing is a government-wide strategy on recognizing, developing, and tapping into these capabilities to use them to inform policy work and make smart, strategic investments. What follows could be elements of such a strategy.

Create a Network: Visualization, Data, and Experts

Pockets of visualization expertise are already dispersed across government depart-ments and agencies and there is evidence of growing curiosity, if not demand, from elected leaders, some executives, and citizens about its use. Many consulting firms are now staking their reputations on supplying such expertise. Galvanizing public servants, representatives of agencies responsible for collecting and ware-housing data, and policy experts with interest or expertise in visualization tech-niques into a network would be an important element of a low-cost strategy to leverage experience, foster learning, and build a constituency for developing new analytic, engagement, and advising tools. Workshops, conferences, and visualiza-tion fairs similar to the academic VisWeek model would provide opportunities to explore the visual design process.

Lever a Network of Distributed Capabilities

With an established network and opportunities for meeting to exchange informa-tion (and presumably some sort of Web portal or intranet capability), governments can lever diverse capabilities across the public sector. Developing a center-of-excellence and clearinghouse in a central or lead agency would facilitate the exchange of practice and techniques and the movement or secondment of visual experts to other organizations as needed. Such an approach would promote deeper capabilities on a distributed basis within departments and agencies where visualization competence already exists, but spread the expert wealth as required.[9] An established network would lead to cost-sharing when purchasing proprietary software, hiring specialized staff, and investing in visualization literacy and train-ing. Because much visualization innovation occurs outside government, this strat-egy should extend to animating and monitoring university and private sector labs and innovation hubs in broader epistemic networks.

Invest in Visualization Literacy and Training

Simply linking policy analysis and visualization expertise will prove insufficient; policy experts should be encouraged to increase visual literacy through training and professional development courses. Although policy and operations units in larger departments and agencies may have sufficient resources to hire and retain certain kinds of visualization expertise, the range of demands for different kinds of visualization and the degree of technical expertise required may rapidly exceed those capabilities. Learning could proceed on a collaborative basis with visualiza-tion programs and researchers associated with universities. Governments might encourage professional public policy and management programs at universities to address visualization in core courses pertaining to policy communications and statistical analysis, linking to visualization experts working at universities (Dawes, Helbig, & Nampoothiri, 2014).

Keep an Eye on the Quality of Data and Provenance of Visualizations

It is easy to focus on the beauty and quality of visualization, and such aesthetics are critical when engaging policy makers, citizens, and stakeholders. However, equally important is gaining access to useful data, ensuring proper analysis and transformation of that data, and developing good response repertoires to policy-driven demands. There is an increasing interest in information visualization and various info-graphics fields to develop repertoires for conveying sources of data and how visualizations were developed (Cox, 2011), including offering outsiders the ability to replicate representations derived from datasets. This is critical in contexts where the opportunity for misinterpretation could be significant.

The short-term costs of investing in visualization capabilities may seem significant—particularly for information visualization, visual analytics, and info-graphics—and the direct benefits to improved decision-making are difficult to demonstrate. However, these costs will undoubtedly be small compared to other IT-related outlays, and increased familiarity and facility with visualization for specific purposes will rapidly improve. Indeed, in certain areas a more defined sense and range of visualization needs may well emerge, which can be factored into analytic and briefing routines.

The extent to which these strategic directions for building and leveraging visualization capabilities across a public sector are successful will depend on the scale of a government and its public sector system. A more circumscribed mix of strategies for investing in and coordinating visualization capabilities might be best advised for public agencies with specific mandates and fewer programs. Such agencies might be inclined to pursue strategies that are bottom-up in nature, collaborating with cognate agencies and other actors in external networks.

Concluding Remarks

Visualization is a diverse, multi-faceted, practice field with considerable promise for assisting policy makers and advisors as they grapple with governance challenges. There is clearly scope for sharing insights about the use of diverse visualization methods for different purposes, their benefits and limitations, and how the investments of scarce public resources can be optimally leveraged. Moreover, finding better ways to share insight from data and dialogue and provide perspective on complex challenges should have important yields for accountability, reporting on the progress of government, and engaging policy makers, citizens, and other stakeholders in more productive dialogue on how to improve policy and service delivery regimes. The demand for visualizations can only continue to increase as more citizens and leaders take an interest in the possibilities such techniques afford, as the computing costs for visual techniques continue to drop, and a new generation of users expect more sophisticated visualizations to accompany presentations and briefings. As more visualization technology connects with decision-makers, additional demands will emerge for quick, capable response times from advisory units.

Even if governments previously under-invested in visualization techniques, this has not precluded bottom-up experimentation and selective investments across the public sector. This chapter has set out a general strategy for cultivating and managing a network of distributed capabilities in visualization across public sector systems. The mutually reinforcing elements of such a strategy include: galvanizing network visualization expertise and interested consumers to exchange insight and best practices, leveraging the network to move expertise around the system and invest in strategic capabilities, investing in improving visualization literacy and technique competencies, and developing quality control and provenance protocols. Such a strategy would leverage existing resources, allow for specialized expertise to develop and be shared, and distribute the front-end costs of making new investments in visualization when longer-term pay-offs can be opaque.

There is much that is new and exciting in visualization, but also much that is old: advisors and analysts have always relied on diagrams, maps, pictures, and data to illustrate patterns, and inform and persuade decision-makers. These tools are part of a greater interest in exploring applications in policy making. The next step is to develop more detailed case studies of how decision-makers, advisors, and analysts use visual techniques in different circumstances and ascertain how improvements can be made, with particular attention to developing quick-response capabilities across the range of visualization techniques, ensuring quality in the design and use of visualizations, and demonstrating their provenance and verisimilitude for the purposes at hand.

Notes

1 This paper draws on two papers (Lindquist, 2011a, 2011b) that informed a series of dialogues with public sector executives proceeding under the auspices of Australian National University's HC Coombs Policy Forum. The author gratefully acknowledges funding from the Coombs Policy Forum and the Australia and New Zealand School of Government.
2 For more detail, see Lindquist (2011a).
3 Shneiderman (1996) identified seven kinds of data: one-dimensional, two-dimensional, three-dimensional, temporal, multi-dimensional, tree, and network data. Ward, Grinstein, and Keim (2010) divide visualization techniques into six broad categories: spatial data, geospatial data, multivariate data, trees/graphs/networks, text, and documents.
4 http://www.ifvp.org
5 http://www.vizthink.com. See also http://infosthetics.com/ and http://www.visual complexity.com/vc/, among many others.
6 See McCandless (2009), a book full of such renderings; but scour the Web too!
7 Participants in the ANU Crawford School Coombs Policy Forum roundtable suggested that professional guidelines be developed inside government for this purpose.

8 Some experimentation with, take-up of, and insistence on visualization across the public sector should not be confused with a corporate strategic investment. When compared to the huge outlays of governments for IT systems, communications, and public relations—in the hundreds of millions of dollars every year—the investments in modern visualization for government policy-development and design challenges are surely relatively small.

9 For example, some governments first developed central-distributed "foresight" capabilities, then encouraged departments to develop their own capabilities, while maintaining connections among them (Singapore, 2010).

References

Althaus, C., Bridgman, P., & Davis, G. (2007). *The Australian policy handbook* (4th ed.). Sydney: Allen & Unwin.

Arnstein, S. (1969). A ladder of citizen participation. *Journal of American Planning Association, 35*(4), 216–224.

Atkinson, C. (2008). *Beyond bullet points*. Redmond, WA: Microsoft.

Baer, K. (2010). *Information design workbook: Graphic approaches, solutions, and inspiration + 30 case studies*. Beverly, MA: Rockport.

Bardach, E. (2005). *A practical guide to policy analysis: The eightfold path to more effective problem solving*. Washington, DC: CQ Press.

Bederson, B.B., & Shneiderman, B. (2003). *The craft of information visualization: Readings and reflections*. San Francisco: Morgan Kaufman.

Bishop, P., & Davis, G. (2002). Mapping public participation in policy choices. *Australian Journal of Public Administration, 61*(1), 14–29.

Blackwell, A. F., Phaal, R., Eppler, M. J., & Crilly, N. (2008). Strategy roadmaps: New forms, new practices. In G. Stapledon, J. Howse, & J. Lee (Eds.), *Diagrams 2008* (pp. 127–140). Heidelberg: Springer-Verlag.

Bresciani, S., & Eppler, M.J. (2010). Choosing knowledge visualizations to augment cognition: The managers' view. In *Proceedings from 14th International Conference on Information Visualization*. Washington, DC: IEEE Computer Society.

Chapman, J. (2004). *System failure: Why governments must learn to think differently* (2nd ed.). London: Demos.

Chapman, J., Edwards, C., & Hampson, S. (2009). *Connecting the dots*. London: Demos.

Checkland, P. (1999). *Systems thinking, systems practice*. Chichester, UK: John Wiley & Sons.

Checkland, P., & Poulter, J. (2010). Soft systems methodology. In M. Reynolds & S. Howell (Eds.), *Systems approaches to managing change* (pp. 191–242). London: Open University, Springer-Verlag.

Chen, C. (2005). Top 10 Unsolved Information Visualization Problems. *IEEE Computer Graphics and Applications* (July/August), pp. 12–16.

Chen, C. (2006). *Information visualization: Beyond the horizon* (2nd ed.). London: Springer-Verlag.

Cox, A. (2011). *How editing & design changes news graphics*. Closing keynote presentation, Vis Week 2011, 28 October 2011, Providence, RI.

Dawes, S. S., Helbig, N., & Nampoothiri, S. (2014). *Workshop report: Exploring the integration of data-intensive analytical skills in public affairs education*. Albany: Center for Technology in Government Research Foundation of State University of New York, http://www.ctg.albany.edu/publications/reports/egovpolinet_workshopreport/egovpolinet_workshopreport.pdf

Duarte, N. (2010). *Resonate: Present visual stories that transform audiences*. Hoboken, NJ: John Wiley & Sons.

Eppler, M.J., & Pfister, R.A. (2010). Drawing conclusions: Supporting decision making through collaborative graphic annotations. In *Proceedings from 14th International Conference on Information Visualization.* Washington, DC: IEEE Computer Society.

Feldman, M.S., & March, J.G. (1981). Information in organizations as signal and symbol. *Administrative Science Quarterly, 26,* 171–186.

Fisher, D. (2010). Animation for visualization: Opportunities and drawbacks. In J. Steele & N. Iliinsky (Eds.), *Beautiful visualization* (pp. 329–352). Cambridge: O'Reilly.

Friendly, M. (2008). A brief history of data visualization. In C. Chen, W. Hardle, & A. Unwin (Eds.), *Handbook of computational statistics* (pp. 15–56). Berlin: Springer.

Geyer, R., & Rihani, S. (2010). *Complexity and public policy: A new approach to 21st century politics, policy and society.* New York: Routledge.

Grammel, L., Tory, M., & Storey, M. (2010). How information visualization novices construct visualizations. *IEEE Transactions on Visualization and Computer Graphics, 16*(6), 943–952.

Heer, J., & Robertson, G.G. (2007). Animated transitions in statistical data graphics. *IEEE Transactions on Visualization and Computer Graphics, 13*(6), 1240–1247.

Horn, R.E. (1998). *Visual language: Global communication for the 21st century.* Bainbridge Island, WA: MacroVU.

Horn, R.E. (2001). *Knowledge mapping for complex social messes.* Presentation at the Foundations in the Knowledge Economy, David and Lucile Packard Foundation. Retrieved from http://www.stanford.edu/~rhorn/SpchPackard.html

Hullman, J., & Diakopoulos, N. (2011). Visualization rhetoric: Framing effects in narrative visualization. *IEEE Transactions on Visualization and Computer Graphics, 17*(2), 2231–40.

Hyerle, D., (2009). *Visual Tools for Transforming Information into Knowledge* (2nd ed.). Thousand Oaks, CA: Corwin.

Lengler, R., & Eppler, M.J. (2007). *Towards a periodic table of visualization methods for management.* Retrieved from http://www.visual-literacy.org/periodic_table/periodic_table.pdf

Lindquist, E. (1988). What do decision models tell us about information use? *Knowledge in Society: An International Journal of Knowledge Transfer, 1*(2), 86–111.

Lindquist, E. (2009). *There's more to policy than alignment.* Ottawa: Canadian Policy Research Networks. Retrieved from http://www.cprn.org/doc.cfm?doc=2040&l=en

Lindquist, E. (2011a). Surveying the world of visualization. *HC Coombs Policy Forum Background Paper.* Canberra: Australian National University. Retrieved from https://crawford.anu.edu.au/public_policy_community/research/visualisation/Visualisation_roundtable_2_Background_Paper.pdf

Lindquist, E. (2011b). Grappling with complex policy challenges: Exploring the potential of visualization for analysis, advising and engagement. *HC Coombs Policy Forum Discussion Paper.* Canberra: Australian National University. Retrieved from https://crawford.anu.edu.au/public_policy_community/research/visualisation/Visualisation_roundtable_2_Discussion_Paper.pdf

Lindquist, E. (2012). Presentation notes for symposium on *Visualization and Policy Development.* Victoria, BC: University of Victoria.

March, J.G., & Sevon, G. (1988). Gossip, information and decision making. In J.G. March (Ed.), *Decisions and organizations* (pp. 429–442). Oxford: Basil Blackwell.

Margulies, N., & Valenza, C. (2005). *Visual thinking: Tools for mapping Your ideas.* Bethel, CT: Crown House.

Mayer, I., Bots, P., & Daalen, E.V. (2004). Perspectives on policy analysis: A framework for understanding and design. *International Journal of Technology, Policy and Management, 4*(1), 169–191.

McCandless, D. (2009). *Information is beautiful.* London: Collins.

McDavid, J.C., & Hawthorn, L. (2006). *Program evaluation and performance measurement: An introduction to practice*. London: Sage.

Nutley, S.M., Walter, I., & Davies, H.T.O. (2007). *Using evidence: How research can inform public services*. Bristol, UK: Policy Press.

OECD. (2009). *Focus on citizens: Public engagement for better policy and services*. Paris: OECD.

Ringland, G. (1998). *Scenario planning: Managing for the future*. Hoboken, NJ: John Wiley & Sons.

Rittel, H.W.J., & Webber, M.M. (1973). Dilemmas in a general theory of planning. *Policy Sciences, 4*(2), 155–169.

Roe, E. (1994). *Narrative policy analysis*. Durham: Duke University Press.

Rosell, S.A. (Ed.). (1995). *Changing maps: Governing in a world of rapid change*. Ottawa: Carleton University Press.

Sabatier, P. (1987). Knowledge, policy-oriented learning, and policy change: An advocacy coalition framework. *Knowledge: Creation, Diffusion, Innovation, 8*(4), 649–692.

Scott, C., & Baehler, K. (2010). *Adding value to policy analysis and advice*. Sydney: University of New South Wales Press.

Segel, E., & Heer, J. (2010). Narrative visualization: Telling stories with data. *IEEE Transactions on Visualization and Computer Graphics, 16*(6), 1139–48.

Senge, P. M. (1990). *The fifth discipline: The art and practice of the learning organization*. New York: Doubleday.

Shneiderman, B. (1996, September). The eyes have it: A task by data type taxonomy for information visualizations. In *Proceedings, IEEE Symposium on Visual Languages, 1996* (pp. 336–343). Washington, DC: IEEE Computer Society.

Sibbet, D. (2010). *Visual meetings: How graphics, sticky notes & idea mapping can transform group productivity*. Hoboken, NJ: John Wiley & Sons.

Sibbet, D. (2012). *Visual leaders: New tools for visioning, management, and organization change*. Hoboken, NJ: John Wiley & Sons.

Singapore, Prime Minister's Office, Public Service Division, Centre for Strategic Futures. (2010). *Strategic planning in Singapore*. Draft paper for New Synthesis Roundtable, Singapore.

Spence, R. (2007). *Information visualization* (2nd ed.). London: Addison-Wesley.

Steele, J., & Iliinsky, N. (Eds.). (2010). *Beautiful visualization: Looking at data through the eyes of experts*. Cambridge: O'Reilly.

Thomas, J. J., & Cook, K. A. (Eds.). (2005). *Illuminating the path: The research and development agenda for visual analytics*. Washington, DC: IEEE Computer Society.

Tufte, E.R. (1990). *Envisioning information*. Cheshire, CT: Graphics Press.

Walny, J., Lee, B., Johns, P., Henry Riche, N., & Carpendale, S. (2011). Visual thinking in action: Visualizations as used on whiteboards. *IEEE Transactions on Visualization and Computer Graphics, 17*(12), 2508–2517.

Ward, M., Grinstein, G., & Keim, D. (2010). *Interactive data visualization: Foundations, techniques, and applications*. Natick, MA: AK Peters.

Ware, C. (2004). *Information visualization: Perception for design* (2nd ed.). San Francisco: Morgan Kaufmann.

Weiss, C. H. (1977). Research for policy's sake: The enlightenment function of social research. *Policy Analysis, 3*(4), 531–545.

Weiss, C. H. (1980). Knowledge creep and decision accretion. *Knowledge, l*(3), 381–404.

Wilson, J.Q. (1989). *Bureaucracy*. New York: Basic Books.

PART III
Analysis

5

THE ENDOGENOUS POINT OF VIEW IN POLICY INFORMATICS

George P. Richardson

Introduction

Several analytic approaches that fall under the label *policy informatics* derive at least a portion of their power from a particular underlying perspective, identified in this chapter as the *endogenous point of view*.[1] Though almost always crucial for the policy implications of a policy informatics study, the underlying endogenous perspective is seldom highlighted, often so deeply embedded in the analyses that it is easily missed. It is the purpose of this chapter to bring that essential core perspective vividly to light, with the goal of enhancing practice in the fields that link to policy informatics and clarifying for others the important contributions it can make to policy and theory-building.

In one policy informatics field, system dynamics, the endogenous point of view has been present, both implicitly and explicitly, throughout its 60-year history. The field arguably provides the most accessible entry into understanding the nature of an endogenous perspective and its importance in policy informatics.[2] So we begin with the early history of the field of system dynamics and the writings of its founder, Jay W. Forrester, and proceed through a number of examples selected to expose the richness the perspective can bring to public and corporate policy.

Endogeneity at the Foundation of System Dynamics

In his seminal article, Forrester (1958) laid out his initial statement of the approach that would become known as system dynamics. He founded the approach on what were then four recent developments, developments we now see as consistent and ongoing: advances in computing technology, growing experience with computer simulation, improved understanding of strategic decision-making, and developments in the understanding of the role of feed-back in complex systems.

Subsequently, Forrester phrased the four foundations differently, but the emphasis was much the same (1961, p. 14):

- the theory of information-feedback systems;
- a knowledge of decision-making processes;
- the experimental model approach to complex systems; and
- the digital computer as a means to simulate realistic mathematical models.

The slight change in this list emphasizes an *experimental* or *iterative approach* to understanding the dynamics of social organizations that would presumably be enabled by repeated computer simulation. Thus, we have the familiar and endur-ing cornerstones of the system dynamics approach as they were expressed at the founding of the field. But within 10 years, Forrester expressed the foundation quite differently.

The Foundation 10 Years Later

Forrester (1968a) focused on a small but deeply insightful model now known throughout the field as the *market growth model*. At that time he was working with the former mayor of Boston, John Collins, and others on urban problems. He began by presenting the structure of the approach he used, not as the four threads outlined above, but as a four-tiered structural hierarchy (Forrester, 1968a, p. 83; Forrester, 1969, p. 12; Forrester, 1968b, pp. 4–17):

- Closed boundary around the system
  - Feedback loops as the basic structural elements within the boundary
    - Level (state) variables representing accumulations within the feedback loops
    - Rate (flow) variables representing activity within the feedback loops
    - Goal
      - Observed condition
      - Detection of discrepancy
      - Action based on discrepancy

In this hierarchy, it is easy to miss the item that appears at the top of the list: the *closed boundary around the system*. But that phrase signals what may be the most significant part of the system dynamics approach for understanding complex systems—Forrester's endogenous point of view.[3]

The Endogenous Point of View

Forrester initially stated that formulating a model of a system should start with the question, "Where is the boundary, that encompasses the smallest number of components, within which the dynamic behavior under study is generated?" (1968b, §4, p. 2). This idea was summarized in Forrester's first principle of system structure, the closed boundary: "In concept, a feedback system is a closed system.[4] Its dynamic behavior arises within its internal structure. Any action that is essential to the behavior of the model being investigated must be included inside the system boundary" (Forrester 1968b, §4, p. 2). Specifically, Forrester notes, "The closed-boundary concept implies that the system behavior of interest is not imposed from the outside but created within the boundary" (Forrester 1969, §4, pp. 1, 2).

Consider the reach of these statements. They tell us to build models that are capable of deriving the dynamic behavior of interest solely from variables and interactions within some appropriately chosen system boundary. They tell us not to depend upon any exogenous forces to produce the dynamics of interest. Moreover, they suggest guidelines for *thinking*: try to think of dynamics that way; try to *understand* system dynamics as generated from *within* some conceptual mental boundary. That effort, seeking the sources of change *within* a conceptual boundary, brings a particular power to policy informatics.

Endogeneity and Feedback

The notion of endogeneity, the generation of change from within, is intimately linked to the notion of circular causality, or *feedback*. It is likely that most fields of human investigation have observed closed loops of information and action and made use of the concepts of balancing and reinforcing feedback processes. Feedback is everywhere in scholarly literature and informal conversation (Richardson, 1991). Policy studies in particular have begun to emphasize the ways in which naturally occurring feedback loops can defeat policy initiatives that otherwise look highly promising, or can reinforce small beginnings to create explosive change. In this chapter on endogeneity in policy informatics, it is vital to realize that feedback loops are really a *consequence* of the endogenous point of view.

Figure 5.1 illustrates the idea. On the left is a simple causal system with causal elements tracing ultimately outside the system boundary. The dynamics of variables A through E are generated partly by interactions among them inside the system boundary, but stem mainly from variables P, Q, R, and S outside the boundary. The dynamics of this system are generated *exogenously* by forces outside the system boundary.

On the right is an endogenous view in which the dynamics of variables A through E are generated solely from interactions among those variables themselves within the system boundary.[5] The figure suggests that taking an endogenous point of view forces the causal influences to form *loops*. Without loops, all causal influences would

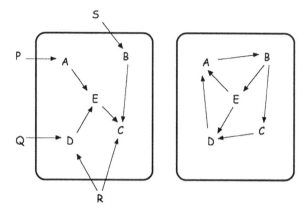

FIGURE 5.1 Left: An *exogenous* view of system structure—causality traces to external influences outside the system boundary. Right: An *endogenous* view—causality remains within the system boundary; causal loops (feedback) must result.

trace to dynamic forces outside the system boundary. Feedback loops thus enable the endogenous point of view and give it structure (Richardson, 1991, p. 298).

Using Endogeneity to Solve Complex Policy Problems

Two examples from Forrester's early writings serve to illustrate vividly what the endogenous point of view looks like in practice and why it is so crucial in thinking and modeling for policy analysis.

Market Growth as Influenced by Capital Investment

The first example (Figure 5.2) is the market growth model on the left (Forrester, 1968a) and vividly shows Forrester's endogenous point of view. The left side of Figure 5.2 shows a structure with no dynamic influences coming from outside the system boundary. The dynamics shown on the right side of Figure 5.2 stem solely from interactions among the variables shown and the balancing and reinforcing feedback loops they form together.

Forrester developed this model to illustrate a potential source of poor corporate performance; the internal operating policies of the company itself. To make the story most vivid, he set the company represented by the model in a potentially infinite market. No external market cap exists in the model to potentially limit corporate growth, production capacity, the size of the sales force, or the number of orders booked per month. Yet the dynamics shown in Figure 5.2 show a company in which production capacity is declining, booked orders have a pronounced oscillatory and eventually declining pattern, and the size of the sales force appears to peak and decline, although there is no external limit to the size of the market that salesmen might reach. In an unlimited market, the company is going out of business.

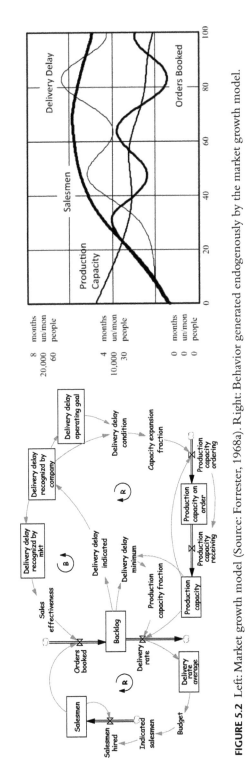

FIGURE 5.2 Left: Market growth model (Source: Forrester, 1968a). Right: Behavior generated endogenously by the market growth model.

The key to these dynamics is the policy the company is using to determine when to add production capacity. The policy requires that additional capacity be added when the delivery delay for the company's product exceeds a target, the Delivery Delay Operating Goal. If that goal is based on past performance, as is the case in this simulation, then it tends to slide upward as the actual delivery delay rises. As booked orders increase, the delivery delay initially increases as well, exerting pressure to expand capacity. But that pressure depends on a comparison of the delivery delay recognized by the company and its target delivery delay. Since that target is based on past performance, it too is rising slightly, somewhat relieving the pressure to expand capacity. In this simulation, the sliding delivery delay target never generates enough pressure on the delivery delay condition to actually expand capacity, and the company therefore loses market share. It is likely that corporate decision-makers in such a setting would believe the causes of the declining sales trace to an overall declining market, when in fact (using this model) that market is potentially infinite. It is the internal operating policies of the company that are causing decline.

Three key insights should inform our understandings of the endogenous point of view here. First, we see in Figure 5.2 (left) Forrester's unmistakable "closed [causal] boundary around the system." There are no causal links coming from outside. There is a hint of "roundness" to the picture that would be characteristic of such a closed causal boundary and the feedback structure it forces.

Second, the dynamics generated by the model come from the interactions of the variables inside the model boundary. There is no declining market cap from outside that would inhibit growth. We see that the self-contained loop structure in the model is sufficient by itself to generate the observed dynamics.

Third, we see that Forrester designed the model to tell the endogenous story of declining sales in the most vivid way possible: He put the company in a potentially infinite market. One might well argue no company exists in such a market and so the model is unrealistic and invalid. But the potentially infinite market takes away all possibility that the declining sales can trace to anything other than an endogenous source. This is modeling for endogenous insight and understanding.

Think how extraordinarily difficult, if not impossible, it would be to reach these potentially crucial policy-related conclusions without an endogenous perspective or, to put it more forcefully, without an endogenous *bias* in one's point of view. This endogenous perspective is neither well understood nor accepted by many policy practitioners and scholars. Forrester's second classical example provides a dramatic illustration.

Urban Dynamics

The 1960s was a period of urban renewal in the United States. Many policies were implemented in efforts to halt central city decay and return old cities to the economic vibrancy of their early days (Forrester, 1969, pp. vii–x). Forrester chose to view an archetypical city in a large and dynamically uninteresting environment.

Significantly, the perspective he took and the model he built did not include sub-urbs around the city or the transportation networks linking it to its environment and was criticized for such glaring omissions. In fact, the environment around the city generated no dynamic influences whatsoever on the modeled city, which was presumed to be small enough in the world setting not to affect its environment. In reality there might be external economic cycles, political and social movements, and cultural changes affecting the city and its environment, but Forrester chose to explicitly ignore those exogenous dynamics. The dynamics of interest were urban dynamics *relative to* the environment outside the city. They were to be insistently and undeniably *endogenous* dynamics generated by the city itself.

To illustrate Forrester's more complex Urban Dynamics model structure, we show in Figure 5.3 the much simpler structure developed by Alfeld and Graham (1976) in the Urban1 model. For each of the stocks shown in Figure 5.3, Forrester had used three stocks, disaggregating housing and businesses into new structures, mature structures, and aging structures, and separating population into underem-ployed, skilled workers, and managerial-professionals. Forrester's model had about 150 equations; Urban1 has 21.

Despite the simplifications, the dynamics of Urban1 match the Urban Dynam-ics model and it provides a marvelous introduction to the insights Forrester obtained. As shown in Figure 5.3 (right), the model demonstrates a growth phase with low unemployment (indicated by the labor-to-job ratio), which turns rather quickly into a no-growth stagnation phase followed by a long period of decline in population and business structures. During the stagnation phase, unemployment rises and remains high for the rest of the simulation.

Like the larger Urban Dynamics model, Urban1 endogenously captures the dynamics of mature cities. The emergence of urban stagnation, the rise in unem-ployment, and the onset of urban decay come about without any external influences.

In addition, the behavior of these urban models illustrates a deep understand-ing about complex dynamic systems: the models are astonishingly resistant to parameter changes. In Urban1 there are simple constants for "jobs per business structure," the average lifetimes of business and housing structures, the normal rate of business or housing construction in the city, the normal rates of immi-gration and emigration affecting the urban population, and even the fraction of the population participating in the urban work force. One can change any of these various parameters, representing realistic policy changes in a city, and the model behaves realistically, but nothing works to prevent the stagnation an decay shown in the graphs on the right of Figure 5.3! Some parameter changes, like increasing jobs in the city, do good things in the short run, but fail in the long run to prevent high unemployment. The policy parameter changes are naturally compensated for by the relevant endogenous feedback pressures at work on the city.

Once again, we see a number of crucial insights stemming from an endog-enous modeling approach and an endogenous perspective on socio-economic dynamics. First, the dynamics of the city in these urban models do not depend

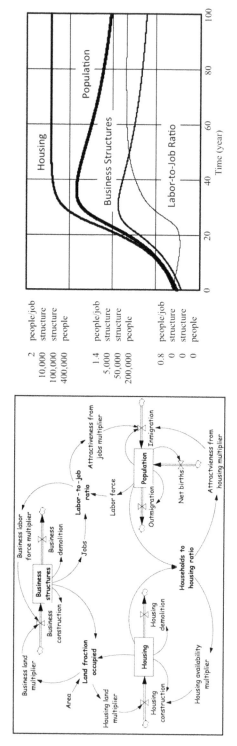

FIGURE 5.3 Left: The stock-and-flow feedback structure of URBAN1 (Alfeld and Graham, 1976). Note no exogenous dynamic influences. Right: Dynamic behavior of URBAN1, showing growth, stagnation, and decay of population and businesses, and the dramatic rise of unemployment in the built-up city.

upon dynamics outside the system boundary, in what Forrester (1969) conceptualized as a "limitless environment." All dynamics are generated from within. Contrary to the dominant perspective at the time, this model shows that suburbs are not causing urban decay; the cities in these models are doing it to themselves.

Second, the possibility that the sources for Urban growth, stagnation, and decay trace to perceptions, interactions, and forces inside the city itself would never have emerged had Forrester expanded the system boundary to include suburbs, national urban policy, or national economic dynamics. The endogenous point of view was crucial for the insights we can derive from Urban1 and Forrester's (1969) study on urban dynamics. In fact choosing a more narrow system boundary than most would have done proved to be the key to deep urban dynamics insights.

Third, forcing dynamics to be endogenous actually forces causality in these models to become circular. Feedback loops in these models are consequences of the assumption of a closed causal boundary. Thus, the endogenous point of view provides the essential perspective capable of illuminating crucial compensating feedback effects that can conspire to defeat favorite policies.

Fourth, feedback loops become crucially important for understanding urban policy. For example, the feedback loops containing the labor-to-job ratio completely neutralize the simulated jobs program and render it useless in the long run.

How might thinking exogenously have affected conclusions emerging from these urban models? Consider just one: suppose, as some suggest, we chose to use time series data for urban population projections. Suppose the data-based projections were very carefully developed by talented data analysts using sophisticated statistical tools and econometric methods, and suppose those projections were fed into Urban1 in place of the endogenous stock of population. Sadly, population would not show a bump because the sophisticated exogenous time series data would not be influenced in the slightest by changing conditions in the model. We would not see the compensating urban migration effect and would miss the crucial conclusion that population and business construction dynamics would naturally offset the effects of the jobs program. We may think it was a long-term policy success, but in reality we would be dramatically misled.

Perhaps that conclusion is obvious here. But it may come as a surprise to learn that some experienced modelers and knowledgeable critics have been seduced by sophisticated statistical methods and believe that time series data improves validity of dynamic models of social systems. Yet such exogenous time series in these urban models would destroy the potential policy insights these models can illuminate.

Endogeneity and Agency

To this point, we have focused on aggregate endogenous forces in dynamic systems without thinking in detail about the actors involved. We turn now to the question of agency in complex policy domains. Who are the actors in the dynamics of a complex system and how do their perceptions, pressures, and policies interact? Are you and I, or the groups we represent, part of the endogenous system structure responsible for the system behavior we perceive? Are we parts of the problem, or parts of the solution, or merely bystanders watching difficult dynamics play out over time? To what extent and how are we agents of our own problematic policy dynamics?

The following examples probe exogenous and endogenous points of view on problems in our time and the question of exogenous or endogenous agency in the problem dynamics. We conclude with an important framework for viewing such complex issues and a rich understanding of what endogenous *systems thinking* in policy informatics can contribute. For illustrations, we consider just two: global climate and floods.

Global Climate

Various groups in the United States, Europe, and the rest of the world are engaged in occasionally heated conversations about global climate change. The data about the average global temperature seems not to be at issue. Global average temperature has been rising steadily throughout the twentieth century, particularly more sharply since 1970. That sharp rise coincides with a sharp rise in the global average atmospheric concentration of carbon dioxide, a significant heat-trapping gas.

That is where the controversy begins. Some argue that human activity is at least partially responsible for the rise in global temperature because it is human production and use of fossil fuels that has pushed CO_2 levels to dramatic heights. Others argue that current changes in global CO_2 concentrations and temperature are simply part of a natural cyclic phenomenon that has been going on for millennia. Thus, there is an exogenous point of view on global warming, maintaining that current conditions are simply part of an exogenously generated cycle, and an endogenous view, holding that human activity in the form of the burning of fossil fuels is exacerbating temperature increases.

The exogenous view has some support in the mechanisms known as Milankovitch cycles,[6] the theory that structural aspects of the earth's orbit and variations in the intensity of solar radiation are responsible for long-term temperature cycles. Milankovitch cycles produce very long-term oscillatory patterns in global temperature that are viewed as stages of glaciation.

Both the exogenous and endogenous views on global climate change have a feedback structure underlying at least some of the dynamics. The water and carbon cycles are influenced by atmospheric and ocean temperatures and they, in turn, influence global temperatures from the reflective effects of clouds and ice and the heat trapping effects of water vapor and CO_2 concentrations. Figure 5.4 shows four external influences on interactions in the global climate system. Those who argue that there is a human role in rising global temperatures emphasize fossil fuel use and human production of aerosols and greenhouse gases. Those who argue that there is a long-term natural cycle similar to the Milankovitch theory would de-emphasize those external forces and emphasize variations in the solar energy reaching earth.

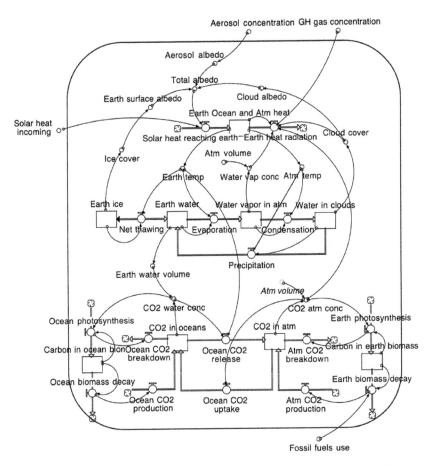

FIGURE 5.4 An overview of the water and carbon cycles and their feedback effects on global temperature, showing four potentially exogenous influences. (Source: Bernstein, Richardson, & Stewart, 1994, and related work of the author; Fiddaman, 2002)

Thus, while balancing and reinforcing feedback interactions exist in all serious views of global climate dynamics, the global warming debate comes down to questions of agency and the implications of human action. Table 5.1 summarizes the points of view.

Floods

The destruction caused by devastating floods along rivers and coastal waters provides another vivid contrast of exogenous and endogenous agency. Costs from flood damage in the United States have increased over the course of the twentieth century (Deegan, 2007; NOAA, n.d.). One obvious perspective on these increasing costs states that over the twentieth century, floods in the United States became more frequent, or more severe, or both (see Figure 5.5 [left]).

TABLE 5.1 Summary of exogenous and endogenous agency perspectives on global climate change

Global Climate	Perspective	Policy Implication
Exogenous View	We are in the warm phase of a 100,000 year cycle caused by exogenous, structural characteristics.	Adapt to the inevitable.
Endogenous View	Human activity is exacerbating the natural cycle.	Alter human habits to minimize the coming tragedies.

However, Deegan's (2007) extensive analysis suggests another explanation, an endogenous view of the dynamics of flood damage that takes into account the human role in creating property vulnerable to flood damage. He points out that "disaster occurs when hazard meets vulnerability." He then traces the dynamics of vulnerability to the interacting actions of the capacity of the local environment to withstand floods, development pressure, property tax needs, perceived risks of development, moral hazard, policy entrepreneurs, and other people pressures. See Figure 5.5 (right) for an overview of the feedback structures that he hypothesizes are at work in the dynamics of flood-related property damage. Deegan's (2007) simulation experiments are telling. In the base run, he simulates five identical floods ten years apart. The important observation is that the property damage is not the same after every flood. The causes of damages over time trace to endogenous feedback structures that generate the dynamics of vulnerable property.

Again, we have open-loop and feedback perspectives on the dynamics of flood damage. Table 5.2 summarizes these exogenous and endogenous points of view.

We should observe that only in the endogenous point of view with explicit human agents is there the empowering perspective that human behavior can have a significant policy role that could minimize future flood-related damage. It is the identification and endogenous inclusion of various stakeholders and policy entrepreneurs in Deegan's (2007) model that gives it the power to aid reliable policy analysis and decision-making. Those actors pressured by policies respond endogenously in the model, as they would likely do in reality, and we can anticipate the dynamics they can create. Thinking endogenously in policy informatics can make a real-world difference.

Implications for Policy Informatics: Choosing Points of View

As we think about the way natural and social systems play out over time, we have a choice of perspectives. We can view dynamics as arising largely from exogenous forces outside our purview and control, or we can see ourselves and the decisions we make as part and parcel of the ebb and flow of things. It also seems reasonable that reality has a similar *choice*: the dynamics we think about might materialize into reality from largely exogenous forces or from predominantly endogenous

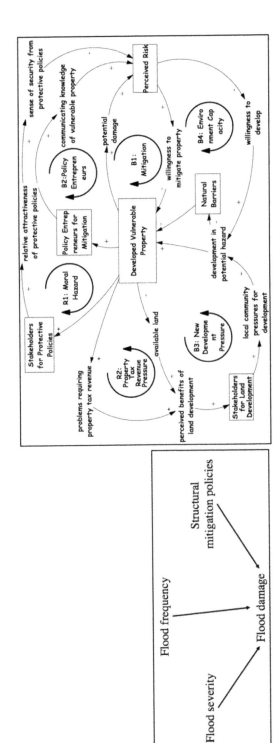

FIGURE 5.5 Left: An exogenous perspective on floods and flood damage, suggesting the reasonable conclusion that the main causes of flood damage are the random severity and frequency of floods. Right: Simplified endogenous view of the dynamics of flood-related damage (Deegan, 2007).

TABLE 5.2 Summary of exogenous and endogenous agency perspectives on flood damage

Flood Damage	Perspective	Policy Implication
Exogenous View	Floods sometimes occur; the greater the flood, the worse the damage.	When floods occur, recover and rebuild.
Endogenous View	Damage occurs when hazard meets vulnerability; vulnerability is a result of people policies.	Recognize the human role in damage. Work with stakeholders to minimize vulnerabilities.

interactions. Thus we have a two-by-two table, the X/N matrix in Figure 5.6, a distant cousin of the Taylor-Russell diagram.

Figure 5.6 suggests four possibilities. In two of the cells, we are *correct*. The cell in the lower left of the table represents those situations in which decision-makers see the phenomena as largely exogenously produced and in reality that is essentially the true situation. In the upper right cell, decision-makers are taking an endogenous point of view of what are essentially endogenously generated dynamics. The other two cells represent situations in which our perspective does not match reality, analogous to Type I and Type II errors in statistics. Perspectives and reality probably do not fall neatly into such discrete boxes, but the simple characterizations help to dramatize the implications of the exogenous and endogenous points of view.

In the lower left cell (exogenous view of exogenous phenomena), we see correctly that there is little that can be done to alter the future state of affairs, so our best course is to accept our fate—predict and prepare for whatever we believe is coming. The expressions on the faces in the table suggest that even though this is a correct perception, we probably are not very happy about it.

In the upper left cell (endogenous view of exogenous phenomena), we are trying to discover endogenous understandings, but we will fail in such attempts because the situation is predominantly determined externally. In this scenario, we are wrong, and our efforts at endogenous explanation and understanding are misguided. Yet, for this author, and perhaps most of us, the effort to find some endogenous aspects here give us hope for understanding and control (or at least the *illusion* of control) and probably makes us feel better than we do in the lower left. So the faces are still mixed, but happier.

In the upper right (endogenous view of endogenous phenomena), we are taking the correct perspective for the situation and we are potentially empowered by it. The situation is best understood in feedback terms, with the possibility that human agency contributes to problems and solutions. We see the endogenous aspects of corporate conditions, urban dynamics, global climate, or terrorism from an endogenous perspective and we have the chance of wisely influencing how things will play out over time. That cell merits three very happy faces.

In contrast, the cell on the lower right (exogenous perspective on endogenous phenomena) is a dismal prospect, characterized by three very unhappy faces. Here we take the point of view that circumstances are largely caused by external forces,

		Exogenous	Endogenous
Predominant Mode of Analysis	Endogenous	Striving for understanding and leverage, but failing ☺☹	Achieving understanding and leverage ☺☺☺
	Exogenous	Accepting fate, Predicting, Preparing ☺☹	Confused, Misguided, Misguiding ☹☹☹
		Exogenous	Endogenous

True (Predominant) State of Affairs

FIGURE 5.6 The X/N matrix—eXogenous and eNdogenous perspectives contrasted with their corresponding "true" states of affairs that are mainly exogenous or endogenous, showing four possible (idealized) perception/reality combinations and their implications

when in fact there are essentially correct endogenous, empowering explanations. In this cell, we are destined to be confused and misguided and we will misguide others.

If one were to recast the table in Figure 5.6 as a decision tree, the decision to take an endogenous point of view in all circumstances would have the highest net payoff, at least in terms of happy faces and the real feelings they represent. An endogenous point of view is potentially empowering and that feels good to us.

Conclusion

We began this exploration with the early thoughts of Forrester on the essential nature of the system dynamics field he created. Through those early writings and a number of wide-ranging examples, we have explored the nature of exogenous and endogenous points of view.

Our first implication is straightforward, though somewhat narrow: *the endogenous point of view is a crucial foundation of the field of system dynamics*. In this interpretation, it is fair to say that system dynamics is the use of informal maps and formal models with computer simulation to uncover and understand endogenous sources of system behavior. This characterization is true to the writings of the founder of the field and is reflected in the best work of its practitioners.[7]

The endogenous point of view is probably most developed and exercised in the system dynamics approach. But a view of *system as cause* is no less crucial in other policy informatics perspectives and to policy analysis in general. It is only such an inward-looking view that can reveal the many ways complex systems react naturally to defeat favorite policy initiatives, to compensate internally for externally imposed strategies. Ideally, the endogenous point of view will become central in some way to all policy informatics approaches as we seek sustainable policy improvements for the common good.

Notes

This chapter is based on Richardson, G. P. (2011). Reflections on the foundations of system dynamics. *System Dynamics Review, 23*(3), 219–243. Reprinted with permission.

1 "Endogenous" (from the Greek fragments "endo" and "gen," meaning "inside" and "production") refers to an action or object coming from within a system. It is the opposite of "exogenous," something generated from outside the system (http://en.wikipedia.org/wiki/Exogeny).
2 While system dynamics may give us the easiest entry, the notion of endogeneity has a historic and deeply significant presence in the social and natural sciences and even informal common discourse (Richardson, 1991).
3 In more recent literature in the broad area of systems thinking, the notion of endogeneity is often encapsulated as *system as cause.*
4 Forrester's use of the term *closed* means *causally* closed. His use is different from the notion of a closed system in general systems theory, which refers to a system that is *materially closed*, that is, it does not exchange material or information with anything outside the system boundary. Forrester's *closed boundary* systems are, in general systems theory terms, *open systems* because they include little clouds representing sources and sinks of material outside the system boundary (Richardson, 1991, p. 298).
5 In practice, confronted with a view like the left side of Figure 5.1, one strives to expand the system boundary to draw in the influences that initially seemed exogenous and create a rich endogenous view.
6 See http://en.wikipedia.org/wiki/Milankovitch_cycles.
7 For example, see http://www.systemdynamics.org/AwardRecipients.htm for a list of winners of the Forrester Award.

References

Alfeld, L.E., & Graham, A.K. (1976). *Introduction to urban dynamics.* Waltham, MA: Pegasus.
Bernstein, D.S., Richardson, G.P., & Stewart, T.R. (1994). A pocket model of global warming for policy and scientific debate. In *Proceedings from International System Dynamics Conference.* Stirling, Scotland.
Deegan, M. (2007). *Exploring U.S. flood mitigation policies: A feedback view of system behavior.* (PhD dissertation). Albany, NY: Rockefeller College of Public Affairs and Policy, University at Albany—SUNY.
Fiddaman, T.S. (2002). Exploring policy options with a behavioral climate-economy model. *System Dynamics Review, 18*(2), 243–267.
Forrester, J.W. (1958). Industrial dynamics: A major breakthrough for decision makers. *Harvard Business Review, 36*(4), 37–66.
Forrester, J.W. (1961). *Industrial dynamics* (Vol. 2). Cambridge, MA: MIT Press.
Forrester, J.W. (1968a). Market growth as influenced by capital investment. *Industrial Management Review, 9*(2), 83–105.
Forrester, J.W. (1968b). *Principles of systems.* Waltham, MA: Pegasus.
Forrester, J.W. (1969). *Urban dynamics.* Waltham, MA: Pegasus.
NOAA. (n.d.). *National Weather Service Hydrologic Information Center.* Retrieved from http://www.nws.noaa.gov/oh/hic/flood_stats
Richardson, G.P. (1991). *Feedback thought in social science and systems theory.* Philadelphia, PA: University of Pennsylvania Press.
Richardson, G.P. (2011). Reflections on the foundations of system dynamics. *System Dynamics Review, 23*(3), 219–243.

6

MODEL-BASED POLICY DESIGN THAT TAKES IMPLEMENTATION SERIOUSLY

David Wheat

Introduction

Low-cost and low-risk policy experimentation is an acknowledged virtue of simulation models of complex public issues (Ghaffarzadegan, Lyneis, & Richardson, 2011). But the virtue can become the vice. It is often *too* easy to ask, "What if this parameter value could be changed?" and get a quick quantitative answer. Model-based analysis that relies exclusively on parameter sensitivity testing may ignore how parameter changes in a computer model can be implemented by public organizations in the real world. Of course, testing a model's sensitivity to variations in policy parameters is an important exploratory step in model-based policy analysis. Too often, however, there is no next step. A content analysis of three decades of articles published in the *System Dynamics Review* found that policy analysis has been limited to parameter sensitivity testing in nearly 75 percent of models of public issues (Wheat, 2010). This situation has developed despite admonitions from experienced modelers. Richardson and Pugh (1989) warned that "policies represented as parameter changes frequently tend not to be very effective in system dynamics analysis" (p. 332) and Sterman (2000) reminded modelers that "policy design is much more than changing the values of parameters" (p. 104).

To illustrate the limits of parameter-based policy analysis, suppose simulation results show that a flu epidemic might be avoided by vaccinating 20 percent of an at-risk population monthly (i.e., the monthly vaccination parameter equals 0.20). What assumptions about institutional arrangements, organizational capacity, and public cooperation are implicit in that simulation experiment? When is it advisable to *model* some of those assumptions explicitly—to reveal the "plumbing" of a policy and how it functions in the model? And what modeling tools, policy concepts, and literature guidance would be helpful? To answer such questions, we encourage model-based implementation planning and analysis during the policy design stage of system dynamics simulation modeling. This continues our quest for a synthesis of system dynamics and the implementation paradigm in the public policy literature (Wheat, 2010). The goal is to encourage more *operational* thinking during model-based policy design and produce models that are less reliant on wishful thinking and more useful to policy makers.

We begin with a brief description of system dynamics *explanatory* modeling as a prerequisite for *policy* modeling. The focus then shifts to implementation planning and the importance of operational thinking during the policy modeling stage. The entire process is illustrated with models of two public health issues: a flu epidemic and automobile pollution. The final section highlights the application of useful modeling insights gleaned from implementation literature.

The System Dynamics Modeling Process

It is useful to think of system dynamics (SD) modeling in terms of two high-level tasks: *problem explanation* and *policy design*. More than 40 years ago, Forrester (1969) emphasized the practical value of this distinction and recently reiterated it (Forrester, 2009). The goal of *problem explanation* modeling is to identify the historical systemic reasons for a pattern of behavior widely viewed as a serious issue (e.g., increasing traffic congestion or declining employment). The *policy design* task is to explore and evaluate ways to alleviate the problem; that is, to improve the dynamic performance of the model system in ways that suggest feasible, cost-effective policies in the real world system that the model represents.

Problem Explanation

The first task is to build and test an *explanatory model*. Although SD scholars emphasize different details of the modeling process, there is a consensus about the key steps when developing the explanatory model: (1) specify the symptomatic problem in terms of its dynamic behavior, (2) develop a hypothesis in the form of a model that offers a structural explanation of the problematic dynamics, and (3) analyze the model with various validation tests aimed at discovering whether it provides a robust endogenous explanation of the problematic behavior (Randers, 1980; Richardson & Pugh, 1989; Coyle, 1996; Sterman, 2000; Barlas, 2002; Morecroft, 2007; Roberts, 2007; Ford, 2010).

An explanatory SD model uses functionally interdependent variables to simulate problematic behavior patterns (e.g., undesirable trends in crime or pollution) that have been observed in the past. Bardach (2005) concludes that policy analysts will find a "good causal model . . . especially [useful] . . . when the problem is embedded in a complex system of interacting forces, incentives, and constraints—which is usually the case" (p. 17). Forrester (2009) states it more bluntly: "Only by clearly understanding what is causing the problem can one begin to see where [policy] attention should be focused." Moreover, a policy has a better chance of acceptance and adherence when its underlying causal theory

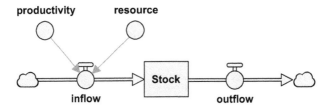

FIGURE 6.1 Example of a simple system dynamics stock-and-flow model

resonates with those whose cooperation is needed for adoption or implementation (Mazmanian & Sabatier, 1981; Hogwood & Gunn, 1984).

A simple example of an explanatory model is displayed in Figure 6.1, which shows two central concepts in an SD model: a stock and its flows.[1] In this example, productivity and resource are parameters, the values of which are determined out side the boundary of a model. The inflow could be defined as the product of those parameters. A stock accumulates (or dissipates) over time as the net flow determines its rate of change. Examples of stocks include patients in a hospital, pollution in the air, and housing in a community. Corresponding inflows would be the rate of patient admissions (patients/month), pollution emissions (pollutants/hour), and housing construction (houses/year), while outflows would be the rate of patient discharge, pollution dispersion, and housing demolition.

When revising a simple explanatory model so that it reflects more realistic complexity, the modeler reformulates some of the exogenous *parameters* as *variables*. Those variables, in turn, would be functionally dependent on a combination of other parameters and variables along chains of causality. The process of *modeling backwards* from a flow along a particular causal chain continues until (a) it ends with an exogenous parameter, or (b) it traces a circular path all the way back to the flow, thus forming a feedback loop—the third central SD concept—that provides an endogenous explanation of the observed dynamics.

Policy Design

The next task is to build and test a *policy model*. Policy modeling requires representing a real-world policy option in the form of new structure added to the explanatory model. The potential for alleviating problematic behavior is then evaluated with simulation analysis. The SD literature provides examples of adding policy structure (Richardson & Pugh, 1989; Sterman, 2000; Ford, 2010), but more conceptual and technical guidance is needed. That is our purpose in this chapter. We build on the feedback tradition in SD policy design in a way that integrates parameter and structure analysis, prompts operational and politically insightful questions when implementation planning is needed, and facilitates adding implementation constraints to a policy model when it is useful to do so.

The Feedback Approach to Policy Design

Integrating Parameter and Structure Analysis

Parameter testing is necessary and useful, even if it is insufficient as a sole policy analysis method. Making an exogenous adjustment to a parameter value and then simulating is a quick and easy way for a modeler to estimate the potential impact of a general strategy for influencing key feedback loops in a problematic model system, and find what Richardson and Pugh (1989) call "leverage points" in the system (p. 322). What is needed, however, is a way to use parameter testing as a springboard to structural design of policy options, i.e., to integrate parameter and structure analysis.

The behavior of the model in Figure 6.1 would depend, in part, on the values estimated for the parameters: *resource* and its *productivity*. Simulating with alternative values for the *resource* would enable preliminary testing of the impact of a policy initiative aimed at changing the acquisition and utilization of the *resource*. Figure 6.2 displays a simple policy model that extends the Figure 6.1 model and specifies a dynamic feedback decision process, in which changes in the *resource* depend on a *goal* for the *stock*. For our purposes, it highlights potential constraints on the impact of a policy due to implementation requirements.

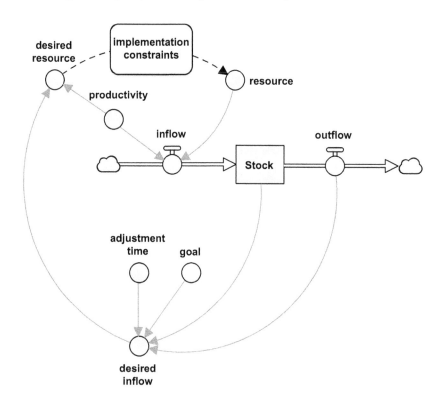

FIGURE 6.2 Simple policy model highlighting implementation constraints

The policy modeling process begins with specification of a *goal* for the *stock* and a realistic and politically acceptable *adjustment time* for closing the gap between desired and actual conditions. For example, a fiscal policy goal could be a level of government debt that is considered sustainable. An environmental policy goal might be an air pollution concentration level that is deemed safe to breathe. A social policy goal could be the number of families receiving welfare assistance that corresponds to a politically acceptable fraction of all families in a society.

It is also necessary to choose a basic strategy for achieving the policy goal. Managing a stock toward a goal requires regulating at least one of its flows. Thus, generating a list of strategic options means specifying whether the inflow or outflow side of a stock will be the target of the policy initiative. For example, a business manager faced with declining demand and rising inventories might adopt an inflow strategy (lay off workers to cut production) or an outflow strategy (cut prices to spur sales) or some combination of both. A public health official fearing an epidemic might contemplate draining the stock of infected persons through an isolation strategy that reduces their contact with susceptible persons (Wheat, 2010). A vaccination strategy, on the other hand, seeks to drain the stock of susceptible persons before many of them have contact with infected persons (Wheat & Shi, 2011). In Figure 6.2, the strategy focuses on regulating the *inflow* by modifying the *resource* level. The *resource* and its *productivity* will drive the *inflow* and, after some delay, the *stock* will have a feedback effect—increasing or decreasing the *desired resource*, as needed to move toward the policy goal.[2]

Next, it is necessary to test the policy feedback loop to see *if* the model performs as expected and *why*. Does the stock adjust to its goal during the simulation? If not, the "desired" equations are either inaccurate or incomplete and must be corrected before beginning any other task. Of course, the right result could be achieved for the wrong reason. That is why behavior analysis is only one of several validation tests in the SD modeling process. Each equation must be scrutinized theoretically and empirically, parametric *stress tests* should be conducted to enable assessment of a model's behavior under extreme conditions, and feedback loops should be analyzed to understand the source of endogenous dynamics.

Operational Questions for Implementation Planning

A big assumption along the policy feedback loop is that a *desired* resource can be transformed into an *actual* resource. Sterman (2000) emphasizes the importance of distinguishing between *desired* and *actual* quantities in a model:

> Modelers should separate the desired rates of change in system states from the actual rates of change. Decision makers determine the desired rates of change in system states, but the actual rates of change often differ [from desired rates] due to time delays, resource shortages, and other physical constraints.
>
> *(Sterman, 2000, p. 519)*

In Figure 6.2, the dashed link signifies that transforming the *desired resource* into an *actual resource* is not an automatic, self-executing step. It requires further actions, necessitates use of real resources, and takes time even under ideal conditions. Moreover, policy initiatives usually generate resistance during the implementation phase, and carrying out real-world policies is often constrained by institutional arrangements and organizational and social capacities for change (Hill & Hupe, 2009; Howlett, Ramesh, & Perl, 2009; Knoepfel, Larrue, Varone, & Hill, 2007). Therefore, we refer to the dashed link as a *wishful thinking* link (Wheat & Shi, 2011) because it ignores the operational requirements for transforming desires into reality. Computer-generated policy results are unlikely to be achieved in real life with the same effectiveness, at the same cost, and in the same timely manner.

In Figure 6.2, the oval-shaped box overlaying the wishful thinking link represents a set of implementation constraints that can be expected to impede progress toward a policy goal. As explained more fully in the next section, it is desirable to formulate the set of constraints as a sub-model (i.e., the box should contain additional stock-and-flow structure to represent how the constraints might work). The input to the sub-model would be the *desired resource* and the output would be the *actual resource*.

Anticipating and planning for policy constraints requires what Richmond called *operational thinking*—thinking about how things actually work in the plumbing of the problematic system (Richmond, Peterson, & Vescuso, 1987; Richmond, 1993,

TABLE 6.1 A checklist of feasibility-relevant questions

1	Does the policy require future legislative action by elected officials?
2	Do responsible agencies lack the capacity (budgetary or human resources, technology, or legal authority) to do what the policy requires?
3	Do responsible agencies have discretionary authority to decide what to do and when to do it?
4	Does the policy require a high level of coordination among different agencies within a single government or across jurisdictions?
5	Does the policy require responsible agencies to perform new tasks, develop new procedures, or hire and train new personnel?
6	Do key personnel in the responsible agencies reject the causal theory implicit in the policy, the moral or social justification for the policy, or the basic goals of the policy?
7	Are key officials within responsible agencies distracted by other pressing issues and unlikely to give adequate attention to this policy?
8	Is the policy opposed by organized interest groups with access to the media, the courts, or the responsible agencies?
9	Is the general public sharply divided over the policy?
10	Does the policy conflict with traditional cultural norms and values held by a politically significant segment of the population?

"Yes" answers indicate potential constraints on policy implementation

1994, 2000). As modelers, we cannot know all the answers regarding a policy's feasibility; thus, it is necessary to consult policy domain experts during the modeling process. Insights gleaned from public policy implementation literature can help modelers pose useful questions to the experts (Wheat, 2010). Therefore, before attempting to model the constraints, we should express them as answers to questions that highlight the implementation challenges inherent in a particular policy option.

The questioning process begins by asking where our implementation analysis is located on the *policy timeline*. Has legislation already been passed establishing general policy goals and authorizing (and funding) a government agency to design and execute a specific program of action? If so, implementation means "carrying out a basic policy decision" that has been made by government officials exercising formal authority (Mazmanian & Sabatier, 1983, p. 1). On the other hand, if a policy has yet to be authorized, implementation planning means anticipating what needs to happen (Ferman, 1990). Placing our policy analysis on a timeline reveals the nature of the constraints (legislative, budgetary, managerial, technical, social, cultural, etc.) a particular policy is likely to encounter. The expected benefits of any policy option should be discounted if formal authorization (or adequate funding) is uncertain.

Other questions shift attention to distant points on our policy timeline and require working back toward the present. Elmore (1979, p. 604) urges analysts to ask about "specific behavior at the lowest level of the implementation process that generates the need for a policy," and from there urges analysts to use *backward mapping*—"to back up through the structure of implementing agencies" with questions about requisite organizational capacity along the implementation path. Using the model in Figure 6.2 as an example, we would say that the inflow is the behavior targeted by the policy, and that changing the *resource* quantity applied to that inflow is the strategy to be analyzed. We then need to envision implementing agencies that are likely to be involved, the extent to which their operating procedures could generate the *desired resource*, the decisions and actions needed to activate those procedures, and the capacity and incentives for those decisions and actions. There may be a hierarchy of key agencies and actions, and mapping backwards requires sequential attention to each.

Elmore's (1979) "backward mapping" provides an operational approach to questioning how government agencies function in particular issue settings—a prerequisite for informed policy feasibility estimates. A complementary, if less operational, approach is suggested in Table 6.1, which contains a checklist of feasibility-relevant questions, compiled by surveying the works of Allison (1969, 1971), Mazmanian and Sabatier (1981, 1983), Hogwood and Gunn (1984), Linder and Peters (1989), and Bardach (2005). For particular types of policies, some questions will be more significant and the answers more critical. In general, however, each "Yes" answer provides a reason to challenge the feasibility of a policy option.

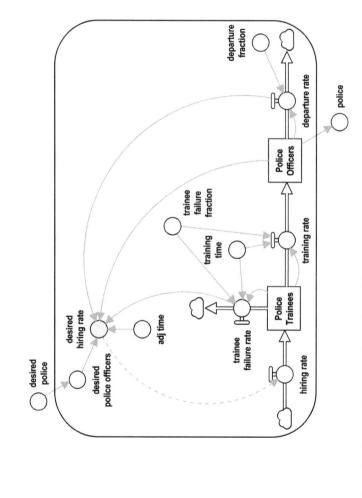

FIGURE 6.3 Inside an implementation constraints sub-model

Answers to both operational and feasibility questions help modelers flag potential constraints on a particular policy option. The answers also facilitate distinguishing between policy options that otherwise suggest similar benefits and costs during wishful thinking simulation experiment. Moreover, both types of questions are helpful when building a sub-model of policy constraints, a simple illustration of which is presented in the next section.

Modeling Constraints on Policy Implementation

Policy design requires more than testing parameters, constructing wishful thinking links, and raising questions about the feasibility of a policy option. It requires adding stock-and-flow structure that transforms wishful thinking links into operational links. When Forrester discussed policy modeling in *Urban Dynamics* (1969), he emphasized the need to ". . . restructure the system so that the internal processes lead in a different direction" (p. 113). Richardson and Pugh (1989) explain that "policy improvement . . . involves the addition of new feedback links" and their examples reflect a feedback perspective on policy design (pp. 332, 337–359). Sterman (2000) emphasizes that "policy design . . . includes the creation of entirely new . . . structures and decision rules" and he provides detailed feedback-based formulation guidelines for effective decision-rule design (pp. 104, 513–550). Ford (2010) encourages "constructing a stock-and-flow diagram to describe the details of policy implementation" and provides numerous examples (p. 158).

Figure 6.3 displays a simple example of modeling implementation constraints when the requisite resource is a force of police officers. The input to the sub-model is *desired police*. The actual number of *police officers* is determined within the sub-model and is the output to the *actual police*.

The sub-model in Figure 6.3 operationalizes at least part of the process of changing an actual resource to its desired level. Operationalizing the constraints enables a more realistic simulation assessment of the expected impact of a policy that requires changing the number of police officers. At a minimum, the constraints in the sub-model delay the impact; it will take longer to achieve the desired number of officers than the wishful thinking link would suggest. The stock-and-flow structure reflects the need to train new recruits for some time before they become officers and operate with the expected level of productivity. Additional delays are due to continuous recruitment and training as officers depart the police force and some trainees fail to become officers. Moreover, if the failure rate were high, the constraints in this example could generate oscillations—too few officers in one time period followed by too many officers later on.[3]

Note that a new wishful thinking link appears in Figure 6.3: a change in the *desired hiring rate* would be instantly implemented in the actual *hiring rate*. Only a little armchair brainstorming is needed to recognize that nothing has been said about the budget for hiring new officers or whether there is a ready supply of recruits. More implementation modeling could be done and, in this case, should be done. In the public health modeling examples discussed below, more is done. One reason for leaving the example in Figure 6.3 unfinished is to keep it simple for explanatory purposes. Another reason is to underscore a key point: There is a limit to what can (or should) be modeled explicitly. Even large, complex finished models will contain some wishful thinking links (whether modelers acknowledge them or not). Taking implementation seriously does not mean modeling minutia or modeling past the point of diminishing returns for the user of the model. Rather, it means being vigilant when closing feedback loops—always aware of unstated assumptions implicit in a link and conscious of the next step: whether to add more stock-and-flow structure, raise more red flags with insightful questions, or both.

Public Health Issue Examples

The use of the policy design framework will be illustrated with two examples. The first draws on a classroom exercise that introduces the modeling process to students in the international system dynamics master's degree program at the University of Bergen in Norway.[4] A physical simulation game generates data that, along with the rules of the game, provide sufficient information to build an explanatory model that reproduces the behavior observed in the game. All of this is done during the first lecture to provide students with an overview of the skills they will develop during the course. During the second lecture, we demonstrate designing a policy model to control the epidemic.

The second example concerns auto pollution in Zimbabwe. The model has been adapted from a student research project in which Madoma (2011) developed an SD model to explain the growth of auto pollution in Zimbabwe's urban centers from 1990 to 2010. She also developed a policy model to test the impact of inspecting cars and impounding those that emit pollutants at a higher rate than allowed by law.

Although contrived, the epidemic model is a simplified version of SD epidemic models that have been developed for real-world situations (Dangerfield, Fang, & Roberts, 2001). Here, its simplicity makes it accessible to a wide range of readers. The auto pollution example, although adapted from a complex model of a messy real-world situation, has also been simplified to facilitate exposition.

The Epidemic Game[5]

Before the game begins, one anonymous student is discreetly designated as "infected." During the game, each student has one random "daily" contact (handshake) with another student. When an infected student and uninfected student make contact, there is a chance of a new infection based on the outcome of a coin toss. Eventually, an "epidemic" occurs. The cumulative number of infected persons grows at different rates over time: slowly during the first few days, then rapidly, then slowly again, and finally no growth at all after everyone becomes infected.

Explanatory Model of the Epidemic

We conceptualize the epidemic game data in terms of two stocks—*Susceptible Persons* and *Infected Persons*, and a flow called the *infection rate* that is the daily rate of new infections. As infections occur during the game, people flow from a *Susceptible Persons* stock to an *Infected Persons* stock. The epidemic grows due to the positive feedback loop between *Infected Persons* and the *infection rate*, and slows due to the negative feedback loop between *Susceptible Persons* and the *infection rate*. When graphed, the stock of *Infected Persons* follows an s-shaped growth pattern.

Policy Design for the Epidemic Game Model

After the dynamics of the problem are understood, we explore policy options to combat an epidemic. We begin by selecting a broad strategic option for analysis, in this case a vaccination strategy. Figure 6.4 displays a sub-model of constraints that could impede a vaccination strategy. To keep the focus on the policy model (left side of the diagram), the explanatory model (bottom right) is highly simplified with only the feedback loop effects on the infection rate shown.

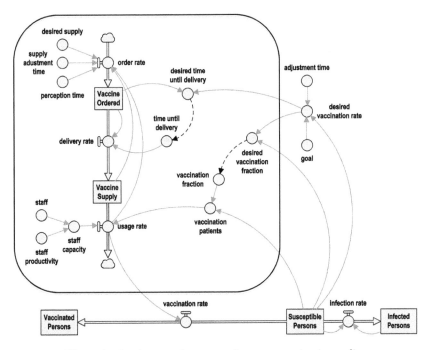

FIGURE 6.4 Illustrative implementation constraints on a vaccination policy

Susceptible Persons is the stock to be managed and, in Figure 6.4, a new outflow—*vaccination rate*—has been added to the explanatory model. The strategy is to drain the stock of *Susceptible Persons* before they have contact with *Infected Persons*. The *vaccination rate* is the flow to be regulated and in the absence of constraints on policy implementation would be instantly equal to the *desired vaccination rate*. But that is wishful thinking.

Modeling Implementation Constraints on the Vaccination Policy

Vaccinations occur when there is a demand for vaccine, a supply of vaccine, and adequate human resources to administer the vaccine. These conditions determine the vaccine *usage rate* and, therefore, the *vaccination rate*. The demand for vaccine is the number of persons who show up to get vaccinated, whether voluntarily or through coercion. The *vaccine supply* on-site and ordered—often problematic under real-world epidemic conditions—is represented as a supply chain. The human resource capacity constraint is represented by parameter values for *medical staff* and *staff productivity*.

Various simulation scenarios, each designed to reflect different assumptions about the constraints, are used to test the impact of the vaccination policy. The best-case scenario is that the number of infections is reduced by nearly half if the vaccination policy can be carried out as desired, i.e., without constraints. If fewer-than-desired people step up to be vaccinated, if there is insufficient medical staff capacity, or if the vaccine is not produced and delivered quickly enough to be useful, the policy will not be as effective as the best-case scenario suggests.

One of the wishful thinking links is a reminder that a vaccination strategy must include a program designed to encourage susceptible persons to get vaccinated. Another highlights the need for pharmaceutical firms to deliver sufficient quantities of vaccine in a timely manner. Neither requirement is a foregone conclusion. Whether to continue building this model (i.e., replace these two wishful thinking links with operational links) is a judgment call. The alternative is to highlight both issues in discussions with policy makers or program administrators and be sure that remaining implementation obstacles are not overlooked. No matter where we stop, there will always remain some wishful thinking links; not everything will be modeled. There is value, however, in surfacing hidden assumptions and highlighting constraints that remain.

Auto Pollution in Zimbabwe

In contrast with the contrived epidemic example, the auto pollution example is based on Madoma's (2011) study of actual conditions in Zimbabwe. From 1990 to 2010, vehicles in urban centers increased by 150 percent and air pollution attributed to vehicles rose even more, indicating that emissions per vehicle were increasing.

Most are poorly maintained low-cost second hand vehicles that do not meet strict emission standards of the countries of their origin. Zimbabwe has become a dumping ground for used cars because people rely heavily on importing second-hand vehicles as they cannot afford new vehicles sold in the country. As a result vehicle emissions . . . have become a major source of pollution in the . . . cities.

(Madoma, 2011, p. 5)

In SD terms, auto emissions add to the stock of air pollution while the dispersion rate subtracts, and the pollution level grows when the inflow is greater than the outflow. Specifically, the auto emissions rate is due to the number of cars and the average emissions per car. The dispersion rate is a function of the auto pollution level and the time required to disperse the pollutants under local atmospheric conditions.

Policy Design for Auto Pollution Model

To slow the growth in the auto pollution stock, its net inflow must be decreased. The outflow is assumed to be beyond the control of public authorities, because it depends on the properties of the pollutants in the air and the dispersing influence of local atmospheric conditions. The proximate cause of the inflow, the *emissions rate*, is the combined effect of the number of cars and the emissions per car, either or both of which might be the target of policy initiatives.

Madoma evaluated the strategy of impounding *imported cars* that violate Zimbabwe's emissions regulations. The strategy aims to influence the emission rate in two ways. First, it could reduce the total number of cars because some drivers who lose their used imports in the impoundment process could not afford more expensive replacement cars. Secondly, with the reduction in the number of used imports, the weighted average emissions per car would decline.

Figure 6.5 displays a simplified version of Madoma's implementation constraints sub-model, which receives an input from *desired impoundment rate* and generates the *actual impoundment rate* output. Within the sub-model, organizational, technological, and social constraints hinder achievement of the policy goals.[6]

The sub-model structure focuses on the funding, hiring, training, and equipping of inspectors who would be responsible for impounding used imports that are in violation of Zimbabwe's auto emissions standards. A stock-and-flow feedback approach is used to model the dynamics of the *inspectors* stock and the *wage budget* on which it depends. A key question is how much of the desired *wage budget* might actually be funded. Instead of attempting to model the budgetary process in Zimbabwe, Madoma uses a parameter, *fraction funded*, to test the sensitivity of the model to various assumptions about program funding. Another important parameter, *probability of corruption*, has a negative influence on *inspector productivity*, and reduces the actual *impoundment rate*.

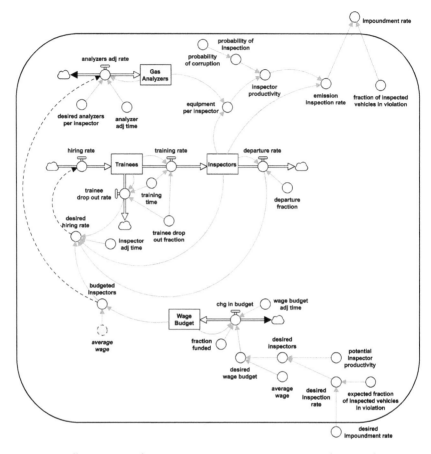

FIGURE 6.5 Illustrative implementation constraints on an impoundment policy

Simulation scenarios are used to facilitate an assessment of the expected impact of an impoundment policy in the auto pollution model.[7] Wishing away all constraints in the first scenario, the policy performs in its desired manner and stops the growth in auto emissions. The next scenario activates the implementation constraints sub-model and assumes a fully funded *Wage Budget* and no lack of inspections due to corruption. Compared to the best-case scenario, the results are not as good because delays occur within the budget and hiring processes and some trainees drop out. Although emissions slow dramatically, they continue rising because the policy sub-model is unable to impound the polluting cars at a rate that keeps pace with their importation and use.

The final scenarios reflect trouble at the government level and/or the street level. The third scenario assumes only half the budget requirements are funded but corruption is not a problem, while the last scenario makes the same budgetary assumption but assumes that inspections are cut by half due to corruption. The impact of the impoundment policy is weak in both cases. Lack of full support at either the top or the bottom of the policy chain undermines the policy, and the pollution level continues to rise near the business-as-usual rate. Should that occur, money spent on the program would be perceived as wasted. The high likelihood of a disappointing outcome means that weak links in the implementation structure must be strengthened or other policy options must be considered.

Discussion

Model-based policy design that relies exclusively on parameter sensitivity analysis raises questions about how parameter changes in a computer model could be implemented in real-world settings. In this chapter, we turn that question around—asking how implementation constraints could be embedded in a computer model—and provide an answer with a framework for model-based policy design that takes implementation seriously.

The framework builds on the traditional SD distinction between problem explanation and policy design. Policy design begins with a high-level perspective that ignores operational details, seeks the logical connection between the problem and the solution, and tests the potential impact of broad strategies. The result is a simulation model containing a series of feedback loops that regulate target flows of the stocks being managed. The loops contain wishful thinking links between desired and actual results, but motivate operationally and politically insightful questions about the feasibility of specific policy options.

Replacing wishful thinking links with stock-and-flow structure that represents operational thinking is the final task in the policy design process. Modeling the implementation process for a policy requires a street-level perspective, questions operational detail and feasibility, adds model structure to represent real-world constraints, and evaluates the policy in light of those constraints. But modeling everything is not possible. There will always be a need for wishful thinking links to highlight implementation constraints that are better addressed the old fashioned way—by constructive questioning and discussion.

To illustrate how the framework can be applied, we modeled two dynamic problems—a simple, contrived flu epidemic and a more complex, real-world auto pollution problem—complete with policy sectors containing stock-and-flow representation of practical details inherent in policy implementation. For both models, the value of the framework has been enhanced by reliance on key elements of the implementation paradigm in the public policy literature.[8] In this final section, we highlight pieces of that literature that can be useful for almost any policy modeling project and note their contribution to the particular examples illustrated in this chapter.

Foremost is Elmore's (1979) backwards mapping approach, which guides modeling for both the epidemic and pollution cases. Importantly, it is not limited to policy modeling; it also aids our *modeling backwards* approach to explanatory modeling.

When thinking about a vaccination strategy, it is useful to think in terms of policy outputs emerging from standard organizational processes (Model II perspective in Allison, 1969, 1971), both within public or quasi-public health institutions and within private pharmaceutical companies. In that context, feasibility of the vaccination strategy ranked high, with the main uncertainty relating to government's authority and determination to speed up production and delivery of vaccine. The impoundment strategy for the auto pollution problem, on the other hand, requires a new program with new budgetary claims in an unstable political culture. The losers under this policy will be identifiable and vocal; winners will be anonymous and silent. That will increase pressures for repeal within the government and pressures for evasive action, non-compliance, and corruption at the street level where the policy would be enforced. The bureaucratic politics perspective (Model III in Allison, 1969, 1971) and the policy feedback concept in the policy dynamics literature (Pierson, 1993; Baumgartner & Jones, 2002) are useful for thinking about the program's prospects for both passage and survival. These insights translate into low feasibility estimates reflected in the parameter assumptions about program funding and the probability of corruption.

Thinking about how governments function leads to thinking about the scope and limits of their power to deal with public issues. Linder and Peters (1989) show how the choice of the "instruments of government" facilitates or impedes implementation. Brainstorming policy options is more likely to remain hitched to reality if SD modelers have ready access to Bardach's (2005) list of "what governments do." In the contrived epidemic case, we have no reference to political culture; thus, the scope of formal authority is unknown. Nevertheless, a vaccination program is one of several standard ways that public health authorities in most countries respond to epidemics; thus, there is little doubt that the instrument of choice could be activated. An auto impoundment strategy, on the other hand, raises issues that would be insurmountable obstacles in some political cultures. Even in Zimbabwe, the environmental laws passed in 2002 authorized fines (that were unenforced) but did not authorize impoundment. Legislative action is needed to bring the impoundment club out of the closet; thus, the importance of examining at least one simulation scenario that reflects a low conditional estimate of appropriations probability (given a low estimate on the chances for authorization).

Building bridges between the SD and public policy communities promises mutual benefits. The public policy literature is a valuable source of conceptual and empirical support for SD representation of delays, capacity limits, feedback effects, and nonlinearities that characterize complex policy systems notoriously resistant to change. And simulation modeling that takes implementation seriously can be a valuable addition to the policy analyst's toolkit. The framework presented in this chapter aims to encourage more operational thinking and less wishful thinking during model-based policy design. Ultimately, it aims to encourage building models that are more useful to policy makers.

Notes

This chapter is a merger and revision of papers presented at the 2011 conferences of the International System Dynamics Society (Wheat & Shi, 2011) and the Association of Public

Policy and Management (Wheat, 2011). Suggestions from anonymous reviewers of the original papers, from graduate students at the University of Bergen, and from the editors were helpful during the revision process. Several models serve as illustrations but discussion of equations is kept to a minimum and full explanatory models are not shown. Readers with questions about specific equations or any other topic should contact the author at david.wheat@uib.no.

1   Models have been created with Stella® Professional Modeling and Simulation Software, available from www.iseesystems.com.
2   If we assume that the *inflow* is the product of the *resource* and its *productivity*, then in re-arranged terms, *resource = inflow/productivity*. Expressed in goal-directed terms, *desired resource = desired inflow/productivity*.
3   The potential for oscillations is due to multiple delays along the negative feedback loop that seeks to close the gap between the desired and actual number of police officers.
4   "Systems Education at Bergen" is available at *Systems* 2014, 2(2), 159–167; http://www.mdpi.com/2079-8954/2/2/159.
5   The Epidemic Game described here has been adapted from the original version (Glass-Husain, 1991).
6   The sub-model in Madoma's (2011) original work includes additional constraints, primarily relating to the funding, acquisition, and use of the gasoline analyzers needed for the inspections. Those are aggregated and simplified in Figure 6.5.
7   Madoma (2011) developed the pollution index for Zimbabwe's urban centers by combining measures of five pollutants: sulfur dioxide, carbon monoxide, carbon dioxide, nitrogen oxides, and particulate matter.
8   Essential roadmaps to the implementation literature are contained in Saetren (2005) and Hill and Hupe (2009).

References

Allison, G.T. (1969). Conceptual models and the Cuban Missile Crisis. *American Political Science Review, 63*(3), 689–718.

Allison, G.T. (1971). *Essence of decision: Explaining the Cuban Missile Crisis.* Boston, MA: Little, Brown and Company.

Bardach, E. (2005). *A practical guide for policy analysis: The eightfold path to more effective problem solving.* Washington, DC: CQ Press.

Barlas, Y. (2002). System dynamics: Systemic feedback modeling for policy analysis. In *Knowledge for Sustainable Development: An Insight Into the Encyclopedia of Life Support Systems.* Oxford: UNESCO Publishing.

Baumgartner, F.R., & Jones, B.J. (2002). *Policy dynamics.* Chicago: University of Chicago Press.

Coyle, R.G. (1996). *System dynamics modelling: A practical approach.* London: Chapman & Hall/CRC.

Dangerfield, B., Fang, Y., & Roberts, C. (2001). Model-based scenarios for the epidemiology of HIV/AIDS: The consequences of highly active antiretroviral therapy. *System Dynamics Review, 17*(2), 119–150.

Elmore, R.F. (1979). Backward mapping: Implementation research and policy decisions. *Political Science Quarterly, 94*(4), 601–616.

Ferman, B. (1990). When failure is success: Implementation and Madisonian government. In D.J. Palumbo & D.J. Calista (Eds.), *Implementation and the policy process: Opening up the black box* (pp. 306–334). New York: Greenwood Press.

Ford, A. (2010). *Modeling the environment* (2nd ed.). Washington, DC: Island Press.

Forrester, J.W. (1969). *Urban dynamics.* Waltham, MA: Pegasus Communications.

Forrester, J.W. (2009). Email communication from Jay Forrester to System Dynamics K-12 Discussion listserv, December 6, 2009, 12:40 a.m. GMT. Used with permission.

Ghaffarzadegan, N., Lyneis, J., & Richardson, G.P. (2011). How small system dynamics models can help the public policy process. *System Dynamics Review, 27*(1), 22–44.

Glass-Husain, W. (1991). *Teaching system dynamics: Looking at epidemics.* System Dynamics in Education Project. Massachusetts Institute of Technology. Retrieved from http://clexchange.org/ftp/documents/Roadmaps/RM5/D-4243-3.pdf

Hill, M.J., & Hupe, P.L. (2009). *Implementing public policy: An introduction to the study of operational governance* (2nd ed.). London: Sage Publications.

Hogwood, B., & Gunn, L. (1984). *Policy analysis for the real world.* Oxford: Oxford University Press.

Howlett, M., Ramesh, M., & Perl, A. (2009). *Studying public policy: Policy cycles and policy subsystems* (3rd ed.). Oxford: Oxford University Press.

Knoepfel, P., Larrue, C., Varone, F., & Hill, M. (2007). *Public policy analysis.* Bristol, UK: Policy Press.

Linder, S.H., & Peters, B.G. (1989). Instruments of government: Perceptions and contexts. *Journal of Public Policy, 9*(1), 35–58.

Madoma, P.L. (2011). Auto pollution policy in Zimbabwe. A system dynamics perspective. (Unpublished master's thesis). Bergen: University of Bergen.

Mazmanian, D., & Sabatier, P. (Eds.). (1981). *Effective policy implementation.* Lexington, MA: Lexington Books.

Mazmanian, D., & Sabatier, P. (1983). *Implementation and public policy.* Chicago: Scott Foresman.

Morecroft, J. (2007). *Strategic modelling and business dynamics: A feedback systems approach.* Chichester, UK: John Wiley & Sons.

Pierson, P. (1993). When effect becomes cause: Policy feedback and political change. *World Politics, 45*(4), 595–628.

Randers, J. (1980). *Elements of the system dynamics method.* Cambridge, MA: MIT Press.

Richardson, G.P., & Pugh, A.L. (1989). *Introduction to system dynamics modeling.* Waltham, MA: Pegasus Communications.

Richmond, B. (1993). Systems thinking: Critical thinking skills for the 1990s and beyond. *System Dynamics Review, 9*(2), 113–133.

Richmond, B. (1994). Systems thinking/system dynamics: Let's just get on with it. *System Dynamics Review, 10*(2–3), 135–157.

Richmond, B. (2000). *The "thinking" in systems thinking.* Waltham, MA: Pegasus.

Richmond, B., Peterson, S., & Vescuso, P. (1987). *An academic user's guide to STELLA.* Hanover, NH: High Performance Systems.

Roberts, E.B. (2007). Making system dynamics useful. *System Dynamics Review, 23*(2–3), 119–136.

Saetren, H. (2005). Facts and myths about research on public policy implementation: Out-of-fashion, allegedly dead, but still very much alive and relevant. *The Policy Studies Journal, 33*(4), 559–582.

Sterman, J. (2000). *Business dynamics: Systems thinking and modeling for a complex world.* Boston: McGraw-Hill.

Wheat, I.D. (2010). What can system dynamics learn from the public policy implementation literature? *Systems Research and Behavioral Science, 27*(4), 425–442.

Wheat, I.D. (2011). *Implementation modeling.* Paper presented at the 2011 Fall Research Conference of the Association of Public Policy and Management. Washington, DC.

Wheat, I.D., & Shi, L. (2011). Exploratory policy design. In *Proceedings from International Conference of the System Dynamics Society.* Washington, DC.

7

PUBLIC-PRIVATE PARTNERSHIPS

A Study of Risk Allocation Design Envelopes

David N. Ford, Ivan Damnjanovic, and Scott T. Johnson

Introduction

The system dynamics modeling methodology can be useful for policy informatics research and policy development by application to public-private partnerships (PPP). Many PPP have been used for large transportation infrastructure development. In the United States (U.S.) transportation infrastructure has traditionally been owned and operated by government agencies. However, as the result of tax shortfalls and depletion of the federal highway trust fund, less funding is available. Unavailability of resources in underfunded agencies responsible for highway infrastructure results in increased congestion and deferred maintenance. To overcome this problem, agencies have turned to alternative project delivery methods such as PPPs. According to the National Council for Public-Private Partnerships,

> A Public-Private Partnership is a contractual agreement between a public agency (federal, state or local) and a private sector entity. Through this agreement, the skills and assets of each sector (public and private) are shared in delivering a service or facility for the use of the general public. In addition to the sharing of resources, each party shares in the risks and rewards potential in the delivery of the service and/or facility.
>
> *(NCPPP, 2011)*

PPP for the private provision of public infrastructure, such as transportation, water supply and waste water treatment, and energy production and distribution, is becoming a more common method of delivering societal lifeline services. Although there are many other forms of PPP, this study involves both infrastructure development and operations over a fixed period before transfer of the infrastructure to the public agency at the end of the project. These are sometimes called concession agreements (Yescombe, 2002).[1] Potential advantages of concession agreements include leveraging the strengths of both the public and private sectors, reducing development risk and public capital investment, faster development, improved cost-effectiveness, and improved services to communities. Key features of a typical PPP for a toll road include: a limited liability project company to develop the infrastructure in accordance with public agency requirements on right-of-way (ROW) typically provided by the public agency; a concession agreement for the right to operate and manage the toll road for profit for a fixed period of time; and contracts that define the allocation of risks to address uncertainties, such as ROW acquisition, cost overruns, and shortfall in revenue. Concession agreement contracts are complex as they involve multiple stakeholders with different objectives who seek to maximize their interests. This makes successful agreements difficult to design and manage. As a primary project participant the decisions and policies of the government agency are critical to PPP project success. Therefore policy informatics can play a valuable role in PPP design and management.

Contractual risk allocation among project stakeholders is at the core of every PPP project. This agreement assigns risks to responsible parties, and is largely what defines the PPP. Achieving a balanced and fair allocation of risk between the public and private partners is critical to project success. If too much risk is borne by the private partner the cost becomes higher than necessary and the public agency may be forced to take premature ownership from a failed developer/operator; if too little is borne by the private partner public funds are wasted. Risk allocation theory suggests that risks should be allocated to the parties that are best able to understand, control, and manage or diversify the risk. This minimizes total costs, as a lack of understanding and control tends to generate inflated prices. In addition, bearing risk should be adequately rewarded. Here the public and private sectors differ because their missions differ. The mission of the public agency is to meet the needs of the public, whereas the mission of the private sector is to maximize return on employed capital. Hence, before proper risk allocation can occur the reward measures for both the public and private partners should be measured using the same units to allow adequate comparison of potential risk allocations.

One of the most important risks in toll road transportation projects is a potential shortfall in revenue. In PPP projects private partners borrow funds for development and therefore require a steady revenue stream to repay loans. As will be described, lenders also impose revenue support requirements on PPP projects as conditions for providing development funds. A guaranteed revenue stream is considered critical to PPP success by professional organizations (NCPPP, 2011) and practitioners (Hebert, 2011). Multiple sources of revenue may be needed to provide a stable revenue stream, including tolls, tax districts, land development opportunities, and subsidies by the public sector partner. A primary means of providing part of these funds and reducing the risk of PPP for private developers is the provision of a guaranteed revenue stream during operations (i.e., government subsidies if revenues fall below debt service costs).

Tipping Points

Tipping points can describe some important PPP project behaviors. A tipping point is a set of conditions that separate two very different, internally driven behavior modes. Gladwell (2000, p. 9) defined one tipping point as "that one dramatic moment in an epidemic when everything can change all at once." Tipping points can be identified in time series data by inflection points, maximum or minimum values in behavior patterns. Sociologists have used tipping points to explain the increase of participants in a riot (Granovetter, 1978; Granovetter & Soong, 1983), school desegregation (Clotfelter, 1976), societal segregation (Schelling, 1971), and social problems associated with ghettos (Crane, 1991). Recently the term *tipping point* and the concept have grown into widespread use, typically used to describe conditions at any significant change in behavior. For example, Gay (2014) describes the current conditions in professional (American) football playoffs as potentially being at a tipping point, and that expanding those playoffs may initiate a decline in live (vs. television) attendance at games. Others have used tipping points to describe a coming onslaught of knuckleball pitchers in major league baseball (Cavanaugh, 2013) and (satirically) the popularity of names for newborn children (Fry & Lewis, 2006). More ominously, former UN secretary-general Kofi Annan called the conditions between the government and protesters in Syria at the end of May 2012 as being "at a tipping point" (British Broadcasting Corporation, 2012) and soon thereafter, U.S. secretary of state Hillary Clinton warned that Syria was "spiraling towards civil war" (Crabtree, 2012).

But behavior-based descriptions have no explanatory power. Some researchers have used the concept of a "critical mass" within a system (e.g., of infected persons or violent protesters) to partially explain how tipping points occur. This explanation uses an analogy to the critical mass required to sustain or expand a nuclear chain reaction within systems. System dynamics researchers have added more rigor and explanatory power to the tipping point concept by formally modeling causal feedback structures that create tipping point conditions and the resulting behavior modes and system performance. This allows system dynamicists to both describe and explain tipping points. From a feedback perspective, systems tend to remain stable as long as the systems' controlling (balancing) feedback loops dominate (Sterman, 2000). But if dominance shifts to the systems' reinforcing feedback loops, behavior can become unstable and drive the system to extremes. The pairing of controlling and reinforcing loops in a tipping point structure creates conditions in which small changes can shift feedback loop dominance, change behavior modes, and drive projects to success or destroy project value and drive them to failure. The conditions that separate these two realms, the tipping points conditions, exist when the controlling and reinforcing loops in the tipping point structure are balanced. The current study uses tipping point structures to explain complex PPP behavior and performance.

Problem Description

Toll road transportation infrastructure projects are inherently risky due to the multiple large uncertainties that can severely impact project performance. Although the PPP framework offers a broad range of approaches and tools to produce successful outcomes, significant challenges remain with the development of equitable risk allocation strategies that effectively address project uncertainties. Anecdotal evidence of this problem can be seen in the controversies that often surround these agreements. Some agreements are believed to unfairly benefit the private party because the public bears too much risk and the private developer captures too much profit. In other instances, projects have failed because the public sector exercised its sovereign power and effectively nationalized the projects. This power is implemented directly using executive power or indirectly via creeping nationalization when a project environment has become significantly altered so that the private party's only option is to abandon the project.

More direct manifestation of the risk allocation problem is seen in the wide variance associated with the cost of project delivery. The forecasts of both construction costs and the traffic volumes associated with operations that can determine project success or failure can be very flawed. Standard & Poor's Ratings Services study of traffic forecasts in 2004 detected optimism bias in traffic revenue studies for PPP projects, suggesting that revenue forecasts are often overestimated by 20 percent to 30 percent (Standard & Poor, 2004). Flyvbjerg, Holm, and Buhl (2005) report that half of the toll road projects from a sample of 210 averaged 20 percent traffic forecast error. As a specific example, the Dulles Greenway toll road project in the U.S. assumed that the traffic demand would start at 34,000 vehicles per day and increase at an annual rate of 14 percent per year for the first six years (Garvin & Cheah, 2004). But actual average traffic was only 11,500 vehicles per day in the first six months (Fishbein & Babbar, 1996).

Developing more reliable forecasts to predict project performance is complicated by multiple factors, including externalities such as new regulations, the development of competing infrastructures, and financing crises. Consider a scenario of a state department of transportation deciding to improve the public road network near a project. This can result in a decrease in traffic volumes on the toll road, resulting in reduced revenue and financial viability of the project (Damnjanovic, Duthie, & Waller, 2008). However, externalities can also be positive, such as roadway improvements on system feeder links that can funnel traffic to the toll road and thereby stabilize or increase project revenue (Vajdic, Damnjanovic, Suescan, & Waller, 2011). Consequently, private partners in PPP seek to reduce risks associated with negative externalities through traditional risk management tools, such as non-compete clauses in agreements with their public partners (Hulsizer, 2011). A non-compete clause could restrict the public sector's right to add capacity within a specified distance from the toll road (Ortiz & Buxbaum, 2008), thus reducing the demand risk for the private partner. Positive externalities (e.g., profits from or increased tax revenue due to development adjacent to the toll road exits) are often ignored or, at best, addressed using only intuition without the aid of formal models.

Although there are many examples of successful PPP projects,[2] the complex risk allocation problem described earlier appears to be widespread, chronic, and costly. Several root causes related to risk allocation may contribute to the failure of PPP projects that could otherwise succeed: (1) forecasts of project behaviors (e.g., demand growth and impacts of raising tolls on demand) may not be well understood and incorporated into project planning; (2) different performance measures are used by PPP project participants, and therefore they may not adequately understand how risk allocation impacts each participant differently; and (3) participants may overrely on informal mental models to bridge significant differences between stakeholder understanding of risk characteristics and potential solutions that can protect project performance.

This study focuses on two types of risk allocation: (1) provision of subsidies by public agencies to developers if the project fails to generate adequate revenue to support loan repayment, and (2) toll adjustments with upper limits to address both revenue needs and user protection. In particular, we seek to advance the understanding of the effects of risk allocation designs defined by contractual agreements between the stakeholders. The formal model used in this study includes traffic and infrastructure maintenance and repair, project debt, loans, and credit risk in feedback loops that mimic operations and two forms of debt financing. By activating reinforcing feedback loops that reflect continuous refinancing of the project, we create a simulated environment in which lenders take no risk. This mental model of project risk could be used to stress-test projects for their robustness against financial failure. We use this feedback structure to investigate the sensitivity of project performance measures to project risk allocation designs. More specifically, we take an envelope or boundary perspective, where primary risks are fully assumed by one of the three primary stakeholders (public agency, private developer, or toll road users). By simulating project behavior under these boundary conditions, we develop insights about how risk allocation impacts PPP project performance for different stakeholders.

Previous research provides a foundation for the use of tipping points to explain PPP project behavior. The effects of tipping points on development projects have been investigated. Repenning (2001) showed that a tipping point in a series of projects can cause *fire-fighting*, which is the phenomenon of projects being over-whelmed with work to the point of failure. Black and Repenning (2001) inves-tigated the effectiveness of several managerial policies to counteract fire-fighting, including adding resources to the project, releasing lower-quality work, and resource allocation policies. In contrast, Taylor and Ford (2006, 2008) investigated tipping points in single development projects. They showed that tipping point dynamics can lead to runaway project backlogs and explain some forms of single development project failure (Taylor & Ford, 2006). In subsequent work (Taylor & Ford, 2008), they modeled operational solutions to tipping point vulnerability in single projects, including using an experienced workforce to minimize rework, slipping the deadline, increasing detail in planning to decouple work and thereby reducing ripple effects, reducing scope, and staffing engineers at night to speed responses to construction issues. Although useful within the contexts in which they were applied, these works described tipping point structures as a single pair of feedback loops, one balancing and one reinforcing. PPP projects, and the model of a PPP project described here, are more complex and therefore require an expanded description of tipping point structures. Therefore, in addition to using tipping points to explain PPP behavior, the current work extends the theory of tipping point dynamics. This is done by describing and explaining the tipping points in the PPP with multiple interacting structures that each consist of sets of reinforcing and balancing loops.

Overview of Chapter

The next section describes the research approach, including the methodology, model of the toll road project, and typical model behavior and model testing. This is followed by a description of the modeling results for three specific purposes: (1) to evaluate impacts of several extreme risk allocations on project performance, (2) to evaluate how financial stress impacts the effects of risk allocation design on project performance, and (3) to explain PPP project success and failure with the model's tipping point structures. The importance of modeling different perspec-tives is then explained, and concluding thoughts are provided.

Research Approach

This section describes a system dynamics model of a PPP toll road project that was developed and tested for its usefulness as a tool for investigating risk alloca-tion strategies. Included is a discussion of the foundational concepts of system dynamics and why it is appropriate for studying risk allocation. Then we define the performance measures, project types, and strategies used to assess project per-formance for each stakeholder. We finish with a more detailed description of the various model sectors and the model behavior and testing used to validate that the model is appropriate for the task.

Methodology

The system dynamics methodology was applied to model a PPP toll road project. System dynamics is one of several established and successful approaches to systems analysis and design (Flood & Jackson, 1991; Jackson, 2003; Lane & Jackson, 1995). System dynamics shares many fundamental systems concepts with other systems approaches, including emergence, control, and layered structures. Therefore, system dynamics can address issues such as risk in large, complex systems (Lane, Gröbler, & Milling, 2004), such as PPP projects. The system dynamics methodology applies a control theory perspective to the design and management of complex engineered systems. The perspective focuses on how the internal structure of a system impacts managerial behavior and performance over time. The system dynamics approach is unique in its integrated use of causal feedback, stocks and flows, time delays, and adaptive decision-making to model physical and financial structures and management strategies and policies. Conceptual modeling uses causal loop diagramming in which system components are linked with arrows that indicate the direction of influence and create feedback loops. Formal modeling expands causal loop diagramming into computer simulation models. Additional variables link system conditions to decision-making algorithms, thereby changing system conditions through feedback loops. Forrester (1961) developed the methodology's philosophy, and Sterman (2000) specified the modeling process with examples and described numerous applications.

System dynamics has been extensively used for system analysis and design, including the study of several aspects of projects, including schedule risk management (Ford & Bhargav, 2006), technology development risk (Ford & Sobek, 2005), and tipping points (Taylor & Ford, 2006, 2008). See Lyneis and Ford (2007) for a summary and review. The methodology's ability to endogenously model many diverse system components (e.g., vehicular traffic, stakeholders, money, information), processes (e.g., toll road operations, repair, and maintenance), and managerial decision-making and actions (e.g., adjusting toll rates) makes it useful for investigating PPP for built infrastructure.

Modeling risks and rewards in PPP projects requires understanding project performance measures. In this study we use the net present value (NPV), a widely used metric of project performance, from the perspectives of the developer and the public. Developer cash flows used to estimate NPV include initial investment (equity), debt service payments, operating maintenance and repair costs, toll revenue, and subsidies received from the government agency. The net cash flow to the public—that is, public net benefits less government agency subsidy payments—was used to estimate the project NPV from the public's perspective. Public participation in projects is often associated with economic evaluation measures to justify issuing public financing instruments (e.g., municipal bonds, direct tax expenditures) and thereby fund infrastructure projects. From the viewpoint of the traveling public (i.e., users), the key performance measure is the reduction in their travel time and its monetary equivalent. The difference between monetary value of the time saved by toll road users and the tolls those users paid for the use of the toll road reflects the net benefits to users. The government agency's subsidy expenditures are its primary performance measure; a project that can be executed with less subsidy support is preferred over a project that needs more, *ceteris paribus*.

Model variables for loan period, subsidies, and toll rates were used to describe different project types and stylized risk allocations that define an envelope of the potential risk allocation strategies. Three extreme risk strategies were used to describe the envelope of potential risk allocation strategies. Each risk strategy assumes one of the three primary project stakeholders (developer, agency, users) holds the majority of the project risk. When the developer holds most of the project risk, the loan period is relatively short (15 years) and the agency pays no subsidies and does not allow tolls to change. Under this risk allocation strategy the developer is dependent on adequate toll road traffic for project performance, the agency has no subsidy costs, and users are protected from toll increases. When the agency holds most of the project risk, the loan period is relatively long (25 years), subsidies are adjusted to cover operating deficits, and tolls are fixed. Under this risk allocation, the agency takes the traffic risk by covering deficits with subsidies, thereby protecting the developer (and, in turn, the lenders) and not requiring toll increases. When users hold most of the project risk, the loan period is relatively long (25 years), the agency pays no subsidies, but the developer increases tolls to a relatively high contractual limit in response to operating deficits. Under this risk allocation strategy, users bear the traffic risk, potentially protecting the developer's return through toll adjustments. Table 7.1 summarizes the envelope risk allocation strategies.

Project types were described with the initial daily traffic demand and the rate at which traffic would increase without the impacts of tolls. Simulations of these

TABLE 7.1 Envelope risk allocation strategies in PPP project model

		Risk Allocation Descriptor		
		Loan Period	*Subsidy Covers Deficit?*	*Adjustable Toll Rates?*
Primary	Developer	15 years	No	No
Holder of	Agency	25 years	Yes	No
Risk	Public	25 years	No	Yes (X2.5 max. increase)

project types with different risk allocation strategies forecasted the performance of the project for the developer, agency, and users. Simulation results were then used to evaluate the attractiveness of different risk allocation strategies on different project types to stakeholders. Further, the model was altered to reflect projects experiencing financial stress by adding a debt services coverage ratio (DSCR) feedback loop. The simulated performance of stressed projects was compared with that of unstressed projects to reveal additional insights about the effects of risk allocation in PPP on project performance. The model structure that best explains project performance is described using tipping point structures. The extreme risk allocations are used to activate these structures and thereby explain simulation results and the drivers of PPP project performance.

A Model of a Public-Private Partnership for a Toll Road

The model captures the important components and interactions that drive PPP project performance based on a specific case study. Model sectors and major interactions are shown in Figure 7.1. The case study toll road provides an alternate route around a metropolitan area. The model is purposely simple relative to the actual project to expose the relationships relative to risk allocation among the primary project stakeholders and performance. Therefore, only those features

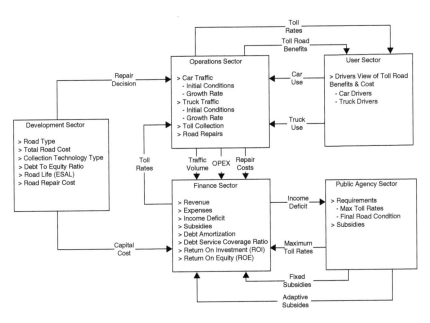

FIGURE 7.1 Sector diagram for toll road public–private partnership model 3

that describe the relevant characteristics of the case study are included. This makes the model valuable for developing insights about risk allocation in PPP projects, but not an effective tool for designing specific PPP projects. The primary project stakeholders in the case study are the private developer and operator of the project (collectively, the "developer"), the state highway authority seeking to meet a public good (the "public agency"), and the users of the toll road. The model has four primary sectors: operations, public agency, development, and financing.

The operations sector models car and truck traffic (modeled separately) on the toll road and how traffic volumes change over time in response to regional development (that increases traffic) and changes in tolls (higher tolls decrease growth rates). Traffic flows and toll rates impact toll income to reflect volume and price risk, the two components of revenue risk (Yescombe, 2002). Toll road operations use the project income deficits (modeled in the finance sector) and constraints imposed by the public agency to adjust tolls charged. Toll road operations create fixed expenses, such as maintaining toll collection infrastructure. Collecting tolls creates variable operating expenses. Repair costs also contribute to operating costs. Repair costs are based on the roadway surface material chosen during development (concrete or asphalt) and traffic flows, measured in equivalent single axle loadings, ESALs (AASHTO, 1993). Cumulative simulated traffic volumes are compared to the number of ESALs that the road can absorb before requiring repair (higher for concrete surfaces than for asphalt). This formulation generates a short-term spike in operating costs during repairs. Congestion on the toll road and adjacent system roads is described with the fraction of the capacity used by traffic. This impacts travel times and thereby the time saved compared to travel times without the toll road.[3]

The public agency sector describes the requirements imposed and supports provided by the agency. Requirements can include minimum roadway conditions required at the end of the project when it is turned over to the public agency. Including this requirement often forces the developer/operator to repair the roadway just before the transfer of ownership, increasing expenses. The public agency can also set maximum toll rates (Yescombe, 2002). Minimum toll revenue guarantees are possible, but monetary subsidies paid to the developer/operator are the primary form of support by the public agency. The model can simulate two subsidy payment structures: fixed amounts paid over a specified period of time or adaptive subsidies that fluctuate with the income deficit (from finance sector). Adaptive subsidies are used in the current work.

The development sector captures critical decisions that are made before the operations stage, including whether to build the road with asphalt or concrete and the fraction of the development cost that is provided by developer equity versus debt. The material decision impacts the initial development cost, traffic loads (in ESALs) that the road can support before repairs are required, and repair costs.

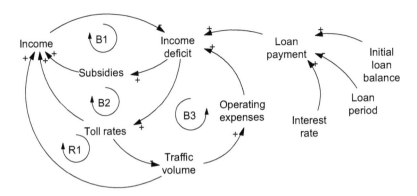

FIGURE 7.2 Toll road public-private partnership conceptual model feedback control of income

Unit transaction costs for collections reflect technology decisions made during development, such as whether to use open-flow electronic toll systems or human-based collection systems.

The finance sector primarily reflects the viability of the project throughout the 50-year project life. The feedback structure of a portion of the finance sector is shown in Figure 7.2 as a causal loop diagram.[4] Reflecting typical project financing, the development cost is disaggregated into equity and debt and the debt is amortized over a loan period based on the interest rate paid to determine the loan payment (Figure 7.2, right side). To measure the financial performance of the project, revenue and expenses are combined using standard definitions to determine monthly and cumulative returns on investment and equity. Subsidies and the products of toll rates and traffic volumes generate income (Figure 7.2, center and left side). Toll rates also influence the amount of traffic, as users assess the value of using the toll road based on their benefits (time saved) and costs (tolls). Traffic also generates operating expenses. The size of the loan payment and operating expenses relative to income determine the project's income deficit, which is used as the basis for increasing tolls or requesting higher subsidies. These basic project finances create three balancing feedback loops, which use subsidies (loop B1) and toll rates (loop B2) to increase income and traffic volume to decrease operating expenses (loop B3), thereby reducing income deficits.

Figure 7.2 also describes the traffic income loop (loop R1), a particularly important feature of PPP toll roads. This feedback loop describes the increase in toll rates due to income deficits. Users respond to increased toll rates by reducing use of the toll road, which reduces income and increases the deficit further. This feedback loop will be used later to explain project performance.

Model Testing and Typical Behavior

The model was tested using standard test methods for system dynamics models (Sterman, 2000). Basing the model on previously tested project model structures and the literature improved the model's structural similarity to PPP project processes and practices. For example, target tolls are based on the need to cover deficits, loaned funds are conserved, elasticities of demand to tolls adjust traffic flows, and standard economic models for calculating loan payments and discounting cash flows were used. Tests for unit consistency in the equations improved confidence in model formulation. Extreme conditions tests were performed by setting model inputs, such as initial car and truck traffic, to extreme values and simulating project behavior. Model behavior remained reasonable. Through a continuous process of model behavior comparison to realistic system behavior the model's structure was evaluated, improved, and found capable of generating the same behavior patterns created by actual PPP projects for the same reasons. Comparison of simulated behavior of the complete model to realistic PPP project behavior also supported model validity.

The complete model was calibrated to reflect the case study project, including assumptions for some variable values (e.g., fraction of development cost borrowed). Calibration sets model parameter values to reflect a specific case that can be used to compare simulated and actual system behaviors. Typical model behavior for the calibrated case is illustrated by four parameters (see Figure 7.3): traffic, roadway condition, operating expenses, and cumulative net income. Traffic is reflected in the total (car and truck) annual average daily traffic (Car AADT, upper left in Figure 7.3). Roadway condition is reflected in the cumulative number of ESALs (Cumulative ESAL, upper right in Figure 7.3), a standard measure of combined car and truck traffic. Total operating expenses include fixed annual costs, toll collection, and roadway rehabilitation costs (Cumulative OPEX, lower left in Figure 7.3). Cumulative net income combines revenues, subsidies, and expenses due to operations and repairs (Cumulative Net Income, lower right in Figure 7.3). As shown in Figure 7.3, traffic initially increases exponentially and adjusts in response to tolls over the project life under other conditions. The roadway condition deteriorates (cumulative ESALs applied increases) due to the growth of the total AADT until it reaches a predetermined level for the chosen road surface (concrete or asphalt), which indicates that the roadway requires rehabilitation to return it to its original condition. Rehabilitation incurs significant short-term costs, generating discontinuous increases in cost and decreases in cumulative ESAL and cumulative net income at weeks 1,132, 1,919, and 2,522. As AADT grows exponentially, the frequency between road repairs decreases due to the faster accumulation of ESAL, from 1,132 months to 787 months to 603 months. In addition, cumulative operations and repair costs increase gradually between road rehabilitations, reflecting repair costs.

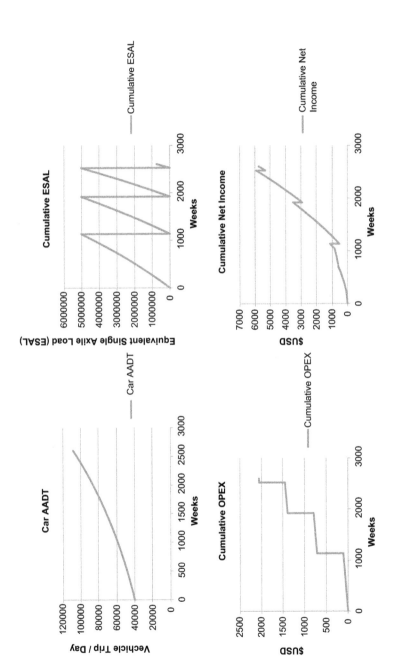

FIGURE 7.3 Typical toll road public–private partnership model behavior traffic, roadway conditions, operations expenses, and net income

Results

Impacts of Extreme Risk Allocation on Project Performance

Extreme risk allocations were modeled using the policies shown in Table 7.1. Possible traffic patterns were modeled with the initial traffic and growth rate in traffic using high and low initial traffic and fast and slow traffic growth. This generated 12 simulations (three risk allocations and four traffic patterns). The project was first simulated for these 12 scenarios to find the developer's and public's project NPV.

The simulation results (not shown for brevity but available from the authors) demonstrate some expected results. Increasing traffic increases NPV. Scenarios with high initial traffic provide attractive returns to the developer and public, and vice versa. Developer returns increase if the public agency or users bear the project risk. The public's return decreases if it bears the risk. In addition NPVs for the public were found to have a much wider range than for the developer, reflecting the impact on thousands of toll road users. The results also indicate the risks involved in PPP. For example, one of the critical challenges of PPP for toll roads is the forecasting and management of traffic volume. The difficulty of forecasting traffic volumes could be reflected in an inability to accurately predict which project conditions a project will experience. Therefore, the PPP project developer and the public might consider the average project NPV across traffic conditions in addition to individual traffic conditions. Average developer NPV values when the developer bears the project risk are negative, suggesting that, if only extreme risk allocations are used, the developer would enter into the project only if the public agency bears the project risk. In contrast, the public benefits regardless of which party bears the risk.

The results also indicate that 4 of the 12 traffic condition risk allocation scenarios generate negative developer NPV. This occurs with low initial traffic when the developer or users bear the risk. Developers would not participate in such projects if they knew those conditions would occur. When the developer bears the risk, this occurs because the developer has no remedies for low revenue (through higher tolls or subsidies) and cannot cover expenses and the loan payment. When users bear the risk, tolls are raised but traffic volumes fall, trapping the project in the traffic income loop (Figure 7.2) as explained previously.

Impacts of Financial Stress on Risk Allocation Design and Project Performance

Projects such as the toll road case study can perform significantly differently if stressed by external or internal factors. Analysts at lender institutions typically simulate project performance under a set of extreme conditions that can impact project cash flow. These stresses can have large impacts on project performance due to the large portion of development costs that are typically borrowed and the resulting high loan payments. If the project's performance metrics withstand extreme conditions, the project's borrowing cost is reduced. One extreme financial stress test is the continuous adjustment of interest rates to reflect the level of project risk. In such a stress test, the developer continuously refinances the project at current rates and even borrows additional funds to cover operating losses if they occur. In this extreme condition, the lending institution bears no risk because it is always being compensated through interest rates that reflect current project risk.

Lenders typically use the debt services coverage ratio to reflect a project's financial risk and to determine corresponding interest rates on projects. The DSCR is the ratio of the project's net operating income to the loan payment (Figure 7.4, bottom and right). A DSCR of one represents a project that can just pay its expenses and loan obligation, but has no funds remaining for profit. Intuitively, the DSCR describes the safety factor indicating the ability of a developer to pay its debt to the lender.[5] A decrease in the DSCR reflects increased risk of default on the loan and vice versa. A decrease in a project's DSCR can initiate toll increases or higher subsidies to improve the DSCR (Figure 7.4, loop B4 and B5), aided by decreased traffic expenses (Figure 7.4, loop B6). A project's financial risk and DSCR can vary significantly during a project because the DSCR is impacted by everything that influences the project's net operating income (e.g., traffic, tolls, subsidies, operating expense) and changes in financing that influence the loan payment (e.g., interest rate, debt balance, amortization period).

In this study, we use the project's DSCR to model the mental model of lenders using the extreme stress test described earlier by increasing the interest rate charged on the remaining loan balance when the DSCR decreases and vice versa (Figure 7.4, right). Projects with relatively large safety factors (high DSCR) are

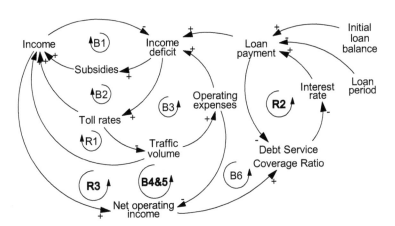

FIGURE 7.4 Toll road public–private partnership conceptual model feedback control of income and debt service risk[6]

TABLE 7.2 Impacts of stress tests on PPP project net present values

Traffic conditions			When the primary holder of the project risk is the …			When the primary holder of the project risk is the …		
			Developer	Public agency	Toll road users	Developer	Public agency	Toll road users
			… the change in the developer's NPV (X$1,000,000) is …			… the change in the public's NPV (X$1,000,000) is …		
Initial traffic	Traffic growth	R/C	a	b	c	d	e	f
Low	Slow	1	-866	37	-748	0	-355	-89
Low	Fast	2	-866	37	-746	0	-356	-100
High	Slow	3	-858	35	-339	0	-97	-2,270
High	Fast	4	-857	103	-309	0	-96	-2,288
Average across traffic conditions		5	-862	53	-536	0	-226	-1,187

"rewarded" with lower interest rates. More dangerously, those with small safety factors (low DSCR) are "punished" with higher interest rates, increasing project expenses and further depressing the DSCR (Figure 7.4, loop R2). Imposing a DSCR loop on a project models an extreme stress test of the project's finances.

The case study was simulated for the 12 scenarios described earlier (four traffic conditions and three extreme risk allocation strategies) under the stressed condition of a DSCR feedback loop. As in the initial set of simulations the developer's and public's average NPV were used as performance measures, in this case to investigate the impacts of financial stresses on the effectiveness of risk allocation strategies.

Some results for the stressed project (not shown for brevity but available from the authors) are similar to those found for the unstressed project. Returns increase with traffic. Developer returns increase if the public agency or users bear project risk. The public NPV range exceeds the developer's NPV range. Uncertain traffic conditions drive developers to want the public agency to bear risk, but the public can benefit regardless of who bears the risk.

However, stressing the project has significant impacts. First, it doubles the scenarios that are unattractive to developers from four to eight and the scenarios in which the public NPV is negative from two to three. Because developers do not enter PPP projects with negative NPV, this indicates that non-extreme risk allocations may be required for a stressed project. The differences between the developer's and public's NPV under stressed and unstressed conditions were calculated to further investigate the impacts of stress testing (see Table 7.2).

As expected, stressing the project generally reduces returns, indicated by the majority of the impacts in Table 7.2 being negative. However, counterintuitively, public return losses due to stressing increase as traffic increases if users bear the risk. This unexpected result is explained in the next section.

Explaining PPP Project Success and Failure With Tipping Point Structures

The simulation model of the toll road PPP project includes two tipping point structures, one operations-based and one based on project finances. In the operations tipping point structure in the unstressed project model (Figure 7.2), the subsidies loop (B1), tolls loop (B2), and traffic cost loop (B3) are the controlling loops that work to counteract deficits by increasing income (B1 and B2) or reducing costs (B3). The traffic income loop (R1) is the reinforcing loop in the operations tipping point structure. This loop can push deficits higher and higher by increasing toll rates and reducing traffic and income. Two of the project conditions and risk allocations modeled (low initial traffic with users bearing risk) generate negative returns for the developer. These negative returns are due to the dominance of the reinforcing feedback loop in the tipping point structure defined by the reinforcing loop (R1) and controlling loops (B1, B2, and B3). Income deficits early in these projects cause a developer to increase toll rates, which decreases traffic,[7] reduces income, and initiates another round of toll increases. This pushes the project across the tipping point from dominance by the controlling loops to dominance by the reinforcing loop. The result is a spiral of increasing toll rates, decreasing traffic, and increasing deficits.

The extreme risk allocation strategies modeled (Table 7.1) activate the tipping point structures in the model. Therefore, the impact of the operations-based tipping point structure on project performance can be seen in the simulation results. The operations-based tipping point structure is activated by allowing tolls to change (Figure 7.4, loops B2 and R1). When this occurs, public NPV decreases compared to when the tipping point structure is not active, partially due to the tipping point structure.[8] If these traffic conditions are expected, risk allocations chosen, and the vulnerability to tipping point dynamics recognized, developers will find such projects unattractive. This demonstrates the importance of understanding how traffic volumes and risk allocation interact with tipping point structures to affect project performance.

The model of the financially stressed PPP project includes a second tipping point structure. In contrast to the operations-based tipping point structure in the unstressed model, the second tipping point is purely financial. As shown in bold in Figure 7.4, the balancing loop structure consists of two similar loops (B4 and B5) that decrease the DSCR when net operating income drops (e.g., by termination of subsidies or reduced traffic), thus increasing the interest rate, loan payment, and income deficit. This deficit increases subsidies and toll rates to increase the net operating income, thereby controlling the DSCR. A third balancing loop (B6) has a similar effect by decreasing operating expenses due to reduced traffic volume caused by increased tolls. In contrast, under the same scenario of a decrease in net operating income, the tipping point structure's reinforcing feedback loop (R2) increases the interest rate and loan payment, which decreases the DSCR further, pushing the interest rate and loan payment higher. The finance-based tipping point structure created by feedback loops B4, B5, B6, and R2 can cause cost escalation by strengthening loop R2 enough to increase loan payments sufficiently to threaten the financial viability of the project. When the project is stressed, it becomes more vulnerable to revenue (i.e., traffic volume) risk because small variations in revenue can decrease the DSCR until it crosses the tipping point, raising the refinancing rate and worsening the positions of both the developer and the public. Therefore, the financial tipping point structure can explain why the number of failed (negative NPV for developer or public) traffic condition/risk allocation scenarios almost doubled from 6 in the unstressed project to 11 when the project is stressed. The simulation results reflect the impact of the financial tipping point structure. The financial tipping point structure is activated by stressing the project with the DSCR structure (Figure 7.4, right) when tolls or subsidies can change (Figure 7.4, loops B1 or B2). If subsidies increase to cover deficits, stressing increases developer NPV (Table 7.2, column b) but decreases public NPV (Table 7.2, column e). If users bear the risk through rising tolls, both the developer and public NPV decrease due to stressing the project (Table 7.2, columns c and f).

Potentially more threatening than either of these tipping point structures alone, the two tipping point structures can interact. For example, a dominant reinforcing DSCR loop in the financial tipping point structure (R2) created by a lender setting a high target DSCR can create a slightly inadequate DSCR in an otherwise healthy project. This can increase the interest rate and loan payment until the project runs a deficit. Toll rates are raised in response, potentially until they decrease traffic sufficiently to push the project over the operations tipping point created by loops R1, B1, B2, and B3, thereby driving the project to failure. Similarly, crossing over the operations tipping point can decrease net operating income enough to push the project over the financial tipping point into dominance by the reinforcing DSCR loop (R2), and drive the project to failure. Interacting tipping point structures can initiate project decline earlier and degrade performance further than single tipping point structures. In the PPP project model, this degrades stressed projects with adjustable tolls (i.e., both tipping point structures are active) and heavy traffic to a greater extent by creating deficits earlier and greater decreases in traffic. The simulations support this explanation, in the increasing losses for the developer and public with increasing traffic when users bear the risk. The interacting tipping point structures create this potential for value-destruction by working synergistically through a third reinforcing loop that links the two tipping point structures. This income-loan loop increases tolls in response to increased loan payments and deficits. Traffic volumes decline, decreasing income (in the operations tipping point structure) and the DSCR, which increases loan payments further (the financial tipping point structure). Whether projects maintain reasonable toll rates and DSCR and therefore succeed or escalate to failure depends upon the relative strengths of the sets of reinforcing and controlling feedback loops within and between tipping point structures.

Discussion: The Importance of Modeling Different Perspectives

In the previous sections, we provided a description of a system dynamics model and how it was used to analyze a large envelope of time-based project performance behaviors. In what follows, we first discuss the role, common practice, and challenges of using computer models such as the one presented here in PPP risk allocation strategy development. We then highlight and explain how key features of the system dynamics model and the approach are useful for overcoming these challenges and agreeing on risk allocation. We compare and contrast these features with current tools and practice while simultaneously positioning them within the policy informatics framework.

Equitable risk allocation for transportation PPP projects requires that partners and their stakeholders develop a shared understanding of how potential project designs will perform under uncertainty. In this context, shared understanding means that mental models are sufficiently similar to create the alignment required to reach agreement. Mental models are a key component of inferential processes used by humans to expand knowledge in the face of uncertainty (Doyle & Ford, 1998). Achieving this is complex because stakeholder diversity and basic human cognition constraints limit our ability to juggle the myriad of factors and interactions that must be considered to produce reliable performance forecasts. Consequently, computer models are routinely relied upon to produce the information that is used by stakeholders to develop and improve the mental models that are ultimately used to make decisions.

Common practice is for the agency and developer to each develop their own (often proprietary) spreadsheet models designed specifically to meet their individual needs. These models are built by partner employees or contractors, typically called analysts, who are skilled at translating problems and associated information into mathematical representations that can be studied and communicated. The early phase of model development involves soliciting user requirements to establish a point of view to guide a complicated process of conceptualizing what components and linkages must be included in the model. The conceptualization process requires making multiple decisions regarding boundaries, level of aggregation, what stakeholder interests to include and how, and so forth. During formal model building, additional project information is gathered from various sources and transformed into an integrated set of formulas and functions, often in a spreadsheet. Model development is a highly iterative process wherein intermediate results must be shared within the organization to allow validation and refinement as required. The end result is at least two tools that provide numerical representations of inputs and outputs for direct inspection, graphing, or other forms of analysis. A key feature of these models is that the underlying structure responsible for the numerical results remains largely invisible to model users, available only in the mind of the model builder. Each party typically uses their respective tool in private to conduct analysis and define negotiating positions before submitting proposals and conducting face-to-face negotiations.

The use of the common practices described earlier to reach an agreement will introduce challenges that are rooted in the use of separate models developed in relative isolation using spreadsheet technology. Because each party has different motivations for pursuing the opportunity, they will likely develop different model requirements that emphasize different aspects of the project. Consequently, each modeler will make different conceptualization decisions. When combined with the differences in modeler skill levels and accepted practices within their respective organizations, there is a strong likelihood that the negotiating parties will be equipped with different information that is biased and incomplete. For example, because of a difference in scope, each model may present only a portion of the potential solution space. This forces negotiators to use other means to fill in the blanks to bridge differences in understanding as they search for common ground.

The system dynamics modeling approach overcomes many of these challenges, mainly by using an endogenous perspective to understand how structures interact to produce behavior (see Richardson in this volume). Consequently, a system dynamics modeler will first work to include and link all relevant structures that have real-world meaning without bias toward the perspective of any one stakeholder. Only then is the model simulated to examine the reasonableness of the output. This is a major difference between the two modeling paradigms: a spreadsheet model provides a numerical view of the behavior, whereas a system dynamics model provides a structural and numerical view of the behavior. This means that a spreadsheet modeler is forced to mentally manipulate structures and their interactions when examining the suitability of the calculation results. Consequently, a system dynamics model provides stakeholders with a more evidence-based, transparent description of the problem and potential solutions.

Although this system dynamics modeling application is focused on a specific case study involving the use of PPP for a transportation toll road, the modeling approach and technology can be used by policy makers and researchers alike for a wide range of problems. The use of system dynamics should be given serious consideration whenever the problem calls for facilitating communication and decision-making across multiple stakeholder groups through increased transparency and causal explanations.

Conclusions

Summary

This study used system dynamics to investigate how risk allocation in a public-private partnership (PPP) project can impact its performance from multiple perspectives. A toll road case study was used to develop and calibrate the feedback-based model. Validation tests indicated that the model can realistically simulate PPP project behavior and performance. Simulations of three extreme risk allocation policies and four traffic growth patterns were used to describe how risk allocation policies and financial stress impact the performance from the perspective of the private developer and the public. Stressing the project largely eliminated risk for lenders but reduced returns to both developers and the public when compared to an unstressed project. Stressing the project also doubled the number of traffic/risk allocation scenarios that were unattractive to developers. Finally, some extreme risk allocation policy/traffic scenarios generated failed projects, for both unstressed and stressed projects.

Insights

After comparing all scenarios, we suspect that projects that share risks provide better returns for all primary participants than do projects with extreme (i.e., envelope) risk allocation strategies. However, the potential for pessimistic traffic scenarios can require significant public subsidies to make projects attractive to developers. The tipping point structures in the system dynamics model proved useful in explaining the project behavior and performance. As some of the simulated projects evolved they changed the relative strength of feedback loops, pushed the project across one of its tipping point conditions, and drove the project to failure. In addition, when both tipping point structures were active, they interacted through a third reinforcing feedback loop to destroy more project value and destroy that value faster. The ability of tipping point structures to explain complex PPP behavior in this way makes them potentially valuable for identifying risk sharing solutions.

Contributions

The current work adds to the understanding of how risk allocation and financial stress impact the performance of PPP projects by objectively modeling the actions and perspectives of the three primary stakeholders. The model was used to forecast their relative performance under different combinations of risk allocation and traffic. Additionally, the study expands system dynamics modeling and project modeling in two ways. First, the model addresses risk allocation in PPP toll road projects, demonstrating how feedback modeling can be used to analyze and guide the design of risk allocation in PPP. Second, the work expands the modeling and understanding of tipping point structures for explaining project behavior and performance. More specifically, tipping point structures are described with sets of three or more feedback loops, two previously un-modeled tipping point structures (one operations-based and one finance-based) are modeled and validated, and the interaction of multiple tipping point structures is described and used to explain project behavior and performance. Finally, the work argues for the use of the system dynamics approach in policy informatics.

The model used here reflects PPP project practice better than many previous models and uses system dynamics to model and analyze the effect of different risk allocation schemes under broad sets of project characteristics. However, the limitations of the model constrain the results and conclusions that can be drawn from them. Primary among these limitations is that we modeled only boundary (extreme) risk allocation policies in which one stakeholder bears all major risks. As expected, these extreme risk allocation policies generate projects that fail. Future work can use the model to study gradations of risk allocation between the extremes modeled here. Modeling intermediate risk allocation policies and model analysis can provide a means to investigate how PPP project teams can avoid and manage tipping point-induced project failure. Future work can also model other factors that may impact PPP project performance.

The results have significant implications for the planning of PPP projects. This study uses a single project representation to study project performance for extreme risk allocation designs. In reality, risk burdens should be, and are, shared at levels between these extremes. Those levels are based on the risk-reward allocations that are acceptable to all stakeholders. However, often not all users sit at the table when PPP contracts are negotiated and risks allocated. In particular, the asset users are often only indirectly represented by the public agency. Poor risk allocation can occur if public agencies work to minimize their own risk, potentially at the expense of users. Planners of PPP projects should be sure that all primary stakeholders have an adequate understanding of the project and that impacts of risk allocation are fairly represented in negotiations. The identification of tipping point structures as drivers of project performance should lead PPP project planners to understand and exploit their role in PPP projects. The use of shared computer models is one means of implementing these recommendations.

In closing, public-private partnership projects are complex systems of physical assets and contracting arrangements among developers, government agencies, and asset users. They require policy informatics tools, such as system dynamics modeling, to explain and understand how risk allocation impacts project performance.

Additional modeling can improve the understanding and use of policy informatics and thereby improve the design and management of public-private partnership projects.

Notes

1 The public agency partner in a concession agreement form of PPP may be a government agency at the national, regional, or state level, a state-owned company, or a special purpose entity (Yescombe, 2002).
2 See http://www.ncppp.org/.
3 Congestion also reduces traffic by discouraging users to enter the roadway. However, we assume that users have no knowledge of roadway congestion *a priori*. Therefore this influence is not included in this model.
4 The polarity of causal arrows linking variables in Figure 7.2 describes the direction of the impact of variable X (at the tail) on variable Y (at the arrowhead). A "+" indicates a direct relationship (if X increases, then Y increases, all other things being equal, and vice versa). A "−" indicates an inverse relationship (if X increases, then Y decreases, all other things being equal, and vice versa). Series of causal links that form closed loops generate feedback effects that can drive system behavior. Feedback loops are labeled as either "B," balancing loops (self-correcting), or "R," reinforcing loops. See Sterman (2000) for a more detailed description of causal loop diagrams.
5 The DSCR as a safety factor plays a similar role as debt service reserve accounts that some lenders require the developer to establish and maintain with adequate funds for the payment of debt obligations for a specified period of time (Yescombe, 2002).
6 **Legend of Feedback Loops:** B1—Subsidies loop: increasing subsidies increases income, thereby reducing deficits and the need for higher subsidies; **B2—Tolls loop:** increasing tolls increases income, thereby reducing deficits and the need for higher tolls; **B3—Traffic loop:** increasing tolls reduces traffic and operating expenses, thereby reducing deficits and the need to increase tolls; **R1—Traffic income loop:** increasing tolls reduces traffic and income, thereby increasing deficits and the need to increase tolls; **B4&5—Net operating income loop:** deficits increase income, net operating income and DSCR, thereby decrease the interest rate and minimum required income and deficits; **B6—Traffic loan loop:** Increased tolls reduce traffic and expenses, increasing net operating income and DSCR; this decreases the interest rate and loan payment, thereby reducing the need to increase income through tolls; **R2—DSCR loop:** Decreasing DSCR increases interest rate and loan payment, thereby decreasing the DSCR further; **R3—Income loan loop:** Increased loan payment raises deficit and tolls, decreasing traffic, income, and DSCR, which increases loan payment farther.
7 This effect is exasperated by the lack of elasticity in the price-demand structure of toll road users and the availability of alternative routes to the toll road.
8 The performance differences are not completely explained by the tipping point structures because of the other impacts of risk allocation.

References

AASHTO. (1993). *AASHTO guide for design of pavement structures*. Washington, DC.
Black, L., & Repenning, N. (2001). Why firefighting is never enough: Preserving high-quality product development. *System Dynamics Review, 17*(1), 33–62.
British Broadcasting Corporation. (2012, May 29). Annan on Syria: "We are at a tipping point." Retrieved from http://www.bbc.co.uk/news/world-middle-east-18257059

Cavanaugh, J. (2013, April). "No-spin zone" *United Hemispheres*. Retrieved from http://www.hemispheresmagazine.com/2013/04/01/no-spin-zone. Accessed October 30, 2014.

Clotfelter, C. (1976). School desegregation, "tipping," and private school enrollment. *Journal of Human Resources, 11*(1), 28–50.

Crabtree, S. (2012, June 13). Russia, U.S. quarrel over aid to Syrian parties. *Washington Times*. Retrieved from http://www.washingtontimes.com/news/2012/jun/13/russia-us-quarrel-over-aid-syrian-parties/

Crane, J. (1991). The epidemic theory of ghettos and neighborhood effects on dropping out and teenage childbearing. *American Journal of Sociology, 96*(5), 1226–1259.

Damnjanovic, I., Duthie, J., & Waller, S.T. (2008). Valuation of strategic network flexibility in development of toll road projects. *Construction Management and Economics, 26*(9), 979–990.

Doyle, J., & Ford, D.N. (1998). Mental model concepts for system dynamics research. *System Dynamics Review, 14*(1), 3–29.

Fishbein, G., & Babbar, S. (1996). Private financing of toll roads. *RMC Discussion Paper Series* (No. 117). Washington, DC: World Bank.

Flood, R.L., & Jackson, M.C. (1991). *Creative problem solving: Total systems intervention*. Chichester: Wiley.

Flyvbjerg, B., Holm, M.K.S., & Buhl, S.L. (2005). How (in)accurate are demand forecasts in public works projects? *Journal of the American Planning Association, 71*(2), 131–146.

Ford, D., & Bhargav, S. (2006). Project management quality and the value of flexible strategies. *Engineering, Construction and Architectural Management, 13*(3), 275–289.

Ford, D., & Sobek, D. (2005). Modeling real options to switch among alternatives in product development. *IEEE Transactions on Engineering Management, 52*(2), 1–11.

Forrester, J.W. (1961). *Industrial dynamics*. Cambridge, MA: MIT Press.

Fry, M., & Lewis, T. (2006, July 16). Over the hedge. Comic strip. *Bryan-College Station Eagle*.

Garvin, M., & Cheah, C.Y.J. (2004). Valuation techniques for infrastructure investment decisions. *Construction Management and Economics, 22*(5), 373–383.

Gay, Jason (2014, January 9) Of Course the Playoffs Will Expand. *Wall Street Journal*. Retrieved from http://www.wsj.com/news/articles/SB100014240527023038481045793088837086683124.

Gladwell, M. (2000). *The tipping point: How little things make a big difference*. Boston, MA: Little, Brown.

Granovetter, M. (1978). Threshold models of collective behavior. *American Journal of Sociology, 83*(6), 1420–1443.

Granovetter, M., & Soong, R. (1983). Threshold models of diffusion and collective behavior. *Journal of Mathematical Sociology, 9*, 165–179.

Hebert, R. (2011). What happened to the Trans-Texas Corridor? *Funding Transportation Projects for Arizona*. Phoenix, AZ: National Council for Public Private Partnerships. Retrieved from http://ncppp.org.previewdns.com/wp-content/uploads/2013/04/Pres-Arizona-Hebert2-0811.pdf

Hulsizer, G. (2011). California's transportation PPP experience. *Funding Transportation Projects for Arizona*. Phoenix, AZ: National Council for Public Private Partnerships. Retrieved from http://ncppp.org.previewdns.com/wp-content/uploads/2013/04/Pres-Arizona-Hulsizer-0811.pdf

Jackson, M.C. (2003). *Systems thinking: Creative holism for managers*. Chichester: Wiley.

Lane, D.C., Gröbler, A., & Milling, P.M. (Eds.) (2004). Rationality in system dynamics: Selected papers from the first European system dynamics workshop, Mannheim

University. Special edition of the *International Journal Systems Research and Behavioral Science, 21*(4).

Lane, D. C., & Jackson, M.C. (1995). Only connect! An annotated bibliography reflecting the breadth and diversity of systems thinking. *Systems Research, 12*, 217–228.

Lyneis, J.M., & Ford, D.N. (2007). System dynamics applied to project management: A survey, assessment, and directions for future research. *System Dynamics Review, 23*(4), 157–189.

National Council for Public-Private Partnerships (NCPPP). (2011). *7 keys to success.* Retrieved from http://www.ncppp.org/ppp-basics/7-keys/

Ortiz, I.N., & Buxbaum, J.N. (2008). Protecting the public interest in long-term concession agreements for transportation infrastructure. *Public Works Management & Policy, 13*, 126–137.

Repenning, N. (2001). Understanding fire fighting in new product development. *Journal of Product Innovation Management, 18*, 265–300.

Schelling, T. (1971). Dynamic models of segregation. *Journal of Mathematical Sociology, 1*(2), 143–186.

Standard & Poor. (2004). *Traffic foresting risk: Study update 2004.* New York: McGraw-Hill.

Sterman, J.D. (2000). *Business dynamics, system thinking and modeling for a complex world.* New York: Irwin McGraw-Hill.

Taylor, T., & Ford, D.N. (2006). Tipping point failure and robustness in single development projects. *System Dynamics Review, 22*(1), 51–71.

Taylor, T.R., & Ford, D.N. (2008). Managing tipping point dynamics in development projects. *ASCE Journal of Construction Engineering and Management, 134*(6), 421–431.

Vajdic, N., Damnjanovic, I., Suescan, D., & Waller, S.T. (2011). Impact of network improvement actions on toll roads revenue performance and bonding costs. *Journal of Transportation Research Record* (No. 2187). Washington, DC: TRB, 8–15.

Yescombe, R.R. (2002). *Principles of project finance.* New York: Academic Press.

8

POLICY INFORMATICS WITH SMALL SYSTEM DYNAMICS MODELS

How Small Models Can Help the Public Policy Process

Navid Ghaffarzadegan, John Lyneis, and George P. Richardson

Introduction

Policy informatics is the use of information technologies, computational modeling, and simulation analysis to address complex policy problems. It is built on the fundamental premise that information can be efficiently and effectively mobilized to identify leverage points and to design, implement, and analyze public policies (Johnston, in this volume). In this chapter, we focus on a specific simulation-based approach to policy informatics, system dynamics, and emphasize the particular potential contributions of small models.

Richardson (in this volume) notes that system dynamics was originally developed in the 1950s by Jay Forrester at MIT based on principles from control systems theory. The method is a simulation approach to modeling complex social systems (Forrester, 1961; Richardson, 1991; Sterman, 2000). The system dynamics approach emphasizes stocks and flows and feedback mechanisms that explicitly represent circular causality. System dynamics models deliberately take a broad boundary so as to capture system behavior endogenously. The endogenous perspective is argued to be a major foundation of the method (Richardson, in this volume).

The system dynamics approach has two main advantages relative to other policy informatics approaches. First, by emphasizing feedback, system dynamics can identify and highlight potential areas of *policy resistance* (Sterman, 2000). Models illustrate how policy actions can trigger reactions—often delayed and unanticipated—that feed back to undermine original policy objectives and even exacerbate original problems. An understanding of the sources of policy resistance is essential for the design of improved public policies.

Second, the feedback approach enables system dynamics models to capture complex dynamics with minimum detail. In contrast to other modeling techniques that generate complexity from detailed depictions of individual agents, the system dynamics approach allows modelers to isolate those dynamics generated by the broader feedback structure of systems. The holistic feedback-based approach can produce models that are small enough to easily communicate core insights to policy makers, yet sophisticated enough to replicate counterintuitive behaviors. We provide an example of such a system in the domain of social welfare later in this chapter.

To show how small system dynamics models can be useful for policy making, we first review five characteristics of public policy problems that make resolution difficult using traditional approaches. These characteristics are: policy resistance, the need for and cost of experimentation, the need to achieve consensus between diverse stakeholders, overconfidence, and the need for an endogenous perspective. We next review an insightful system dynamics model. The model illustrates the usefulness of small system dynamics models for policy making. Most notably, it reveals counterintuitive behavior that is not readily apparent in the absence of an endogenous, aggregate simulation approach. In the last section, we explore these features and develop a set of arguments about how and why small system dynamics models can uniquely address the characteristics of public policy problems identified in the first section and complement other policy informatics approaches. The review sheds light on the factors that modelers should consider when developing effective models for policy makers.

"The Problems" with Public Policy Problems

Public policy problems have several characteristics that impede resolution using traditional approaches. We first explore these characteristics, defining and giving examples for each.

Policy Resistance from the Environment

The first characteristic of public policy problems is the complexity of the environment in which problems arise and in which policies are made. Such complexity leaves policies highly vulnerable to *policy resistance* (Forrester, 1971b; Sterman, 2000). Policy resistance occurs when policy actions trigger feedback from the environment that undermines the policy and at times exacerbates the original problem.

Policy resistance is common in complex systems characterized by many feedback loops with long delays between policy action and results. In such systems, learning is difficult and actors may continually fail to appreciate the full complexity of the systems they are attempting to influence. For example, if a policy increases the standard of living in an urban area, more people will migrate to the area and consume resources (e.g., food, houses, and businesses), thereby causing the standard of living to decline and reversing the effects of the original policy

(Forrester, 1971a). Similarly, when police forces are deployed to control an illegal drug market, drug supplies decrease, leading to higher drug prices, greater profits per sale, and drug dealing becomes a more attractive pursuit. This causes the number of dealers to increase, thereby undermining the original policy (Richardson, 1983). These examples illustrate how attempts to intervene in complex systems often trigger reactions from the environment that compensate for or undermine the original policy action. When policy makers fail to account for important sources of compensating feedback, policies may fail. As a result, traditional tools that lack a feedback approach may fail to anticipate the best policy actions.

Need to Experiment and the Cost of Experimenting

A second characteristic of public policy problems is the importance and cost of experimentation with proposed solutions. Experimentation is important because the stakes of public policy problems are high; experimentation is costly because once implemented, policies are often not reversible. Policy resistance and long delays between actions and their consequences make effective experiential learning extremely difficult (Sterman, 2000; Rahmandad, Repenning, & Sterman, 2009). For example, interest groups may form surrounding a new policy, making a switch to a new approach exceedingly difficult.

Need to Persuade Different Stakeholders

Policy making is not a straightforward process where a decision-maker decides and others immediately implement. Rather, different constituencies, pressure groups, and stakeholders in and outside of government all play important roles in developing policies and influencing their effectiveness throughout society. When effective policies are especially counterintuitive—as is often the case in complex systems—policy makers face the added challenge of generating support from those with diverse and entrenched interests.

An effective means to inform and persuade stakeholders is essential to the development of powerful policy. Otherwise, social pressures from citizens, political opponents, pressure groups, lobbyists, and other constituencies can lead to the enactment of policies focused on short-term gain at the expense of longer-term outcomes.

Overconfident Policy Makers

Effective resolution of public policy problems is also hindered by judgment and decision-making shortfalls at individual levels. Among several different forms of biases, overconfidence (Lichtenstein & Fischhoff, 1977; Lichtenstein, Fischhoff, & Phillips, 1982) is a serious barrier to learning. Studies show that individuals tend to be overconfident in their decisions when dealing with moderately or extremely difficult questions (Bazerman, 1994). Overconfidence is common among naïve as well as expert decision-makers (Henrion & Fischhoff, 1986; Griffin & Tversky, 1992). In complex systems with long delays and a large degree of uncertainty, overconfidence is especially likely given the difficulty that policy makers have learning about their own performance and capabilities.

Overall, individuals' general bias toward their own capabilities, combined with the complexity of the public affairs context, makes overconfidence an important problem in the policy making process. Although overconfidence is not the only bias that exists among decision-makers (Tversky & Kahneman, 1974), research suggests that it has an especially important influence on the ability of policy makers to question their assumptions, models of thinking, and strategies. Overconfidence also makes convincing stakeholders with diverse interests to support policies more difficult.

Need for an Endogenous Perspective

A final characteristic of public policy problems is the tendency of decision-makers to attribute undesirable events to exogenous rather than endogenous sources. In judgment and decision-making literature, such a tendency is usually referred to as *self-serving bias* (Babcock & Loewenstein, 1997). An endogenous perspective is necessary for individual and organizational learning (Richardson, in this volume). Individuals who attribute adverse events to exogenous factors and believe "the enemy is out there" lack the ability to learn from the environment and improve their behavior (Senge, 1990).

Attributing the shortfalls of policies to oppositional parties, international enemies, and other exogenous forces is very common among policy makers and politicians. To illustrate this point, Senge (1990, pp. 69–71) gives the example of the arms race between the Soviet Union and the United States during the Cold War. Rather than viewing actions in the context of the entire feedback system, each party focused only on the link between the threat from the other party and its own need to build arms (threat from the other → need to build own arms). For both, the arms buildup of the other was viewed as an exogenous threat rather than an endogenous consequence of their own earlier actions. The result was an expensive and dangerous escalation.

In summary, public systems and public policy problems have numerous characteristics that inhibit making and implementing effective policies. We argue that small system dynamics models are especially well suited to overcome the above issues and thus can effectively complement other policy informatics approaches. To develop our argument, we first review a small system dynamics model as an example. We then use the example to illustrate the specific characteristics of small system dynamics models that address the problems described above and distinguish the approach from alternatives.

A Small Model

Here, a small system dynamics model that successfully reveals critical public policy insights is reviewed. It was developed to analyze welfare policy in New York State, and is termed the *swamping insight model* (Zagonel, Rohrbaugh, Richardson, & Andersen, 2004).

The Swamping Insight Model

In 1996, then president Bill Clinton signed the Personal Responsibility and Work Opportunity Reconciliation Act to change the role of the federal government in providing support to poor families. The legislation replaced programs that had the potential of providing lifetime federal support to indigent families with Temporary Assistance to Needy Families (TANF). Passing this law shifted responsibility to individuals, states, and counties and made many local government agencies more responsible with welfare issues. For both policy makers and researchers, the condition was new and difficult to fully address (Zagonel et al., 2004; Richardson, 2013).

In January 1997, Aldo Zagonel, John Rohrbaugh, George Richardson, and David Andersen were involved in a simulation project with a coalition of New York State agencies and three county governments to address state level policy making issues regarding TANF.[1] The project led to the creation of several conceptual and simulation models, one of which was a small system dynamics model to capture insight about the effects of investment in different parts of the welfare system (pre-employment and post-employment). This piece, like the other models that were developed, is grounded in the qualitative data extracted through a group model building process. Some insights from the model are reported by Richardson, Andersen, and Wu (2002) and Richardson (2013). This model—later referred to as the *swamping insight* model—can be considered a common archetype of systems that include recidivism.

The model (Figure 8.1) represents the flow of potential recipients of TANF support (i.e., total families at risk). The chain includes two main stock variables, *families on TANF* and *post TANF employed*. The stock *families on TANF* represents people who receive TANF support. The stock *post TANF employed* represents people who have recently found jobs and no longer receive direct support, but are still at risk of losing their jobs. A holistic view of the problem suggests that policy makers should consider all at-risk families, including those that are in the program and those that may return to the program. The number of families on TANF increases as families enter the program and decreases as they find employment and move to the post TANF employed stock. Most individuals in the group of post TANF employed families are employed in low wage, temporary jobs. Thus, these families are still at risk of recidivism and can return to the former stage (i.e., families on TANF) if an individual becomes unemployed.

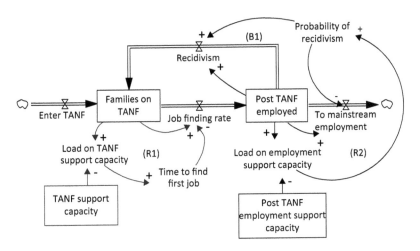

FIGURE 8.1 "Swamping insight" model

The modelers formulated the flow rates based on two variables to represent supportive capacities in the system. *TANF support capacity* influences the *job finding rate*—as support capacity increases, people find jobs more quickly and move to the next stage. A similar effect exists for the downstream capacity (*post TANF employment support capacity*), which captures the economic condition of the region and number of jobs available for post TANF families, who face a high risk of losing employment and returning to a state of need. Families may graduate from post TANF employment into mainstream employment, after which they hold much greater job security. We assume that once families enter mainstream employment they will not need (and will not be eligible for) future TANF assistance. The model captures the recidivism phenomenon by defining a variable named *probability of recidivism*. The probability is a function of the post TANF employment support capacity: As this capacity increases, more people exit the chain of people at risk and enter mainstream employment and fewer return to the TANF program.[2]

One of the main questions that the model addresses is the effect of increasing TANF support capacity. The model demonstrates a counterintuitive result: In contrast to what policy makers intuitively expect, a rise in the upstream capacity makes outcomes worse by increasing the number of families on TANF as well as the total number of families at risk (see Figure 8.2a). Imagine the consternation of policy makers and welfare practitioners seeing that despite increasing the capacity to speed people out of TANF their workload is increasing and they get more and more request for TANF support.

By increasing the upstream capacity, more people flow downstream and the downstream load increases. As a result of the downstream overload, people do

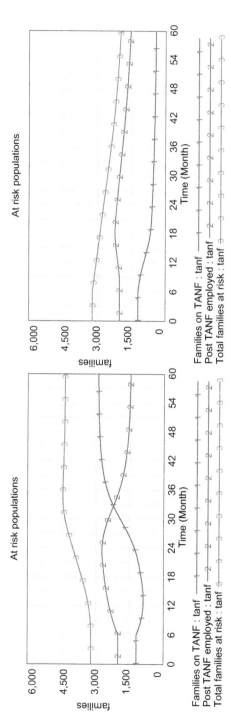

FIGURE 8.2 The effect of 20 percent increase in: (a) upstream (TANF support) capacity; and (b) downstream (Post TANF employment) capacity. (Note: Total families at risk is equal to families on TANF plus post TANF employed).

not receive quality downstream services, causing their economic condition to deteriorate. Ultimately, such families cannot sustain their situation and therefore return to the TANF program, reloading families on TANF.

In contrast, an increase in the downstream capacity (post TANF employment support capacity) has a positive effect on the system by decreasing the number of families on TANF and the total number of families at risk (Figure 8.2b). Such a policy decreases the downstream load as well as the upstream load by decreasing recidivism.

To understand why the system resists a policy of increasing the upstream capacity, we examine the effects of two important feedback loops (Figure 8.1): first, the balancing loop B1 from *families on TANF job finding rate* → *post TANF employed* → *load on employment support capacity* → *probability of recidivism* → *recidivism* → *families on TANF*; and second, the reinforcing loop R2 from *post TANF employed* → *load on employment support capacity* → *probability of recidivism* → *to mainstream employment* → *post TANF employed*. As defined by these feedback loops, the probability of recidivism is an endogenous variable that changes as a function of load on employment support capacity.

The effect of loops B1 and R2 on system behavior can be illustrated by changing the functional relationship between the load on employment support capacity and the probability of recidivism. Figure 8.3 shows three different functional forms listed as scenarios. In the first scenario (the base run), the function is formulated based on data from expert meetings. In the second scenario, we assume that both feedback loops are broken and that the probability of recidivism is a constant number. This scenario illustrates the rationality of decision-makers who assume an absence of feedback through the probability of recidivism. In the third scenario, we reintroduce feedback, but decrease the gain of the loops by changing the sensitivity of *probability of recidivism* to load on employment. The third scenario illustrates how the system will behave if the *probability of recidivism* is not as sensitive to load on employment as in the base run, but still varies endogenously.

Figure 8.3 shows that the above mentioned feedback loops play significant roles in outcomes results. In the second scenario, when we assume a constant probability of recidivism (no feedback through this variable), as policy makers expect, an increase in the upstream capacity results in a decrease in *families on TANF*. Thus, the reason the system resists an increase in upstream capacity in the base run is the increase in *probability of recidivism*, which in turn decreases the outflow from *post TANF employed* to mainstream employment, and increases the outflow from *post TANF employed* to *families on TANF*. Interestingly, in the third scenario, *families on TANF* ends up in a higher equilibrium.

Naturally, policy makers are prone to concentrate on the part of the system for which they are most responsible. If focus is on the increasing number of people at the upstream, in the absence of a holistic view of the system, policy makers may attribute worsening outcomes to exogenous influences. In reality, it is their own policy that has reduced effectiveness of downstream services per individual and raised the level of recidivism.

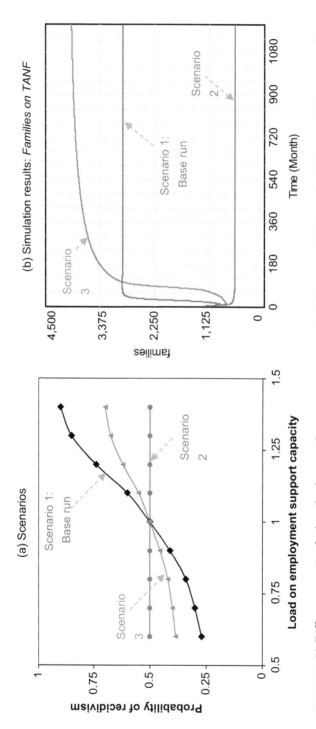

FIGURE 8.3 (a) Different scenarios for how load on employment support capacity influences probability of recidivism and (b) corresponding simulation results for Families on TANF

Overall, this small model shows that adding capacity upstream can swamp downstream resources, increasing the recidivism rate and resulting in increased demand upstream. In other words, adding capacity upstream, by itself, can increase the upstream load and make the entire system worse off. This simple model helps policy makers: (1) develop a holistic view of their problem, (2) better understand counterintuitive lessons from the model, (3) experiment to find better policies for implementation in the real world, and (4) learn about swamping insight and the endogenous causes of policy failure.

The Characteristics of Small System Dynamics Models

As discussed above, public policy problems have several characteristics that make effective policy making especially difficult. Yet, by examining small system dynamics models like the swamping insight model, important insights regarding the source of policy failures can be uncovered. In this section, we highlight characteristics of the model presented and discuss how small system dynamics models can contribute to public policy making more generally. The discussion is summarized in Table 8.1.

We argue that four central characteristics make small system dynamics models especially well-suited for learning about and designing effective policies: (1) the feedback approach and emphasis on endogenous explanations of behavior, (2) the aggregate approach, (3) the simulation approach, and (4) the fact that the models are "small" enough that structure is clear and the link between structure and behavior can be easily discovered through experimentation. In an earlier chapter, Richardson introduces the first three characteristics and describes their historical importance to the system dynamics method (Richardson, in this volume). In this chapter, we explore each of these four characteristics in brief by making connections to the discussed model.

Feedback Approach

The feedback loop is the primary tool used to represent an endogenous source of problem behavior in system dynamics models (Richardson, in this volume). The swamping insight model has a feedback loop approach to modeling that emphasizes endogenous sources of behavior. The model shows how increasing resources allocated to upstream welfare services can result in even *greater* demand for such services. The problem behavior is endogenously created and is highly counterintuitive.

Such counterintuitive behavior is often an example of *policy resistance*. Too often, policies fail due to unanticipated compensating feedback. Emphasizing feedback and an endogenous perspective helps policy makers understand how policy resistance can arise. The model challenges common beliefs about how systems work

TABLE 8.1 The significance of small system dynamics models in addressing public policy problems

Public Policy Problems Characteristics	Small System Dynamics Models Characteristics			
	Feedback Approach	*Aggregate Approach (Stock-Flow)*	*Simulation Approach*	*Small Model Size*
The policy resistance environment	Feedback is the major source of policy resistance.	Accumulations (stocks) are essential to understanding policy resistance.	Simulation can illustrate why some intuitive policies lead to policy resistance and allow for the design and testing of more robust policies.	Small size allows for exhaustive experimentation and sensitivity analysis, wise interpretation of parameters, and parameter changes.
Need to experiment and cost of experimenting	Feedback diagrams and mental simulation (thought) must substitute here for actual policy trials.	Aggregate approach decreases the cost of developing and running models, allowing for more experimentation.	Simulations allow for exhaustive experimentation and games for policy makers without incurring actual social and economic costs.	Small size ensures that the results of experiments can be fully and easily understood by policy makers.
Need to persuade different stakeholders	Feedback diagrams and qualitative analysis can contribute to policy discussions.	Aggregate approach facilitates presentation of lessons to others. Highlights feedback and endogenous sources of problem behavior.	Simulations can help build consensus around difficult policy problems that may otherwise have multiple interpretations.	Small size facilitates presentation of lessons to others. Short exposition and holistic view are made possible.
Overconfident policy makers	Causal loop (feedback) diagrams reveal new insights and challenge policy makers to be wary of overconfidence.	Failure to understand the dynamics of accumulation is a common source of policy error.	Simulations effectively communicate the counterintuitive nature of policy problems to policy makers who otherwise may remain un-persuaded.	Small size ensures that model insights are fully understood, allowing policy makers to appreciate and address their own overconfidence.
Need to have an endogenous perspective	Feedback approach helps policy makers learn what an endogenous view is and why it is necessary to effective policy making.	Aggregate approach leaves more room in an individual's cognitive capacity to concentrate on feedback and develop an endogenous perspective.	Simulations allow policy makers to explore how behaviors are created endogenously through a broad model boundary.	Small size allows individuals to see the feedback structure as a whole and not be frustrated by the need to track many variables and links at once.

by revealing feedback loops that can exacerbate the situation, thereby facilitating learning for even the most overconfident users.

Aggregate Approach

Second, the swamping insight model takes an aggregate approach to modeling. More specifically, it does not track each individual in the population separately. Instead, it models a group of individuals in the aggregate. In keeping with the system dynamics modeling tradition, the building blocks of model structure are stocks and flows rather than individual agents.

Rahmandad and Sterman (2008) argue that differential equation-based models—of which this model is an example—are easier to understand than detailed agent-level models and usually have similar policy implications. In addition, aggregation reduces the size of a model, thereby decreasing the cost of developing and running the model and allowing more experimentation. Given limitations in the cognitive capacity of individuals, aggregation also allows users to focus on feedback ahead of agent-level detail and as a result can develop a more holistic and endogenous perspective of the problem.

Further, recent research has shown that individuals often fail to understand the dynamics of accumulation (Sterman, 2008), creating significant implications for the policies that they will support. By focusing on stocks and flows as the building blocks of model structure, system dynamics models can directly help policy makers build intuition regarding the dynamics of accumulation and thereby overcome one potential source of policy error.

Simulation Approach

Third, mathematical simulations provide a virtual environment in which to conduct counterfactual experiments and train public employees and administrators. Although many lessons can be learned from a paper causal loop diagram and through the process of model building and communicating (Kim, 2009; Black & Andersen, 2012), other more substantial insights require the development and testing of a simulation model (Ghaffarzadegan & Andersen, 2012; Kim, MacDonald, & Andersen, 2013). In the above case, simulation helped illustrate why intendedly rational policies lead to policy resistance.

Furthermore, simulation models provide learning environments where modelers, policy makers, and others can design and test policies. Given the complexity of many policy environments, experimentation is essential for the design of effective policies. Simulations provide a helpful environment where policy makers can experiment and learn about the effects of different policies without any significant social and economic cost to policy makers (Sterman, 2000). These environments can also be used as flight simulators for the purpose of training public employees and decision-makers (Senge, 1990).

In addition, simulations can help build consensus surrounding difficult policy problems. By communicating the counterintuitive nature of policy problems to policy makers, simulations can encourage dialogue and lead to the development of shared interpretations regarding the source of problem behavior. A set of system dynamics models, including the swamping insight model, can communicate counterintuitive lessons and generate consensus among multiple stakeholders in the New York State welfare policy community (Zagonel et al., 2004). In the area of project management, system dynamics models have been used to negotiate disputes and reach consensus regarding the causes of expensive cost and schedule overruns in complex projects (Lyneis & Ford, 2007).

Small Model Size

Finally, the model illustrated here is small. We defined *small* to mean models that consist of a few significant stocks and at most seven or eight major feedback loops. There are two main benefits to a small size. First, a small model size allows for exhaustive experimentation through parameter changes. With lower order models it is much easier to learn from sensitivity analysis and examine the interactions among different parameters (as shown in the swamping insight model, Figure 8.1). Thus, important leverage points in the system can be more easily identified.

Second, a small size ensures that the results of experiments can be fully and easily understood by policy makers. Short exposition makes a holistic view possible; individuals can see the feedback structure as a whole and not be frustrated by the need to track many variables and links at once. In addition, short exposition facilitates presentation of lessons to others and helps bring dynamic lessons to stakeholder meetings. Our emphasis on small models echoes that of Repenning (2003), who argues that, in an academic context as well, small models are necessary to build the intuition of readers who are not accustomed to a dynamic or holistic view of systems.

All told, small system dynamics models bring numerous benefits to the public policy making process. Table 8.1 depicts how each of the characteristics of small system dynamics models can help address the challenges inherent in public policy making.

Conclusion and Discussion

Policy informatics seeks to address current challenges that policy makers face when making decisions in complex systems with abundant data sources. To be successful in that effort, analyses must target the main sources of complex or counterintuitive behavior patterns and help both researchers and policy makers make sense of policy resistance. In this chapter, we argue that small system dynamics models can play such a unique role and thereby serve as a necessary complement to other policy informatics approaches. After listing several common difficulties in policy making, we reviewed an insightful model and used it as an example to examine how small system dynamics models can address some of the most pressing challenges that policy makers face.

We believe that small system dynamics models can contribute significantly to policy making due to four central characteristics: First, they take a feedback approach; second, they are aggregated; third, they present simulation runs; and fourth, they are *small*. Because of these characteristics, small system dynamics models can illustrate the sources of policy resistance in the environment, facilitate learning through extensive experiments, overcome the issues of overconfidence, bring different stakeholders to a shared understanding, and help policy makers learn about the importance of an endogenous perspective to problem solving.

Although we focus on small system dynamics models in this chapter, we should acknowledge that larger system dynamics models are also at times highly effective and necessary for policy analysis. Many successful system dynamics applications combine small models used to communicate insights with larger disaggregated models that represent multiple sectors or agents and provide more fine-grained policy analysis. Very often, core structures, insights, and policy recommendations from larger models are well-captured by simpler models.

Despite these benefits, it is important to mention that small models do also have limitations. First, customers in general, and policy makers in particular, often demand an exclusive model that considers all possible causal links either observed or contemplated. In such a situation, policy makers may lose their trust if they see that their hypothesized link or variable does not exist in the model. Further, in many situations, stakeholders may want to see their own organizations, departments, or communities separately represented, thereby increasing the level of disaggregation. In such a case, having an exclusive version of the model and showing that the final behavior is not qualitatively sensitive to the policy makers' assumed important links or variables can be helpful.

A further limitation is that the use of small models may lead modelers to underestimate the role of some feedback loops, which may in reality be important. It is critical to mention that effective small system dynamics models must not only be simple, but must also include all of the most dominant loops. As a result, the process of building small models may in fact be more difficult than building larger models that include multiple feedback loops. In many cases, small models may emerge only after extensive examination of a larger model to allow for the identification and isolation of only the most dominant feedback loops. Once a large model is developed and the modeler has a clear indication of the dominant loops, he or she can build a smaller version to present to policy makers.

Furthermore, building small models should not impede *operational thinking* and modeling. System dynamics encourages clearly thinking about causalities and how variables are actually connected to produce behavior (Richmond, 2001). Modelers should be clear about how a variable ultimately influences another variable by stating step-by-step the path of the causal link. Being precise in the formulation of causal links and clarifying important capacity constraints are essential aspects of good modeling practice. Although small models may omit some of the details behind causal links, variables can and must remain operational at a high level.

Overall, despite these limitations, we argue that small system dynamics models can greatly aid the policy making process. Small models help policy makers learn about the environment and sources of policy resistance, build learning environments for experimentation, overcome overconfidence, and develop shared understanding among stakeholders. For all of these reasons, we believe that the policy informatics community should do more to help policy makers incorporate the use of small system dynamics models into the policy making process.

Notes

This chapter is based on a previously published paper: Ghaffarzadegan, N., Lyneis, J., & Richardson, G. P. (2011). How small system dynamics models can help the public policy process. *System Dynamics Review, 27*(1), 22–44. Reprinted with permission.

1 The project is reported in several articles, including Zagonel, Rohrbaugh, Richardson, and Andersen (2004), Richardson, Andersen, and Wu (2002), and Richardson (2013). In addition to playing an important role in developing and testing different policies at the state level, the project was also one of the cases used to develop more general processes of group model building (Richardson & Andersen, 1995; Vennix, 1996; Andersen & Richardson, 1997).

2 The outflows from the *post TANF employed* stock are formulated as follows:

> *Recidivism = (post TANF employed/time in post TANF employed)* * *Probability of recidivism*
> *To mainstream employment = (post TANF employed/time in post TANF employed)* * *(1 − probability of recidivism).*

References

Andersen, D.F., & Richardson, G.P. (1997). Scripts for group model building. *System Dynamics Review, 13*(2), 107–129.

Babcock, L., & Loewenstein, G. (1997). Explaining bargaining impasse: The role of self-serving biases. *Journal of Economic Perspectives, 11*(1), 109–126.

Bazerman, M.H. (1994). *Judgment in managerial decision making* (3rd ed.). London: Wiley.

Black, L.J., & Andersen, D.A. (2012). Using visual representations as boundary objects to resolve conflict in collaborative model-building approaches. *Systems Research and Behavioral Science, 29*(2), 194–208.

Forrester, J.W. (1961). *Industrial dynamics*. Cambridge, MA: MIT Press.

Forrester, J.W. (1971a). *World dynamics*. Cambridge, MA: Wright-Allen Press.

Forrester, J.W. (1971b). Counterintuitive behavior of social systems. *Technology Review, 73*(3), 52–68.

Ghaffarzadegan, N., & Andersen, D.F. (2012). Modeling behavioral complexities of warning issuance for domestic security: A simulation approach to develop public management theories. *International Public Management Journal, 15*(3), 337–363.

Ghaffarzadegan, N., Lyneis, J., & Richardson, G. P. (2011). How small system dynamics models can help the public policy process. *System Dynamics Review, 27*(1), 22–44.

Griffin, D., & Tversky, A. (1992). The weighing of evidence and the determinants of confidence. *Cognitive Psychology, 24*(3), 411–535.

Henrion, M., & Fischhoff, B. (1986). Assessing uncertainty in physical constants. *Journal of Physics, 54*(9), 791–798.

Kim, H. (2009). In search of a mental model-like concept for group-level modeling. *System Dynamics Review, 25*(3), 207–223.

Kim, H., MacDonald, R. M., & Andersen, D. F. (2013). Simulation and managerial decision-making: A double-loop learning framework. *Public Administration Review, 73*, 291–300.

Lichtenstein, S., & Fischhoff, B. (1977). Do those who know more also know more about how much they know? *Organizational Behavior and Human Performance, 20*, 159–183.

Lichtenstein, S., Fischhoff, B., & Phillips, L.D. (1982). Calibration of probabilities: The state of the art to 1980. In D. Kahneman, P. Slovic, & A. Tversky (Eds.), *Judgments under uncertainty: Heuristics and biases* (pp. 306–334). New York: Cambridge University Press.

Lyneis, J.M., & Ford, D. (2007). System dynamics applied to project management: A survey, assessment, and directions for future research. *System Dynamics Review, 23*(2/3), 157–189.

Rahmandad, H., Repenning, N.P., & Sterman, J.D. (2009). Effect of feedback delays on learning. *System Dynamics Review, 25*(4), 309–338.

Rahmandad, H., & Sterman, J.D. (2008). Heterogeneity and network structure in the dynamics of diffusion: Comparing agent-based and differential equation models. *Management Science, 54*(5), 998–1014.

Repenning, N.P. (2003). Selling system dynamics to (other) social scientists. *System Dynamics Review, 19*(4), 303–327.

Richardson, G.P. (1983). Heroin addiction and its impact on the community. In N. Roberts, D. F. Andersen, R. Deal, M. S. Garet, & W. A. Shaffer (Eds.), *Introduction to computer simulation, a system dynamics approach*. Reading, MA: Addison-Wesley.

Richardson, G. P. (1991). *Feedback thought in social science and systems theory*. Philadelphia, PA: University of Pennsylvania Press.

Richardson, G. P. (2013). Concept models in group model building. *System Dynamics Review, 29*(1): 42–55.

Richardson, G. P., & Andersen, D. F. (1995). Teamwork in group model building. *System Dynamics Review, 11*(2), 113–137.

Richardson, G. P., Andersen, D. F., & Wu, Y. J. (2002, July 28–August 1). Misattribution in welfare dynamics: the puzzling dynamics of recidivism. In *Proceedings of the 2002 International System Dynamics Conference*, Palermo, Italy.

Richmond, B. (2001). *An introduction to systems thinking*. Watkinsville, GA: High Performance Systems, Inc.

Senge, P. (1990). *The fifth discipline: The art and practice of learning.* New York: Doubleday.

Sterman, J.D. (2000). *Business dynamics: Systems thinking and modeling for a complex world.* Boston, MA: Irwin/McGraw-Hill.

Sterman, J.D. (2008). Risk communication on climate: Mental models and mass balance. *Science, 322*(24 October), 532–533.

Tversky, A., & Kahneman, D. (1974). Judgment under uncertainty: Heuristics and biases. *Science, 185*(4157), 1124–1131.

Vennix, J. (1996). *Group model building: Facilitating team learning.* Chichester, UK: Wiley.

Zagonel, A.A., Rohrbaugh, J.W., Richardson, G.P., & Andersen, D.F. (2004). Using simulation models to address 'what if' questions about welfare reform. *Journal of Policy Analysis and Management, 23*(4), 890–901.

9

QUANTITATIVE MODELING OF VIOLENT GROUP BEHAVIOR USING OPEN-SOURCE INTELLIGENCE

Christopher Bronk and Derek Ruths

Introduction

Although interstate conflict still occurs, violence carried out by transnational actors challenges theory crafted to understand warfare between countries (Finney, 2009). In grappling with the challenges posed by the demands and activities of groups employing violence as their preferred modus of political expression, an important first step is the development of a terror group classification system. This provides a framework within which similarities can be identified and employed to inform evolving theories regarding terrorism and terrorist activities. As a component of irregular warfare (Vacca & Davidson, 2011), theory regarding terror group activity remains fairly immature, but lends itself to data analytics falling under policy informatics. What we have attempted to produce is a computational model for organizing how groups behave based on their actions in a manner that goes beyond valuable, but constrained and scarce, subject matter expertise.

Interest in transnational terror as a topic of study rose in prominence during the last decade largely due to the capacity of non-state organizations to hit targets of importance to the United States and its allies. As Enders and Sandler (2006) observe, prior to September 11, 2001, no terrorist attack took more than 500 lives or produced a direct cost in damage of more than $3 billion. However, the attacks on 9/11 considerably elevated the scale of damage, taking almost 3,000 lives and incurring over $80 billion in financial losses. For the first decade of the 21st century, terrorism, including counterinsurgency operations, was the predominant security concern for the United States and much of the globe. For this reason, we chose to study how terrorism data could be aggregated and machine processed, much as we imagined it might be done in secrecy at the world's intelligence agencies.[1]

Several research teams around the United States had been collecting information on terror events before 9/11. Generally, these teams collected information from open news sources regarding politically motivated attacks that employed means aimed at achieving loss of life. Terror groups often choose to attribute their acts as furtherance of their political goals. Our partners for this project, the Institute for Study of Violent Groups (ISVG) at Sam Houston State University, had been collecting and coding a large repository of terror events, encompassing acts from highly lethal jihadist terror groups to comparatively benign actors in the United States and Europe. We wanted to see how those groups were alike in their behavior—tactics, choice of armaments, targets, and geographic distribution. Our curiosity was shaped by the question as to how terror groups acted in relation to one another.[2]

Where politically oriented group violence is concerned, we sought accurate assessment of capacity and intent in advance of incidents of the sort desired by intelligence and security policy makers and practitioners. Often this information is not readily available to the public and even in intelligence agencies may hold conflicting or missing knowledge of responsible groups (Lowenthal, 2011). Furthermore, research on terror events is typically case-based; more general theoretical models may not be at the same level of maturity as other forms of conflict (i.e., militarized disputes between states or civil conflicts). Although predicting when and where a terror group chooses to strike requires considerable insight into a terror group's communications, aggregating large quantities of information about group behavior is useful for trending and intelligence analysis. We undertook our research as social media technologies were evolving, but operated under the belief echoed by Arquila (2013) that *big data* could be useful in combating terror activity.

We believed computational methods offered a good ability to extrapolate group involvement using various data sources. Here, we formalize two problems that seek to quantify violent group behavior for the purpose of characterizing the behavioral tendencies of the group and then use these tendencies to infer the identity of the actors involved in a given violent event. From these, we considered how terror group analysis might be aided by algorithmic processes to sift through large quantities of fairly well structured but multivariate data.

Our employment of news reporting is predicated on a belief that openly available news sources may be employed to create understanding of political events (Jardines, 2005). The question posed here relates to what larger picture may emerge when algorithmic methods for data analysis derived from computer science are used. Accepting the argument that "Given the increasing threat of terrorism and spread of terrorist organizations, it is of vital importance to understanding the properties of such organizations and to devise successful strategies for destabilizing them or decreasing their efficiency," our research is a preliminary effort in that vein of activity. With the network as a unit of analysis and research tool, this effort follows paths examined by Krebs (2002) and Sparrow (1991).

In this chapter, we considered the *group classification problem*, where we seek to distill the behavior of a group into a set of features that can be used to identify different groups with similar behavioral tendencies. Our approach encodes each group's past behavior in a feature-vector. Phylogenetic methods[3] (Felsenstein, 2003) are then used to build a well-supported classification over all groups for which feature-vectors were built. This activity allowed us to assess similarity across all events and groups.

We then defined the *history-biased event participant inference problem* as the task of identifying the violent groups most likely involved in a given event and presented a methodology that addresses this problem. Our approach trains a naive Bayesian classifier[4] (Russell & Norvig, 2009) for violent group behavior using a dataset of open-source events with known group involvement. This model is then used to identify groups whose behavioral model is most consistent with an event for which all participants (or culpable parties) are unknown.

In order to test our methods, we employed a dataset provided by ISVG that consisted of the attributes and responsible groups for over 28,000 violent events reported in open-source news sources. Collected events spanned a 5-year period from January 1, 2002 to December 31, 2007.

To evaluate our event participant inference method, we performed 10-fold cross-validations on this dataset and determined that our method identifies true participants with over 80 percent accuracy in a violent event participant inference exercise employing Bayesian models trained on open-source information. We also studied performance of our method against the November 2008 Mumbai terror attacks, which killed or wounded nearly 500 people. Comparing the ISVG dataset with attributes of the attacks taken from news reports released directly following the incident, our method correctly identified the ultimately responsible group, Lashkar-e-Tayyiba, as a likely participant in the attacks.

These results both validate the utility of our methods and the ability for human-coded open-source intelligence (Best & Cumming, 2007) to offer new insights into violent and clandestine group behavior.

Inferring a Classification for Violent Groups from Behavioral Features

Constructing a classification system involves building taxonomy (a hierarchical classification) based on differences between groups, which are often drawn from case studies (Wyne, 2005). However, the size, scope, and fluidity of the landscape of violent groups pose a significant challenge. Thousands of groups and sub-groups exist, making the process of devising a global classification system exceedingly tedious.

Furthermore, the incompleteness and asymmetry of information produces an environment in which establishing all necessary parameters that are of importance in the study of each actor constrained (Overgaard, 1994). Also, the manifold differences in culture, ideology, and geography for each group make arriving at meaningful qualitative differences complicated. Finally, the frequency with which violent groups appear, dissolve, and shift positions dramatically reduces the useful lifetime of such taxonomy.

Analogs in other fields such as biology and ecology employ data-centric computational clustering algorithms for inferring classification systems (Kaufman & Rousseeuw, 1990; Mitra & Acharya, 2003). Such approaches use observable attributes of objects to construct a classification system. Computational methods have the ability to rapidly construct classification systems of hundreds or thousands of objects. Furthermore, the data-centric aspect of these methods is particularly applicable to violent groups. Due to their clandestine nature, little is known about their inner workings, but a significant record of their activities exists in the form of news reports because terror organizations employing violence consistently seek media attention (Weimann, 1994). This information, gleaned from openly available sources, presents an alternate view for understanding terror activity that values clandestine collection and sophisticated technical intelligence capabilities (Steele, 2000).

We present a computational method that is fast and non-parametric, producing a classification system that contains all groups specified as input. In an analysis of 108 violent groups, we observe that the inferred classification system contains subclasses that have been previously identified in literature (Sageman, 2004). Furthermore, we find that the inferred ontology supports a connection between group behavior, ideology, and geography. These results indicate that we can produce classifications that both capture existing knowledge as well as discover new relationships among groups.

Method

The objective of a clustering method is to take a set of objects and a pairwise group similarity measure and return a classification system over the objects such that highly similar objects are grouped together. In the case of our area of application, our interest is in clustering a set of violent groups based on their similarity to one another. The use of multi-resolution clustering algorithms was chosen for this study. It can construct many self-consistent, overlapping groupings that simultaneously capture multiple interpretations of *highly similar*.

Hierarchical clustering is a method that constructs a tree where the leaves are the objects to be clustered and each internal node in the tree corresponds to the cluster whose contents are the leaves beneath the node. The significance of a hierarchical clustering is the complete ontology that it produces over the clustered objects. In such a classification system, an object belongs to multiple classes, each class offering a different definition of *highly similar*. As a result, hierarchical clustering by single linkage does not require a threshold parameter, making it well suited to the violent group classification problem.

Computing Group Distances

Distances between the violent groups are required in order to determine which groups are close enough to be placed together under a single classification. By *distance*, we mean the intuitive notion that two groups that are highly similar should have a small distance and two groups that are quite different should have a large distance.

It is worth pointing out that, at the outset, we might also be interested in using more characteristic attributes of a group to characterize it and compute its similarity to another group. By *characteristic attributes*, we refer to the quantification of the character of the group, such as ideology, internal organization, and leadership style.

Having decided to use behavioral attributes, two tasks remain: (1) obtaining a record of different group behaviors and (2) computing meaningful numerical distances from the record.

Obtaining Behavioral Attributes

A record of a group's activity exists in a variety of places, ranging from public news archives, to government reports, to secret intelligence. The quantity, quality, and availability of this information vary considerably from source to source. However, fundamentally, a record of activity will provide a list of incidents, participants, and various details about the incident.

Before meaningful computational processes may be undertaken, the relevant details must be parsed out into a computer-readable format, often into a relational database. Through a variety of methods ranging from automated text mining (Feldman & Sanger, 2007) to manual curation, such content extraction can be performed. In our analysis, we used the ISVG database on violent groups (ISVG, 2008).

We assume, without loss of generality, that it is possible to compile a series of statistics (a *group vector*) from a parsed dataset about a group's behavior. In this example, a large number of behaviors belonging to three general behavioral areas have been quantified: weapon usage, tactic usage, and target types.

Computing Distances from Attributes

Having constructed the group vectors, group distances can be as simply the distance between their positions using the Euclidean norm (Kaufman & Rousseeuw, 1990; Mitra & Acharya, 2003).

The result of this clustering is a tree over all input violent groups.

Computing Classification System Correlations

An important question to ask is how strongly the behavior-based classification system corresponds to various known characteristic attributes such as ideology to determine the extent the behavior-based classification reflects characteristic attributes. It will allow testing for correlations between other variables and the classification system. We quantify the quality of a correlation between the classification system and the variable by determining how likely such a correlation will occur by chance.

Computing the Correlation Score

The parsimony score measures how consistent a leaf labeling is with the topology of the tree. The leaf labels are propagated through all internal nodes in such a way that the total number of labeling changes between a node and its parent is minimal. The parsimony score is the total number of changes required by the tree. Because the parsimony score increases as labeling becomes less consistent with tree structure, a higher parsimony score indicates a poorer correlation between the tree and the labeling.

Results

We applied our classification method to a dataset obtained from the ISVG containing over 26,000 violent incidents attributed to 1,411 distinct violent groups. Of the 1,411 groups, there was only sufficient data to support the clustering of only 108 groups,[5] which were included in the clustering and subsequent analysis. The dataset was assembled from the manual collection and encoding of news stories reporting terrorist activities between 2002 and 2007.

We had two objectives in performing this analysis: (1) to evaluate the performance of our method and (2) to study the kind of insights that can be obtained from a behavioral analysis of violent groups.

We used the 50 behavioral variables that were drawn from weapon usage, bomb usage, tactic usage, or target type selection. Each behavioral variable corresponds to an event attribute stored in the ISVG database extracted by at least one human coder from an article describing the event in question. In general, the variables are either integer- or Boolean-valued, indicating either the number of entities (e.g., people, weapons) involved or the presence/absence of some quantity (e.g., the use of vehicles).

We investigated how the inferred classification system correlates strongly to any variables related to the violent groups, but not used as part of the clustering. Correlations with variables known to characterize types of violent groups support the validity of the classifications inferred by our system. Given that the inferred classification system is valid, any correlation with other variables can provide insights into the properties of groups that best characterize different behavioral classes.

Two group attributes that have been conjectured to significantly influence behavior are ideology and regional affiliation (Sageman, 2004). We compiled the ideological and regional locations of the classified groups and considered the correlation between this attribute labeling as the classification. We observed very strong correlations between ideology and regional affiliation and the classification tree.

Considering the Classification Method

The observation that several existing classification systems are highly consistent with the classification system produced by our method suggests that our approach of classifying groups by their behavior as reported in public news sources does generate meaningful clusters. This point is further supported by the strong correlation between our method's classification system and ideology and regional affiliation.

It is important to emphasize that ideology and geographical location were entirely absent from the group attributes used to generate the classifications. None of the variables can encode geographical location or ideology. Therefore, the observed correlations can be explained only as a true correspondence between the behavior of a group and its ideology or geographical location.

Quantifying this correlation is a capability unique to the data-driven classification methodology we have proposed here (Sageman, 2004). Some discrepancies in the correlations do exist, but offer additional insights into the phenomena of group-based violence. For a group to be placed with other groups, it must behave similarly. In the case of the communist group, this may be an indication that the group's behavior is dominated by its regional affinity rather than ideological bearings; the groups have access to similar weapon supplies or training resources; or that this group shares some other attribute in common with the Islamic groups that dominates their behavior; or perhaps that in the life-cycle of violent groups, similar behaviors may be exhibited at similar points.

Discrepancies offer practical insights as well. Since selection of counter-terrorism tactics are based, in part, on the behavior of a group, this means that groups may respond similarly to the success or failure of similar tactics and is a hypothesis in need of testing (see Desouza & Hensgen, 2007).

We observed that hierarchical classifications captured greater relational information than threshold-based methods. Results of evaluation indicate that the hierarchical clustering approach proposed both captures the classifications discovered by k-means and likely by most other threshold-based clustering methods.

Violent Event Participant Inference Using Bayesian Models Trained on Open-Source Information

With the classification method constructed above, there is still the question of the practical utility of such a model and its application to events playing out in real-time. Although group-specific intelligence may be difficult to obtain, violent groups often seek promotion through attribution (Abrahams, 2006).[6] As long as a group exhibits some degree of behavioral consistency—it is more likely to perform events that resemble events it has performed in the past—this historical information is useful for predictive purposes. The behavioral consistency argument in its most reductive form suggests that, all other variables being equal, if a violent event is perpetrated using a specific tactic, we should suspect groups that have a history of using that tactic. However, when dealing with violent groups that can exhibit significant shifts in leadership, membership, and behavior over time, this assumption may not perfectly model behavior patterns. Nonetheless, in the absence of all other group-specific information, which can often be the case with violent groups, we propose that it is a model worth considering. As our results show, despite the potential for inaccuracy, a method based on this assumption may perform quite well. However, even in a seemingly simple scenario, multiple variables can be recorded to represent the unique event, such as weapon type, target type (public, private, government, etc.), and region. This multivariate nature of the data allows inferences to be made on the basis of a range of observed variables.

Method

Our method consists of two components. The first is an algorithm for building models of each group's behavior. For each group, the algorithm extracts past activity from the set of historical events and encodes it as a set of event-feature probabilities, the *group signature*. The second stage utilizes group signatures, along with information from all historical events, to calculate the probability of a group being involved in the query event. Our method returns the top groups most likely to be involved based on the threshold.

Note that every event implicitly has a set of properties that characterize it. For example, a violent event may have involved a handgun and grenade at a police station. In this case, the event has a handgun property, a grenade property, and a police station property, in addition to a regional property (the geographical region within which the event occurred). The properties for an event are determined by how the event is reported and which details are known. However, setting these issues aside, we establish an event encoding scheme that defines a set of properties that an event may or may not have. For this investigation, we restrict our consideration to binary properties and identify the problem of incorporating non-binary properties as a direction for future work.

Results

In order to evaluate the accuracy of our method, we validated our method against the ISVG open-source dataset described earlier. For validation, we considered only those events for which group associations were known. Following this analysis, we considered a real-world case study: the November 2008 bombings that occurred in Mumbai. In both of these analyses, we used a dataset also derived from the ISVG database.

The Open-Source Dataset

As previously described, the ISVG database is a large, manually curated collection of open-source intelligence gathered from news sources worldwide. Since a reasonable record of prior behavior is central to the training of group signatures, only those groups that had data for at least three event properties were retained. This reduced the number of groups in the analysis to 480 and accounted for 24,962 events. This was the final dataset used in the remainder of the analysis. We considered 61 different event properties (note that most events had a small subset of these properties).

Method Performance

From the outset, we considered it unlikely that any computational method would reliably assign participant groups the highest likelihoods of involvement. The noisy and uneven nature of the data being used combined with the complexity of the phenomena being modeled simply precluded this as a reasonable expectation. Furthermore, the practical use for the proposed method is as a decision support tool for analysts and researchers, which can significantly and quickly reduce the space of groups that must be considered to be possible event participants. In such a use case, practitioners will favor including all of the groups identified as being likely participants in their subsequent analysis. For both these reasons, our validation of the method focused on determining whether the true event participants were ranked higher than other groups—close enough to the top of the rank-ordered list that a practitioner would have high likelihood of including the true participants in their analysis based on the results of our method. In order to do this, we evaluated the frequency with which the true participants were ranked among the top groups scored as most likely to participate in specific events. True participants in an event are identified in the top five groups in more than 65 percent of cases. By the time 10 groups are considered, this accuracy has already risen to nearly 80 percent. After including 50 groups (about 10 percent of all groups considered), true participants are detected in 95 percent of cases.

Case Study: Mumbai

During the attacks of November 2008 in Mumbai, we were able to quickly read through news reports as the events unfolded and pull from them properties that were represented in our dataset. Using only news reports from the event, we obtained information on the following properties: the *weapons* used by the attacking group included bolt action or automatic rifles and grenade/rocket launchers;

the *bombs* were grenades; the *tactics* employed were direct fire and explosives, and raid; the *targets* were building and monument/structure/public land; and the *region* was South Asia. Our method identified the following groups as most likely involved (from most to least likely): Liberation Tigers of Tamil Eelam, Taliba, United Liberation Front of Assam India, Lashkar-e-Tayyiba, Al Qaeda, Communist Maoist Party of India, Chechen Rebel Groups (Russian Federation), and United National Liberation Front (India).

Most noteworthy is the fact that the true culprit behind the attacks, Lashkar-e-Tayyiba, is included in the listing. Furthermore, several other groups included in the list were considered to have the motive and history to have conducted such an attack. In this sense, our method performed well. Although the true culprit was not identified as the top candidate, this can hardly be expected given the complexity of violent group behavior and the incomplete data available. By identifying the true culprit along with other groups that plausibly would have been involved in such an attack, our method provides analysts with a significantly condensed list of groups to closely consider.

Conclusion

Whether operating with the support from or independent of states, terror groups are emerging as significant actors in contemporary political decision-making (Atran, 2004). As a result, gaining a better understanding of how they should be understood within a political framework is imperative. Furthermore, obtaining a framework within which group-specific knowledge is portable among groups is essential to addressing the broader array of international threats posed by many violent groups. Fundamental to both of these objectives is the need for a comprehensive classification system.

The sheer number of violent groups and the complexity and ambiguity of their organization and function makes manually constructed ontologies difficult if not impossible to construct. In this chapter, we have presented a methodology that automates the construction of a classification system over a large number of violent groups using only open-source intelligence limited to public news reports. The classification system produced by our method both coheres with existing knowledge, and provides insight into the phenomenon of international terrorism. As a result, our methodology and open-source intelligence in general show promise in significantly improving our understanding of group-based violence as well as other areas of interest in national security and foreign policy circles, both as an abstract phenomenon as well as a concrete challenge to the function of a stable global structure.

From the outset, it is important to place our work within the modern political context: ever increasing information of varying quality is becoming available in the public domain regarding the behavior of violent groups. Additionally, the landscape of violent sub-national actors is rapidly changing in response to social, political, economic, and environmental forces. Computational techniques can enable experts to keep pace with these changes and efficiently leverage available data.

Our results suggest that there is a place for this and future computational methods in the arsenal of tools used by analysts and academicians to study sub-national violence. Although resolving political and social questions will always require the attention of domain experts, methods such as the one we have presented in this paper can be used to sift through large databases of open-source or secret information, narrow the set of hypotheses that should be closely considered, identify useful pieces of evidence in a case study, and provide support for conclusions eventually reached by the expert. In order for our method to play such a role, it is important that in future work its fundamental model of group behavior be extended. At present, we have used a naive Bayesian model in which all properties are: (1) assumed to be independent and (2) equally weighted. Neither of these conditions will hold in the general case and, in future work, we aim to enrich the underlying Bayesian model to handle both of these current assumptions.

Our work also suggests that open-source intelligence can function as a rich source of information about violent group behavior. Since secret or detailed information about specific violent groups can be difficult to acquire, leveraging open-source intelligence for this purpose may prove a valuable tool to researchers who might otherwise have little information to use for analysis. Although our work suggests that open-source intelligence has promise, much future work is required to establish the conditions and purpose for which it can be reliably used.

It is also worth noting that computational methods like ours have an explicit dependency on data encoding. Although text-mining techniques are rapidly improving, they still cannot rival the accuracy of manual coding of textual data. Thus, the utility of computational tools that depend on encoded textual content is limited by the availability of reliably encoded data. Although this is certainly a short-term limitation of computational methods in this area, we anticipate that continued attention to the problem of text-mining and -coding methods will improve these issues.

There are three major directions for future work that emerge from this chapter. As an initial point of inquiry, our analysis yielded a classification system based on group behavior. To our knowledge, this is perhaps the first such system devised. A closer investigation of this system may offer new insights into the connection between group behavior and identity, structure, and other group attributes.

The event participant inference system proposed, while yielding promising performance, presents a number of opportunities for future work, including boot-strapping the system on different behavioral data and incorporating features that encode geographical and other useful variables.

Finally, and most broadly, we have shown that open-source intelligence can provide novel insights into the activities, organization, and behavior of violent groups. Although open-source data is widely available in raw, textual form, there is a tremendous need for an investment in systems for extracting, curating, orga-nizing, and storing content from these data sources. Such datasets will enable large-scale, quantitative analysis that can test many hypotheses about the nature and behavior of violent groups.

Acknowledgments

We gratefully acknowledge the Institute for the Study of Violent Groups (ISVG) for its help in delivering and working with the ISVG 2002–2007 dataset and appreciate the input of Professor Steve Young at Sam Houston State. This man-uscript also benefited from feedback and comments on our presentation of a preliminary version of this work at the 2008 Networks in Political Science Con-ference at Harvard Kennedy School of Government that we presented with John Miller and Devika Subramanian. Our undergraduate research assistant, Sean Gra-ham, provided important technical input and preliminary runs on the Mumbai attacks as they unfolded over the 2008 United States Thanksgiving holiday. Ric Stoll provided further assistance and input. Finally, the Office of the Secretary of Defense's Strategic Multilayer Assessment (SMA) initiative supported our work.

Notes

1 We wondered exactly what the DARPA Total Information Awareness program might be considering for aggregating data into a useful picture, for example.
2 ISVG approached the development of terrorism knowledge as a human coding exer-cise. Graduate and undergraduate students in criminal justice and other disciplines were assigned parts of the world and actively searched for terror incidents. Some of these were noteworthy on a grand scale during our time of study, including the Bali bombings (2002), Madrid train bombing (2004), and coordinated multiple attacks in Mumbai (2007). But most of the incidents were of much smaller significance and, thankfully, lethality. With this information, we wanted to begin thinking about how models drawn from computational science and statistics might group together the small and major events. Human coders at ISVG identified terror incidents and targets, weapons, tactics, location, and likely attribution among other parameters. The structure constructed in our classification method discovers ideological and geographical rela-tionships that identify strong correlations in the violent behavior of groups.
3 Phylogenetics is a mathematical discipline concerned with the inference of the evo-lutionary relationships among a set of species. The methods have been successfully applied to classification and clustering problems as well.
4 As will be discussed later, a naïve Bayesian classifier models the probability of an object belonging to a specific class as a product of the probabilities of that object's features belonging to other objects known to belong to that class.
5 A data point corresponds to the observation of one property in one incident in the dataset.
6 Whether this is ultimately useful is subject to debate. See Abrahams (2006).

References

Abrahams, M. (2006). Why terrorism does not work. *International Security, 31*(2).
Arquila, J. (2013, April 22). Small cells vs. big data. *Foreign Policy*.

Atran, S. (2004, March 16). A leaner, meaner jihad. *New York Times.*

Best, R., & Cumming, A. (2007). *Open source intelligence (OSINT): Issues for Congress.* Washington, DC: Congressional Research Service.

DeSouza, K.C., & Hensgen, T. (2007). Connectivity among terrorist groups: A two models business maturity approach. *Studies in Conflict and Terrorism, 30*(7), 593–613.

Enders, W., & Sandler, T. (2006). Distribution of transnational terrorism among countries by income class and geography after 9/11. *International Studies Quarterly, 50,* 367–393.

Feldman, R., & Sanger, J. (2007). *The text mining handbook.* Cambridge, UK: Cambridge University Press.

Felsenstein, J. (2003). *Inferring phylogenies.* Sunderland, MA: Sinauer Associates.

Finney, P. (2009). Bridging multiple divides in IT theory: Confronting terrorism, international history, culture and the war on terror. *International Relations, 23*(75).

Institute for the Study of Violent Groups. (2008). Retrieved from http://www.isvg.org

Jardines, E. (2005). *Using open-source information effectively.* Washington, DC: House Committee on Homeland Security, Subcommittee on Intelligence, Information Sharing, and Terrorism Risk Assessment.

Kaufman, L., & Rousseeuw, J. (1990). *Finding groups in data: An introduction to cluster analysis.* New York: Wiley.

Krebs, V. (2002). Mapping networks of terrorist cells. *Connections, 24*(3).

Lowenthal, M. (2011). *Intelligence: From secrets to policy.* Washington, DC: CQ Press.

Mitra, S., & Acharya, T. (2003). *Data mining: Multimedia, soft computing, and bioinformatics.* New York: Wiley-Interscience.

Overgaard, P. (1994). The scale of terrorist attacks as a signal of resources. *Journal of Conflict Resolution, 38*(3).

Russell, S., & Norvig, P. (2009). *Artificial intelligence: A modern approach.* East Saddle River, NJ: Prentice Hall.

Sageman, M. (2004). *Understanding terror networks.* Philadelphia: University of Pennsylvania Press.

Sparrow, M.K. (1991). The application of network analysis to criminal intelligence: An assessment of the prospects. *Social Networks, 13,* 251–274.

Steele, R. (2000). *On intelligence: Spies and secrecy in an open world.* Oakton, VA: OSS International Press.

Vacca, A., & Davidson, M. (2011). The regularity of irregular warfare. *Parameters: Journal of the U.S. Army War College, 41*(1), 18.

Weimann, G. (1994). *Theater of terror: Mass media and international terrorism.* London: Longman.

Wyne, A. (2005). Suicide terrorism as strategy: Case studies of Hamas and the Kurdistan workers party. *Strategic Insights, IV*(7). Retrieved from http://calhoun.nps.edu/handle/10945/11469?show=full

10

MAKING A DIFFERENCE

Kimberly M. Thompson

> "All decisions are based on models . . . and all models are wrong." (Sterman, 2002)
> "All models are wrong but some are useful." (Box, 1979)

Building Useful Models

Modelers often say that the process of working on a model to support policy is at least as important as any final products. Despite this realization, independent of the utility of any policy insights produced, analysts rarely capture the experiential learning that led to the end results in the final publication or other products. For example, we fail to record the strategies used to develop successful policies and models, deal with policy resistance, or overcome barriers, as if no bumps ever occurred on the road. The path to making a real difference using policy models to develop insights and support high-level decisions requires engagement in a collaborative, analytic, and deliberative process. Unfortunately, schools do not teach the skills required to establish and maintain effective collaborations, so these skills often only come with experience and not all modelers fully appreciate their importance.

This chapter uses examples from over a decade of collaborative effort to build integrated policy models to support high-level decisions related to managing the global risks of polioviruses (Thompson, 2013). The collaboration began in 2001, a year after the initial global target date of 2000 for wild poliovirus eradication and the end of poliomyelitis, a devastating paralytic disease that affects the nervous system (Thompson, 2013). By 2000, the Global Polio Eradication Initiative (GPEI) successfully eradicated one of the three wild poliovirus serotypes (i.e., type 2), but many challenges remained in the fight against polio (Thompson, 2013), and relatively little modeling supported the global effort until our collaboration began. The chapter covers issues related to understanding the system, framing the analysis, engaging stakeholders and managing expectations, seeing the paths forward, creating and communicating shared insights, and iterating and learning.

Understanding the System

Large problems inherently involve complexity. We must deal with uncertainty, variability, and time, all of which complicate efforts to develop simple models. Uncertainty results from the realities of imperfect information, knowledge gaps, and our inability to predict the future. While we may feel that we understand everything perfectly, or well enough, in most cases it makes sense to assume that one actually understands very little. Here the wisdom captured by Francis Bacon (1605, Section 5, Part 8) rings true: "If a man will begin with certainties, he shall end in doubts, but if he will be content to begin with doubts, he shall end in certainties." The acceptance of ignorance and uncertainty can prove challenging, but after coming to terms with uncertainty, one can begin the quest required to really understand the system and pursue the goal of trying to integrate as much information as possible. The journey should also begin the process of building a partnership with key collaborators who will invest in shared learning.

Unlike uncertainty, variability implies that we must consider the very real differences that exist between individuals, including different people, places, cultures, products, values, and preferences. Variability leads to the very real prospect that two individuals faced with seemingly the same choice may rationally prefer different options. Variability often means that choosing a policy will lead to "winners" and "losers." Truly understanding the system that one seeks to impact requires identifying the stakeholders (including collaborators) and learning about their knowledge, attitudes, practices, beliefs, and incentives. In many cases, we must also understand the reasons stakeholders may hold different perspectives. This requires respecting differences in individual values and preferences.

In addition to the complications of uncertainty and variability, analysts must also deal with time and the dynamics of the system. Individuals and systems adapt to incentives and their environments, and feedback loops exist. In the context of analyses, we might find it useful to evaluate the situation by looking at a snapshot in time, but we cannot expect the system to remain static. Additionally, policy analyses often focus on improving the future, which implies the need to attempt to understand the system well enough to facilitate reliable predictions about the expected or potential paths.

In the context of modeling poliovirus risks and policies related to their management, we began our collaboration after investing in an effort to develop a retrospective model aimed at exploring how the cost-effectiveness of polio interventions changed over time in the United States (Thompson & Duintjer Tebbens, 2006). Prior to meeting with experts, we reviewed the available literature and built a first version of the dynamic disease model. When we felt we reasonably understood the situation in the United States, we set up a meeting with polio experts at the Centers for Disease Control and Prevention (CDC) to request input. Rather than present our model as a *fait accompli*, we invited the CDC experts to help us identify our mistakes and asked them to tell us about the information, literature, data, and insights we missed. The final product (Thompson & Duintjer Tebbens, 2006) improved significantly as a result of discussions and iteration.

This meeting began a true collaboration and process of asking questions together and of each other to support our collective learning. While our CDC colleagues appreciated the value of our retrospective modeling, they also recognized the opportunity to use similar models to address important prospective policy questions and they asked about our interest in modeling current policy issues. Prior to launching a prospective modeling project, we agreed that we would form a real collaboration and that all involved would invest the time and intellectual capital needed to ask the right questions and develop useful models. Our CDC colleagues introduced us to the other spearheading partners involved in the GPEI and other key stakeholders, which helped us much more quickly understand the global picture.

When we first discussed the project of characterizing the risks, benefits, and costs of potential policies to manage poliovirus risks after the eradication of wild polioviruses (i.e., post-eradication), we found that most stakeholders did not appreciate the full complexity. The first product of our collaboration provided a concise summary of the many choices that national health policy makers faced (Sangrujee, Duintjer Tebbens, Cáceres, & Thompson, 2003). We used decision trees to organize the options, and we identified some key differences between countries (i.e., variability), which explained why countries would consider different options (Sangrujee et al., 2003).

As analysts, we typically want to simplify as much as possible, but we must take care not to overdo the simplification so as to not assume away parts of the problem that would lead to stakeholder concerns about the model ignoring issues of importance to them. Similarly, if we assume some of the difficult complications do not exist, then the model may produce irrelevant, misleading, or even harmful results. For the polio modeling, we recognized the responsibility we faced to develop a truly global model, which required that we consider the real differences that exist between countries of different income levels.

Framing the Analysis

As we develop our understanding of the system, we can start to explore its components and the connections between them in more depth. Throughout the process, it often helps to construct maps of the structure that might facilitate communication with various stakeholders and organize the work. These maps might provide a very high-level view, or they might drill down into the details. Often, developing multiple diagrams provides the best strategy to support discussions with various stakeholders. For example, high-level policy makers may want to see only the big picture, but they will most likely want members of their technical staff to understand the details.

With all stakeholders, prior to jumping into the complexity, success requires that we reach agreement on the big picture because most stakeholders will want to ensure that the analysts "get it" and understand the "real issues." Every conversation with a collaborator or stakeholder offers all parties involved an opportunity for learning. In the context of developing the framework for the analysis, it helps to maintain a running list of assumptions and to frequently discuss the potential implications of choices made related to determining the analytical scope.

Typically, policy analyses start with a question or hypothesis and a specific analytical framework in mind. Some analytical frameworks may inherently ignore uncertainty, variability, and/or time. Consequently, they may compel analysts to simplify and make assumptions that will not work for all collaborators and stakeholders. In such cases, analysts should explore existing opportunities to integrate tools. In addition, during the process of framing the analysis, the collaborators may also realize that the initial question requires some modification. Specifically, building a useful model and developing a helpful policy analysis may require reframing to ask the "right" question(s). This process of iterating on the question(s) can become particularly frustrating for analysts, because it may mean essentially renegotiating the terms of the contract and taking steps backward. It may also frustrate other stakeholders because the timelines for presenting results may shift. However, upon discovering important question(s), the collaborators should recognize that making a difference requires focusing on what matters.

In the context of modeling the global benefits, risks, and costs of post-eradication policies for polio, we developed an initial influence diagram (Thompson et al., 2006a, Figure 2) that eventually evolved to a final high-level diagram (Thompson et al., 2008a, Figure 1) and a more detailed technical version (Duintjer Tebbens et al., 2008a, Figure 1). We also systematically developed each of the components of the analysis, beginning with construction of a dynamic disease model for polioviruses (Duintjer Tebbens et al., 2005) and characterization of the risks (Aylward et al., 2006; Duintjer Tebbens et al., 2006a) and costs (de Gourville, Duintjer Tebbens, Sangrujee, Pallansch, & Thompson, 2006; Duintjer Tebbens, Sangrujee, & Thompson, 2006b) of the policy options (Sangrujee et al., 2003), which we then integrated to produce the final results (Thompson et al., 2008a) and extensive uncertainty and sensitivity analyses (Duintjer Tebbens et al., 2008a).

While quantifying the risks (Duintjer Tebbens et al., 2006a), we recognized the very small but non-zero risk of a potential outbreak of polio after eradication, which led us to ask about the likely response to such an event.

Our questions led to discussions about what might work best, and this led us to perform additional modeling (Thompson, Duintjer Tebbens, & Pallansch, 2006b), which helped us appreciate that "faster is better." This motivated the polio partners to invest in multiple efforts to speed up the process of detecting and responding to outbreaks, which began immediately after we shared the insights and well before publication of the actual paper or the eradication of wild polioviruses (Thompson, Duintjer Tebbens, Pallansch, Kew, et al., 2006b).

We also identified the need for a vaccine stockpile for post-eradication outbreak response (Thompson & Duintjer Tebbens, 2008a; Duintjer Tebbens, Pallansch, Alexander, & Thompson, 2010). Our model (Thompson et al., 2008a) demonstrated that after eradication of wild polioviruses, one vaccine, i.e., oral poliovirus vaccine (OPV), would no longer represent a good option because it would cause more cases and costs than it saved in the absence of wild poliovirus circulation. We further demonstrated that due to the disease dynamics, we would need to globally coordinate OPV cessation (Thompson et al., 2008a; Thompson & Duintjer Tebbens, 2008a). This work supported the development of a global agreement for coordinated OPV cessation (World Health Assembly, 2008).

Engaging Stakeholders, Managing Expectations, and Seeing the Paths Forward

Effective modeling cannot occur in a vacuum. Generating influential research results requires that the people who might use them care about and understand the modeling. This may sound obvious, but surprisingly, many modelers do not speak with stakeholders about their model until they think the model is done. While it may seem easier to do this, ultimately the fate of the model and its utility depend on its acceptance. Admittedly, engaging stakeholders can prove challenging because the process takes time and approaching stakeholders requires respect. The best approach to engaging some stakeholders may come in the form of requesting an opportunity for a brief meeting to learn about their perspectives about what would make a model useful.

In most situations, stakeholders will play a role in determining the ultimate impact of the model and its results in actual policy decisions and/or their implementation. Most policies require a financial and political commitment, and the stakeholders who hold the purse strings and represent the public interest always matter. In addition to these individuals and organizations, other stakeholders may also play key roles (e.g., community leaders, individual physicians or educators, members of the media). The process of developing a policy model must balance the needs and desires of the various stakeholders with the ability of the model to deliver the information needed.

One of the most important elements of success relates to the ability to listen carefully to stakeholders to understand evolution in their thinking and to learn about relevant events that occur. Analysts must understand and identify key questions that they can answer before decisions must be made and be able to provide information before it is needed to support decision-making. Models that produce results after a path has been selected typically offer little impact, although they may help lead to course corrections in some cases. Modelers must engage stakeholders to the extent possible and manage their expectations for what the models can and cannot offer. Sometimes, this may mean reprioritizing work to address urgent questions.

As our collaboration on managing poliovirus risks evolved, we consistently expanded the group of stakeholders, and we began to explore different issues and challenges. One of the most important issues encountered fell outside the scope of our funded work. Specifically, while we were focusing on building models to help inform post-eradication policy choices, some stakeholders began to question whether the global eradication effort should continue at all given the missed deadlines and cumulating expenses. As we listened to the discussion, we recognized that our model could provide useful insights about the question of eradication versus control and we prioritized our efforts to address this question (Thompson & Duintjer Tebbens, 2007).

Using integrated economic, risk, and system dynamic models, we demonstrated the need for intensified immunization efforts in northern India, showed that eradication represented a better humanitarian and economic option than control, and explained the undesirability of a wavering commitment (Thompson & Duintjer Tebbens, 2007). We also provided context about the need to view polio eradication as a major project (Thompson & Duintjer Tebbens, 2008b) and demonstrated challenges associated with managing competing priorities (Duintjer Tebbens & Thompson, 2009). Finally, we addressed the question of the economic value of the GPEI in an analysis that estimated net benefits of $40 billion to $50 billion from the polio prevention alone, with another $17 billion to $90 billion in net benefits realized from Vitamin A supplements given as a part of polio vaccination campaigns (Duintjer Tebbens et al., 2011).

Our close stakeholder interactions help us to appreciate their analytical needs. However, in order to deliver useful insights and results, we need to see these early enough to do all of the work required and to get out in front of the issues to the extent possible. In addition, we often need to integrate tools and extend existing methodologies (Thompson, 2006; Duintjer Tebbens et al., 2008; Rahmandad, Hu, Duintjer Tebbens, & Thompson, 2011) and highlight the important role that modeling can play in supporting decisions (Thompson, 2012; Thompson, 2013).

Despite our collaboration, we faced significant challenges in gaining access to some data. We respect that some stakeholders may not want to share data and we used the approach of running the model with our best guesses and producing results using highly uncertain inputs, which we then shared along with the message that better information would help us to improve the model. Typically, we found that demonstrating the need for data and explaining how we would use them helped significantly, but this means running the model once with best guesses before getting to the final model.

Seeing the Paths Forward

Stakeholder discussions can lead to opportunities to expand the use of a model to answer additional questions. Too often, modelers may view a project as a "one-time" effort, not a long-term commitment. Analysts may build a model to address a question of interest to them, publish the paper, and then move on to the next model. This "hit-and-run" approach may provide some value, but making a meaningful impact on policies often requires much more investment. Students sometimes ask what they need to do to get people interested in a modeling paper that they published, and the simple answer typically boils down to reaching out to and talking with the people who might find the model of interest. Ideally, modelers can and should engage potentially interested stakeholders before they publish because this offers the opportunity to learn about any issues, misperceptions, mistakes, and/or questionable assumptions and address them in the analysis before it becomes final.

A large role that modelers and analysts can play relates to helping various stakeholders ask questions and put these questions into the broader policy context. In order to make a difference, modelers must understand where the system is now and where it might go, viewing their role as an enabler of the change or maintenance of the status quo that will lead to improved outcomes. Developing the ability to forecast well may only occur once modelers and their collaborators understand the system and know what they can expect from each other.

While developing a model to help with post-eradication policies, we recognized that models could play a potential role with respect to accelerating the achievement of eradication. Although some stakeholders discouraged the development of models to explore pre-eradication policies, largely on the premise that by the time the models would be ready, they would not help because eradication would be completed, in reality that did not occur. After years of delay, we began to model pre-eradication policies, beginning with a significant expansion of our dynamic disease model. Specifically, we recognized that we needed to more fully consider the potential immunity states that we included, leading us to engage experts in literature review (Duintjer Tebbens et al., 2013a) and synthesis of the available evidence to develop model inputs (Duintjer Tebbens et al., 2013b).

We also reviewed the literature related to OPV and the risk of circulating vaccine-derived polioviruses. As a result, we built the evolution of OPV into the model (Duintjer Tebbens et al., 2013c). In the process of developing the new model (Duintjer Tebbens et al., 2013d), we recognized the importance of providing a comprehensive review of current national (Thompson et al., 2013a) and global policy options (Thompson & Duintjer Tebbens, 2012) and the literature related to individual and population immunity (Thompson et al., 2013b). We explored the dynamics of population immunity in the United States and the implications of the shift from stand-alone vaccine formulations to combination vaccines (Thompson et al., 2012). We also considered the possibility of unde-tected circulation of wild polioviruses after apparent eradication (Kalkowska, Duintjer Tebbens, & Thompson, 2012). We continue to use the model to address a wide range of policy questions (www.kidrisk.org).

Creating and Communicating Shared Insights

Perhaps the most important role that policy models can play derives from the ability to use the model to communicate about the system and the potential impacts of different actions. Developing a model requires evaluating evidence and making assumptions, and presenting the assumptions can allow stakeholders to offer offer suggestions and challenges. The systematic approach to synthesiz-ing the evidence provides structure and it helps reveal conflicts and inconsistent data that should spur discussion. The model development process can serve as a platform for asking questions and exploration, and often discussions can lead to investments by various stakeholders to perform research that will help resolve important uncertainties.

Models can also help provide insights about opportunities and actions. The experience of "ah-ha" moments occurs along the journey and these can cata-lyze actions. However, developing the insights collectively, such that they become shared and embraced, requires engaging some stakeholders in the process at a technical level. For all stakeholders, policy analysts and modelers need to develop short and effective messages that resonate with decision-makers and derive from the analysis. Ideally, these messages should be simple, but we should also not ignore any relevant details that impact their appropriate interpretation.

Our collaboration developed a significant number of messages to help support poliovirus risk management efforts. For example, when discussing the model results related to responding to outbreaks, we repeated the message that "faster is better" (Thompson et al., 2006b); the stakeholders adopted this concept and worked on multiple opportunities for implementing earlier detection and response. Similarly, we demonstrated the problems with a "wavering commitment" to eradication (Thompson & Duintjer Tebbens, 2007), which led to renewed commitments to complete eradication. We also emphasized the importance of using as much OPV as possible to maximize population immunity and thereby achieve eradication of wild polioviruses as quickly as possible, while reporting that after eradication, globally coordinated OPV ces-sation represented the best option (Thompson et al., 2008a). We recognized the importance of managing population immunity and considering the role of potentially re-infected individuals who might participate in transmission from the beginning of our collaboration (Duintjer Tebbens et al., 2005). However, we realized the need to emphasize that "achieving eradication requires complete prevention," and this led us to emphasize the need to shift from a reactive mode, i.e., a focus on detecting and responding to cases, to a preventive mode, i.e., a focus on managing population immunity to stop transmission so no cases occur (Thompson, 2012; Thompson & Duintjer Tebbens, 2012; Thompson, et al., 2013b). Eradication requires a sustained commitment and this requires that people value the investment (Thompson & Duintjer Tebbens, 2011; Thompson et al., 2011). We recognized the importance of providing rigorous economic analyses that help stakeholders understand the benefits and costs of their actions. Throughout the collaboration, we worked hard to publish the results of our efforts in peer-reviewed literature to attain the benefits with doing so. In addition, we developed pages on our website (www.kidrisk.org) that provide a short summary about each of our papers, with a link to the published paper when possible. The website made our work widely accessible and provided both high-level summaries of the main findings and recommendations and access to the technical details. We also attended many meetings to present our work and to listen and learn from the presentations of others.

Iterating and Learning

Building a useful model inevitably requires some iteration. Analysts need to test their models to assess performance and understand limitations. We need to validate assumptions and model results to the extent possible. While everyone appreciates that "garbage in = garbage out," we must recognize that sometimes the path to progress begins with a simple set of assumptions that we modify as we learn. Additionally, in complex systems, analysts may not know which assumptions will drive the model behavior and this may lead to consideration of some aspects that later prove relatively unimportant. If we see models as opportunities for learning and understanding, then this may offer more promise than seeing models as the means to determine "the answer." In this regard, models may help us as much when defining the questions as when developing insights. Of course, models need to answer the key questions, but the context and learning they provide may in fact serve to motivate the actions required without the need to implement a formal policy. Creating a living model and developing a successful collaboration to support policies inevitably requires a willingness to iterate. On many occasions, good modelers change course.

While modeling polio, our collaborators constantly raise new questions, modify assumptions, and challenge the model, and we maintain flexibility in the process to accommodate the resulting iterations. This means that sometimes the model changes in such a way that we produce results much later than planned and it also means that we consider some questions that only marginally fit within the scope of contracts. The ability to collaborate in this context really requires trust and faith in the process to deliver the information needed. The constant desire to learn and improve our collective understanding continues to motivate our work.

Along the way, we missed a few opportunities that could have significantly impacted the path toward polio eradication. For example, we did not begin modeling pre-eradication issues as early as we could have. If our modeling had motivated more aggressive and intensive efforts to manage population immunity earlier, this might have saved money and prevented some cases of paralytic polio. At the same time, our efforts support a large and complex global policy dialogue and many complicated and high-stakes policy challenges remain.

Remarkably, when we first began working on polio, some academic colleagues suggested focusing on another disease because polio was nearly gone. Fortunately, we saw the opportunity to contribute to the development of policies in the polio "end game" and our decision to invest significant analytical resources and commit to the process proved very rewarding. We thank all of the individuals and organizations involved in the journey, and we hope that our experiences and lessons learned (Thompson & Duintjer Tebbens, 2005; Thompson et al., 2006b) will help others as they work to build useful models.

Acknowledgments

I gratefully acknowledge financial support for the body of work described here in the form of unrestricted gifts to the Harvard Kids Risk Project and grants from the Centers for Disease Control and Prevention (CDC): U50/CCU300860, U01 IP000029, NVPO N37 (FY2005), 200–2010-M-33379, 200–2010-M-33679, 200–2010-M-35172, U66 IP000169, the World Health Organization (WHO) APW200179134, and the Bill & Melinda Gates Foundation: 4533–17492, 4533–18487, 4533–21031. I thank Radboud Duintjer Tebbens for embarking and continuing on the entire journey with me. I also thank collaborators from the CDC, including James Alexander, Lorraine Alexander, Larry Anderson, Gregory Armstrong, Albert Barskey, Brenton Burkholder, Cara Burns, Victor Cáceres, Jason Cecil, Susan Chu, Steve Cochi, Kathleen Gallagher, Howard Gary, John Glasser, Steve Hadler, Karen Hennessey, Hamid Jafari, Julie Jenks, Denise Johnson, Bob Keegan, Olen Kew, Nino Khetsuriani, Robb Linkins, Naile Malakmadze, Rebecca Martin, Eric Mast, Steve McLaughlin, Steve Oberste, Mark Pallansch, Becky Prevots, Hardeep Sandhu, Nalinee Sangrujee, Anne Schuchat, Jean Smith, Philip Smith, Peter Strebel, Linda Venczel, Gregory Wallace, Steve Wassilak, Margie Watkins, and Bruce Weniger, and from WHO, including Bruce Aylward, Fred Caillette, Claire Chauvin, Philippe Duclos, Esther deGourville, Hans Everts, Marta Gacic-Dobo, Tracey Goodman, Ulla Griffiths, David Heymann, Scott Lambert, Asta Lim, Jennifer Linkins, Patrick Lydon, Chris Maher, Linda Muller, Roland Sutter, Rudi Tangermann, Chris Wolff, and David Wood. I also thank the Global Polio Laboratory Network, Harrie van der Avoort, Francois Bompart, Anthony Burton, Konstantin Chumakov, Laurent Coudeville, Walter Dowdle, Paul Fine,

Michael Galway, Shanelle Hall, Neal Halsey, Tapani Hovi, Kun Hu, Dominika Kalkowska, Samuel Katz, Jong-Hoon Kim, Tracy Lieu, Marc Lipsitch, Anton van Loon, Apoorva Mallya, Phil Minor, John Modlin, Van Hung Nguyen, Walter Orenstein, Carol Pandak, Peter Patriarca, Christina Pedreira, Stanley Plotkin, Hazhir Rahmandad, Robert Scott, John Sever, Thomas Sorensen, John Sterman, Robert Weibel, Jay Wenger, and Peter Wright.

References

Aylward, R.B., Sutter, R.W., Cochi, S.L., Thompson, K.M., Jafari, H., & Heymann, D. (2006). Risk management in a polio-free world. *Risk Analysis, 26*(6), 1441–1448.

Bacon, F. (1605). *The advancement of learning* (Section 5, Part 8). London: Everyman. Retrieved from http://pages.uoregon.edu/rbear/adv1.htm

Box, G.E.P. (1979). Robustness in the strategy of scientific model building. In R.L. Launer & G.N. Wilkinson (Eds.), *Robustness in statistics* (p. 202). New York: Academic Press.

de Gourville, E., Duintjer Tebbens, R.J., Sangrujee, N., Pallansch, M.A., & Thompson, K.M. (2006). Global surveillance and the value of information: The case of the global polio laboratory network. *Risk Analysis, 26*(6), 1557–1569.

Duintjer Tebbens, R.J., Pallansch, M.A., Alexander, J., & Thompson, K.M. (2010). Optimal vaccine stockpile design for an eradicated disease: Application to polio. *Vaccine, 28*(26), 4312–4327.

Duintjer Tebbens, R. J., Pallansch, M.A., Chumakov, K.M., Halsey, N.A., Hovi, T., Minor, P.D., . . . Thompson, K.M. (2013a). Expert review on poliovirus immunity and transmission. *Risk Analysis, 33*(4), 544–605.

Duintjer Tebbens, R.J., Pallansch, M.A., Chumakov, K.M., Halsey, N.A., Hovi, T., Minor, P.D., . . . Thompson, K.M. (2013b). Review and assessment of poliovirus immunity and transmission: Synthesis of knowledge gaps and identification of research needs. *Risk Analysis, 33*(4), 606–646.

Duintjer Tebbens, R.J., Pallansch, M.A., Cochi, S.L., Wassilak, S.G.F., Linkins, J., Sutter, R.W., . . . Thompson, K.M. (2011). Economic analysis of the Global Polio Eradication Initiative. *Vaccine, 29*(2), 334–343.

Duintjer Tebbens, R.J., Pallansch, M.A., Kalkowska, D. A., Wassilak, S.G.F., Cochi, S.L., Thompson, K.M. (2013d). Characterizing poliovirus transmission and evolution: Insights from modeling experiences with wild and vaccine-related polioviruses. *Risk Analysis, 33*(4), 703–749.

Duintjer Tebbens, R. J., Pallansch, M. A., Kew, O. M., Cáceres, V. M., Jafari, H., Cochi, S. L., Sutter, R. W., Aylward, R. B., Thompson, K. M. (2006a). M. Risks of paralytic disease due to wild or vaccine-derived poliovirus after eradication. *Risk Analysis 26*(6), 1471–1505.

Duintjer Tebbens, R.J., Pallansch, M.A., Kew, O.M., Cáceres, V.M., Sutter, R.W., & Thompson, K.M. (2005). A dynamic model of poliomyelitis outbreaks: Learning from the past to help inform the future. *American Journal of Epidemiology, 162*(4), 358–372.

Duintjer Tebbens, R.J., Pallansch, M.A., Kew, O.M., Sutter, R.W., Aylward, R.B., Watkins, M., . . . Thompson, K.M. (2008a). Uncertainty and sensitivity analyses of a decision analytic model for post-eradication polio risk management. *Risk Analysis, 28*(4), 855–876.

Duintjer Tebbens, R.J., Pallansch, M.A., Kim, J.H., Burns, C.C., Kew, O.M., Oberste, S.M . . . Thompson K.M. (2013c). Oral poliovirus vaccine evolution and insights relevant to modeling the risks of circulating vaccine-derived polioviruses. *Risk Analysis, 33*(4), 680–702.

Duintjer Tebbens, R.J., Sangrujee, N., & Thompson, K.M. (2006b). The costs of future polio risk management policies. *Risk Analysis, 26*(6), 1507–1531.

Duintjer Tebbens, R.J., & Thompson, K.M. (2009). Priority shifting and the dynamics of managing eradicable infectious diseases. *Management Science, 55*(4), 650–663.

Duintjer Tebbens, R.J., Thompson, K.M., Huninck, M., Mazzuchi, T.M., Lewandowski, D., Kurowicka, D., & Cooke, R.M. (2008). Uncertainty and sensitivity analyses of a dynamic economic evaluation model for vaccination programs. *Medical Decision Making, 28*(2), 182–200.

Kalkowska, D., Duintjer Tebbens, R.J., & Thompson, K.M. (2012). The probability of undetected wild poliovirus circulation after apparent global interruption of transmission. *American Journal of Epidemiology, 175*(9), 936–949.

Rahmandad, H., Hu, K., Duintjer Tebbens, R.J., & Thompson, K.M. (2011). Development of an individual-based model for polioviruses: Implications of the selection of network type and outcome metrics. *Epidemiology and Infection, 139*, 836–848.

Sangrujee, N.K., Duintjer Tebbens, R.J., Cáceres, V.M., & Thompson, K.M. (2003). Policy decision options during the first five years following certification of polio eradication. *Medscape General Medicine, 5*(4). Retrieved from http://www.medscape.com/viewarticle/464841

Sterman, J.D. (2002). All models are wrong: Reflections on becoming a systems scientist. *System Dynamics Review, 18*(4), 501–531.

Thompson, K.M. (2006). Poliomyelitis and the role of risk analysis in global infectious disease policy and management. *Risk Analysis, 26*(6), 1419–1421.

Thompson, K.M. (2012). The role of risk analysis in polio eradication: Modeling possibilities, probabilities, and outcomes to inform choices. *Expert Review of Vaccines, 11*(1), 5–7.

Thompson, K.M. (2013). Modeling poliovirus risks and the legacy of polio eradication. *Risk Analysis, 33*(4), 505–515.

Thompson, K.M., & Duintjer Tebbens, R.J. (2005). *Modeling global policy for managing polioviruses: An analytical journey.* Presented at System Dynamics Society 2005 Annual Meeting, Boston, MA. Retrieved from http://www.systemdynamics.org/conferences/2005/proceed/papers/THOMP452.pdf

Thompson, K.M., & Duintjer Tebbens, R.J. (2006a). Retrospective cost-effectiveness analyses for polio vaccination in the United States. *Risk Analysis, 26*(6), 1423–1440.

Thompson, K.M., & Duintjer Tebbens, R.J. (2007). Eradication versus control for poliomyelitis: An economic analysis. *The Lancet, 369*(9570), 1363–1371.

Thompson, K.M., & Duintjer Tebbens, R.J. (2008a). The case for cooperation in managing and maintaining the end of poliomyelitis: Stockpile needs and coordinated OPV cessation. *The Medscape Journal of Medicine, 10*(8), 190. Retrieved from http://www.medscape.com/viewarticle/578396

Thompson, K.M., & Duintjer Tebbens, R.J. (2008b). Using system dynamics to develop policies that matter: Global management of poliomyelitis and beyond. *System Dynamics Review, 24*(4), 433–449.

Thompson, K.M., & Duintjer Tebbens, R.J. (2011). Challenges related to the economic evaluation of the direct and indirect benefits and the costs of disease elimination and eradication efforts. In S.L. Cochi & W.R. Dowdle (Eds.), *Disease eradication in the 21st century: Implications for global health.* Cambridge, MA: MIT Press.

Thompson, K.M., & Duintjer Tebbens, R.J. (2012). Current polio global eradication and control policy options: Perspectives from modeling and prerequisites for OPV cessation. *Expert Review of Vaccines, 11*(4), 449–459.

Thompson, K.M., Duintjer Tebbens, R.J., & Pallansch, M.A. (2006a). Evaluation of response scenarios to potential polio outbreaks using mathematical models. *Risk Analysis, 26*(6), 1541–1556.

Thompson, K.M., Duintjer Tebbens, R.J., Pallansch, M.A., Kew, O.M., Sutter, R.W., Aylward, R.B., . . . Cochi, S.L. (2006b). Perspective: Development and consideration of global policies for managing the future risks of poliovirus outbreaks—Insights and lessons learned through modeling. *Risk Analysis, 26*(6), 1571–1580.

Thompson, K.M., Duintjer Tebbens, R.J., Pallansch, M.A., Kew, O.M., Sutter, R.W., Aylward, R.B., . . . Cochi, S.L. (2008a). The risks, costs, and benefits of future global policies for managing polioviruses. *American Journal of Public Health, 98*(7), 1322–1330.

Thompson, K.M., Pallansch, M.A., Duintjer Tebbens, R.J., Wassilak, S.G.F., & Cochi, S.L. (2013a). Pre-eradication national vaccine policy options for poliovirus infection and disease control. *Risk Analysis 33*(4), 516–543.

Thompson, K.M., Pallansch, M.A., Duintjer Tebbens, R.J., Wassilak, S.G.F., & Cochi, S.L. (2013b). Modeling population immunity to support efforts to end the transmission of live polioviruses. *Risk Analysis, 33*(4), 647–663.

Thompson, K.M., Rabinovich, R., Conteh, L., Emerson, C.I., Hall, B.F., Singer, P.A., . . . Walker, D.G. (2011). Developing an eradication investment case. In S.L. Cochi & W.R. Dowdle (Eds.), *Disease eradication in the 21st century: Implications for global health*. Cambridge, MA: MIT Press.

Thompson, K.M., Wallace, G.S., Duintjer Tebbens, R.J., Smith, P.J., Barskey, A.E., Pallansch, M.A., . . . Wassilak, S.G.F. (2012). Trends in the risk of U.S. polio outbreaks and poliovirus vaccine availability for response. *Public Health Reports, 127*(1), 23–37.

World Health Assembly. (2008). *Poliomyelitis: Mechanism for management of potential risks to eradication* (resolution 61.1). Geneva: WHO.

PART IV

Administration

11

GOVERNANCE INFORMATICS

Using Computer Simulation Models to Deepen Situational Awareness and Governance Design Considerations

Christopher Koliba and Asim Zia

Introduction

Although complex governance arrangements have always been with us, there is evidence that their complexity is growing with trends toward privatization, devolution, partnerships, and deregulation, impacting how public goods, services, and policies are designed and implemented (Koliba, Meek, & Zia, 2010). These newer governance arrangements call for deeper understanding and knowledge regarding how governance structures impact governance functions. Ultimately, such knowledge may be used to improve performance and accountability. Governance informatics is predicated on the assumption that by building capacity to describe governance processes of heterogeneously interacting agents in complex inter-organizational environments, network managers will be in a better position to adaptively manage wicked problems stemming from accountability and performance of inter-organizational governance networks (Koliba, Zia, & Lee, 2011).

Taking this informatics approach, we posit that the governance knowledge to be culled from informatics platforms can contribute to the situational awareness of stakeholders. Pilots, engineers, emergency management professionals, and military strategists have emphasized the importance that situational awareness brings to understanding complex systems. Situational awareness hinges on a combination of systems thinking, the acquisition and filtering of information, and the application of descriptive patterning that may only be developed through extensive experience built up over time. Endsley (1995) observed that stakeholders with situational awareness seek to classify and understand the situation around them. They relied on "pattern-matching mechanisms to draw on long-term memory structures that allowed them to quickly understand a given situation" (Endsley, 1995, p. 34).

Situational awareness "is the perception of the elements in the environment within a volume of time and space, the comprehension of their meaning, [and] the projection of their status in the near future" (Endsley, 1995, p. 34). Situational awareness should explain dynamic goal selection, while giving attention to appropriate critical cues, to form expectancies regarding future states of the situation (Endsley, 1995). If public managers want to operate within a complex and dynamic system, they must know not only what its current status is, but what its status *could be* in the future, and they must know how certain actions that may be undertaken will influence the situation. For this, they need a *structural* knowledge—how the variables in the system are related and how they influence one another (Radin, 2006, p. 24).

Situational awareness implies a bi-modal approach to viewing structural patterns of governance. Governance structures should be viewed within the context of governance functions and the range of performance goals tied to those functions. Understanding how values and perceptions inform decision-making is as important as the structures that shape how decisions are made. Both of these dimensions are critical for situational awareness.

It is important to develop greater situational awareness of the roles that jurisdictional boundaries, institutional authorities, interest groups, and governing rules and relationships play in shaping the governance and policy dynamics of a given situation and context. In order to do so, we consider the governance network to be a complex adaptive system, "one whose component parts interact with sufficient intricacy that they cannot be predicted by standard linear equations; so many variables are at work in the system that its overall behavior can only be understood as an emergent consequence of the holistic sum of all the myriad behaviors embedded within" (Levy, 1993, p. 34).

This complexity displays certain patterns that may only be picked up through systemic observation.

Employing an informatics approach grounded in the co-construction of computer simulation models with stakeholders pushes us to consider how knowledge of governance arrangements informs decision-making in ways that possess real implications for practice. We may study how stakeholders within a governance network use their existing knowledge of governance arrangements to inform their practice and contribute to the governance of their jurisdiction. We also recognize that actors in governance networks can strategically manipulate information and informatics approaches, leading to the creation of a new set of "information haves" and "information have-nots." Further, perception of governance knowledge could also be filtered through ideology. We must acknowledge these issues up front as these are well-known concerns raised in game theory and other organizational development literature.

The primary objective of the governance informatics approach outlined here lies in the conscious development of governance knowledge and use of this knowledge to develop shared mental models of governance functions and performance. When woven into a process of authentic engagement, governance informatics projects can become spaces where transformative and adaptive changes may be undertaken. In the early stages of a governance informatics project, the tools of research and computer simulation modeling are harnessed to stimulate learning, inform planning and design considerations, and develop adaptive management tools and techniques. The value that governance knowledge (e.g., bearing a conceptual understanding of and an empirical language to describe governance networks) brings to a given situation is considered. Although such knowledge may be employed at the level of smaller problems found at more localized levels (e.g., governance knowledge is likely to be very useful at the level of the individual public administrator or policy analyst), knowledge culled from governance informatics can be used to achieve performance and accountability goals relating to the function of the governance network.

Governance Networks

Within the social sciences, traditional views of governance have hinged on the relatively simple framework of unitary government agencies implementing policy decisions in the most efficient and effective manner possible. It is now widely acknowledged that this simple model does not account for the kind of hybridized governance networks that have arisen as a result of the persistence of wicked problems (Kickert, Klijn, & Koppenjan, 1997; Frederickson, 1999; Milward & Provan, 2006) that lack a definitive formulation, have no stopping rule, and no immediate and ultimate tests of a solution (Rittel & Webber, 1973). The challenges associated with wicked problems are accentuated by the complexity of the multi-stakeholder, networked arrangements that exist to address wicked problems. As the result of a synthesis of the literature pertaining to policy networks (Rhodes, 1997; Kickert et al., 1997), policy systems (Baumgartner & Jones, 1993; Sabatier & Jenkins-Smith, 1993), public management networks (Milward & Provan, 2006; Agranoff, 2007), policy implementation networks (O'Toole, 1990), and governance networks (Sorensen & Torfing, 2005), we conclude that inter-organizational networks may be characterized as:

- Facilitating the coordination of actions and/or exchange of resources between agents within the network;
- Drawing membership from some combination of public, private, and nonprofit sector agents;
- Carrying out one or more policy functions;
- Existing across virtually all policy domains and, often times, existing to integrate policy domains;

- Comprising agents from inter-organizational levels, although they are also described in the context of the individuals, groups, *and* organizations that comprise them; and
- Resulting from the selection of particular policy tools.

(Koliba et al., 2010, p. 46)

To accommodate these characteristics, we define a governance network as:

a relatively stable pattern of coordinated action and resource exchanges involving policy actors crossing different social scales, drawn from the public, private or nonprofit sectors and across geographic levels; who interact through a variety of competitive, command and control, cooperative, and negotiated arrangements; for purposes anchored in one or more facets of the policy stream.

(Koliba et al., 2010, p. 35)

Governance network analysis is informed by resource exchange theory (Rhodes, 1997), vertical and horizontal conceptualization of administrative authority (Agranoff & McGuire, 2003), complex systems dynamics (Haynes, 2003), social network theory (Wasserman & Faust, 1994), and an integrated accountability framework previously developed by members of the research team (Koliba, Mills, & Zia, 2011; Zia & Koliba, 2011).

Rhodes (1997) was one of the first scholars to deeply consider the relationship between governance and inter-organizational networks, arguing that governance occurs as "self-organizing phenomena" shaped by the following characteristics:

1. *Interdependence* between organizations. Governance is broader than government, covering non-state actors;
2. *Continuing interactions* between network members, caused by the need to exchange resources and negotiate shared purposes; and
3. *Game-like interactions*, rooted in strategy and trust, and regulated by rules of the game negotiated and agreed upon by network participants.

Governance is, therefore, characterized by the interdependency of network actors, the resources they exchange, and the joint purposes, norms, and agreements that are negotiated between them. Within the context of democratic systems, these purposes, norms, and agreements need to be "democratically anchored," thereby following a set of norms tied to representative government, citizen participation, and transparency (Sorenson & Torfing, 2005). Thus, the interdependence, continuing interactions, and game-like interactions of governance are rooted in a set of values and practices indelibly linked by democratic forms of government and the public-serving goals derived through it.

Considerations of network governance lead to an inevitable consideration of the bargaining and cooperative systems of more horizontally arranged ties in addition to the traditional vertically oriented command and control systems of mono-centric government systems (Kettl, 2006, p. 491). We argue that mixed-form governance networks may incorporate all forms of administrative authority. Table 11.1 provides an outline of the variables that may be used as the basis to describe and analyze governance networks as complex, adaptive systems. It relies on the basic architecture of networks: nodes, ties, and whole network characteristics. A full explanation of each of these variables is provided at length in Koliba et al. (2010) and summarized in Table 11.1.

TABLE 11.1 Taxonomy of governance networks

Type of Variable	*Variable*	*Descriptor*
Agents (Nodes)	Social scale	Individual, group, organizational/institutional, inter-organizational
	Social sector (organizational level)	Public, private, nonprofit
	Geographic scale	Local, regional, state, national, international
	Role centrality	Central—peripheral, trajectory
	Capital resources actor provides (as inputs)	Financial, physical, natural, human, social, cultural, political, knowledge
	Providing accountabilities to . . .	Elected representatives, citizens and interest groups, courts, owners/shareholders, consumers, bureaucrats/supervisors/principals, professional associations, collaborators/partners/peers
	Receiving accountabilities from . . .	See above
	Performance/outputs and outcomes criteria	Tied to policy function and domain
Ties	Resources exchanged/pooled	Financial, physical, natural, human, social, cultural, political, knowledge
	Strength of tie	Strong to weak
	Formality of tie	Formal to informal
	Administrative authority	Vertical (command and control), diagonal (negotiation and bargaining), horizontal (collaborative and cooperative), competitive
	Accountability relationship	See above

(Continued)

TABLE 11.1 (Continued)

Type of Variable	Variable	Descriptor
Whole Network	Policy tools	Regulations, grants, contracts, vouchers, taxes, loans/loan guarantees, etc.
	Operational functions	Resource exchange/pooling, coordinated action, information sharing, capacity building, learning and knowledge transfer
	Policy functions	Define/frame problem, design policy solution, coordinate policy solution, implement policy (service delivery), evaluate and monitor policy, political alignment
	Policy domain functions	Health, environment, education
	Macro-level governance structures	Lead organization, shared governance, network administration organization
	Network configuration	Inter-governmental relations, interest group coalitions, regulatory subsystems, grant and contract agreements, public-private partnerships
	Properties of network boundaries	Open—closed, permeability
	System dynamics	Systems-level inputs, processes, outputs, outcomes

Using this taxonomy we may observe the emergence of certain patterns of governance that align structures with functions. These patterns are formed by the properties assigned to actors and the nature of the ties that bind them together. These actors will likely exist at one or more levels of social scale: individuals, as groups of individuals, organizations and institutions, and inter-organizational networks. Ties will likely be mediated through any number of different institutional rules, regulations, and laws that are set in place through the selection of policy tools. Actors will be tied together through complex configurations of vertical or horizontal administrative authorities. Across these ties flow resources that can be tracked as financial, social, political, human, natural, or physical capital. We have used this framework to describe the composition of watershed governance networks (Koliba, Scheinert, Reynolds, & Zia, 2013), the governance of energy distribution networks (Koliba et al., 2014), and transportation project prioritization (Zia, Koliba, & Tian, 2012; Zia & Koliba, 2013).

Elements of a Governance Informatics Project

The centerpiece of a governance informatics project is a series of models that begin with early scoping models and culminate in pattern-oriented agent-based models (ABMs). Governance informatics projects are predicated on an ongoing cycle of stakeholder engagement, empirical analysis, and model development. Through this process, decisions regarding boundary conditions, model assumptions, pattern identification, and scenario development are collectively made and owned. The elements of a governance informatics project are illustrated in the flow chart below (see Figure 11.1) that demonstrates a relationship among stakeholder engagement, participatory model development, and decision support (Mitroff, 1972; van den Belt, 2004), as described below.

Element 1: Clarify Initial Boundaries, Conditions, and Problem Conceptualization

Initially, stakeholders in the governance network are approached to participate in the governance informatics project through one of several different avenues: as co-producers of model use and output; as sources of information about the network being modeled; and as checks on the robustness of the model as it pertains to its face validity. Stakeholder engagement occurs at various stages of a project

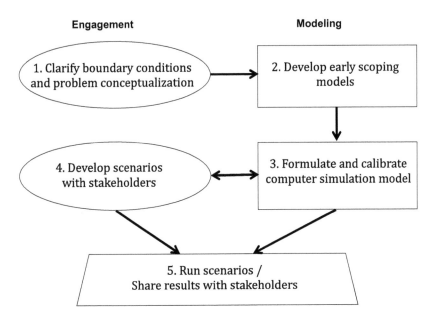

FIGURE 11.1 Elements of a governance informatics project

to build the capacity of modelers and stakeholders to work together to translate empirical and simulated data analysis into useful informatics. Engagement activities include stakeholder interviews, focus groups, surveys, and structured workshops. As model development moves from the conceptual to computational phase, stakeholders are invited to offer feedback and provide substantial input into the range of scenarios generated from working models. This process allows

> knowledge to emerge and be used throughout the course of an interactive analytic process. Consequently, it can provide a bridge for moving from deductive analysis of closed systems, to interactive analytic support for inductive reasoning about open systems where the contextual pragmatic knowledge possessed by users can be integrated with quantitative data residing in the computer.
>
> *(Bankes, 2002, p. 7264)*

Boundaries have been widely recognized as important parameters within policies and governance networks (Kettl, 2006). Determination of the boundaries of a network to be modeled is shaped by conceptual and practical considerations concerning the limitations of theories and methods to model observed patterns. Practical considerations concern limitations that exist around the availability of data and the sheer complexity of a network. These constraints give rise to uncertainty. Determining the boundary conditions of the network to be modeled is of practical importance when scoping out the endogenous and exogenous features of the model. Efforts are made to expand or contract the boundary of the network to capture the optimal level of endogenous features.

The boundaries of the system must also be formed so they capture the imagination of stakeholders. Governance networks are defined by their functionality (Koliba et al., 2010). A governance informatics project will be constructed around a shared common policy domain (such as transportation or water), and along very specific, functional lines (e.g., project prioritization or water quality management needs). The challenges of constructing the boundaries of a system in such a way lie in the tacit processes that unfold—often described by stakeholders as the political elements of the process.

In addition to setting the boundaries of the governance network, the scope of the problems to be addressed must be conceptualized. This process occurs during initial meetings, focus groups, and one-on-one discussions with stakeholders and through extensive literature reviews. Attention to compelling problems that need to be addressed maintains stakeholder engagement and keeps modelers focused.

Element 2: Develop Scoping Models through Qualitative Mapping

Conceptual or scoping models are devised during the very early stages of a project (van den Belt, 2004). These early models are determined as a result of stakeholder engagement (via traditional research collection methods and participatory modeling sessions) and use of theoretical frameworks devised to describe governance network functioning. These models serve as useful boundary objects (Wenger, 1998) around which common assumptions and mental models are formed. Initial scoping models are refined as data is collected and analyzed. Once the initial conditions are set, a prolonged and extensive study of the networks to be modeled is undertaken. The analysis of source documents, existing databases, focus groups, interviews, surveys, legal reviews, and participant observations all feed conceptual, scoping model development.

Some of the following questions may be posed to stakeholders during focus groups, surveys, and interviews to construct initial scoping models:

- What capital resources, types of ties, policy tools, administrative strategies, accountability structures, and performance management systems are in place to ensure that networks function properly?
- Who are the actors involved in the governance network?
- How does a public manager handle accountability trade-offs?
- How is network performance defined? Who establishes the definition? How is network performance measured and managed? Where, within the governance network, is network performance data discussed, used to make decisions, and acted upon?

(Koliba et al., 2010, pp. 287–288)

Koliba et al. (2010) established a step-by-step guide to construct a scoping model of a governance network.

A critical feature of a scoping model is the identification of *action arenas* (Ostrom, 2005) within which critical decisions are made. Action arenas may take the form of working groups, offices or departments, committees, or task forces. Action arenas serve as the veritable "brains" of these networks. The individuals who populate them and the tools and decision heuristics that govern the functionality of these spaces become core features of the pattern–oriented ABMs to follow.

Element 3: Formulate and Calibrate Computer Simulation Model

To cope with the inherent complexity and uncertainty in the social complexity of governance networks, we undertake a variation of pattern-oriented modeling. Note that other variations of computer simulation models, including system dynamics and discrete process modeling, may be used in governance informatics projects. In fact, for the project described later in this chapter, a system dynamics model was constructed to better understand the relationship between local capacity and transportation project funding patterns (Tucker, 2011).

Pattern-oriented ABMs are described by Grimm et al. (2005) as bottom–up models that emphasize the applicability of models to real problem solving (p. 987).

According to Grimm et al. (2005), "Patterns are defining characteristics of a system and often, therefore, indicators of essential underlying processes and structures." By grounding model development in these observed patterns, the "model design directly ties the model's structure to the internal organization of the real system." This is done by asking, "What observed patterns seem to characterize the system and its dynamics, and what variables and processes must be in the model so that these patterns could, in principle, emerge?" (p. 987).

Pattern-oriented approaches are pursued because they help focus and reduce uncertainty in any model of a complex adaptive system. Grimm et al. (2005) add that pursuing a pattern-oriented modeling strategy:

> is a way to focus on the most essential information about a complex system's internal organization. Multiple patterns keep us from building models that are too simple in structure and mechanism, or too complex and uncertain. Using patterns to test and contrast alternative theories for agent behavior or other low-level processes is a way for [modelers] to get beyond clever demonstration models and on to rigorous explanations of how real systems are organized and how they respond to internal and external forces.
>
> *(p. 991)*

The patterns of governance established during the scoping model development stage are refined as computer models are developed. Choices regarding which patterns are most pertinent in addressing the problem at hand must be made. Pattern selection must be grounded in some material substance: real people, institutions, resource exchanges, functionalities, and/or data.

In ABMs, the patterns of financial, knowledge, human, natural, or physical capital flow through a governance network that can be simulated and calibrated to the observed patterns. At various stages of the process, agents respond to the signals coming to them by accessing these resource flows and through the behaviors of other agents. The agents in these models can be individuals, groups, or institutions, as well as discrete tasks, projects, or programs. Discussions of ABMs are carried out extensively in other chapters of this book, and Zia et al. (2012) provide an in-depth discussion on the prospects and limits of the ABM approach for simulating creative, learning-based, and adaptive decision-making agents in inter-organizational governance networks.

Element 4: Develop Problem and Governance Scenarios with Stakeholders

An effective way to envision the uses of governance ABM is through the generation of initial scenarios of alternative future realities. "A scenario is a coherent, internally consistent and plausible description of a possible future state of the world. It is not a forecast; rather, each scenario is one alternative image of how the future can unfold" (IPCC, 2008). Although a wide range of scenario types are offered in the literature, two particular types of scenarios are most relevant to governance informatics projects: *problem* scenarios and *governance* scenarios.

Problem scenarios focus on possible anticipated future states of the system under study. These future states are shaped by the boundary conditions and problem configurations identified earlier in the process. Problem scenarios usually focus on the role that alternating stable and dynamic states of a system play in the existence of a given policy problem (e.g., the future state of water quality, renewable energy, or food production in the region).

Governance scenarios are the alternative institutional or network designs envisioned as a means of responding to problem scenarios, and serve as the critical, distinguishing feature of a *governance* informatics project. The range of governance design considerations that stem from the construction of scoping and computer simulation governance models provide a deeper level of situational awareness about alternate *functionalities* and *structures* of the governance networks to tackle the wicked problems. Governance scenarios may be predicated upon different uses and weighting of performance data, different configurations of actors involved in decision-making, new sources of resource flows, etc.

To aid in developing a deeper situational awareness, both problem and governance scenarios must possess the following characteristics:

- *Relevance*: Scenarios align with the problems and questions of interest to stakeholders and decision-makers.
- *Legitimacy*: Scenario development process includes diverse stakeholder views and beliefs.
- *Plausibility*: Scenarios tell coherent stories that could conceivably occur.
- *Understandable*: Scenarios are understandable to stakeholders.
- *Distinguishable*: Scenarios are sufficiently dissimilar to show contrasting impacts.
- *Scientifically credible*: Scenarios have been developed using scientifically robust, credible methods.
- *Computationally accessible*: Scenarios may be represented in the inputs, through-puts, and/or outputs of the computer simulation models being devised.
- *Comprehensive*: Scenarios consider relevant drivers of change, including both those beyond and within the control of decision-makers.
- *Iterative*: Scenarios are refined over time to incorporate stakeholder feedback, emerging knowledge, trends, and issues.
- *Transparent*: Scenarios are accessible to a wide array of stakeholders. The assumptions embedded within the scenarios are clear.

- *Insightful*: Scenarios challenge assumptions and broaden perspectives about unexpected developments. (Adapted from McKenzie & Rosenthal, 2012)

Approaches for working with stakeholders to develop scenarios can range from loosely guided, open ended discussions of possible alternative future states to tightly designed, protocol-driven processes. By relying on a pattern-oriented approach to model development, the range of possible alternative governance scenarios envisioned to address problem scenarios will be anchored within empirical observations from existing systems and structures, which may or may not be viewed as barriers to effective change. The extent to which existing governance arrangements serve as impediments or enablers of effective change will clearly indicate the approaches to be selected.

Element 5: Run Scenarios to Inform Decision-Making

The final phase of a governance informatics project involves running alternative governance scenarios conceived during earlier stages of the project. In an ABM, these scenarios may be run using sliders or other input tools that allow users to "play" with alternative configurations. Whether to build these sliders into the model should be determined through active engagement with stakeholders during the scenario development phases.

Squazzoni and Boero (2010) suggest that our goal for projects of this nature does not lie in putting ourselves in the position "to predict the future state of a given system, but to understand the system's properties and dissect its generative mechanisms and processes, so that policy decisions can be better informed and embedded within the system's behavior, thus becoming part of it." (p. 3)

The very process of providing feedback concerning a system's dynamics back to the system itself becomes an important component of decision-making and action. "Once viewed as something that is embedded within the system, rather than taking place before and off-line, policy [and governance] starts to be practiced as a crucial component that interacts with other components in a constitutive process" (Squazzoni & Boero, 2010, p. 3). Thus, knowledge and information regarding network governance processes may be fed into the communication systems of the network, facilitating system-wide learning, adaptation, and the emergence of new strategies (Guney & Cresswell, 2010). The types of pattern-oriented ABMs of governance networks presented here can be used "when policy makers need to *learn from science* about the complexity of systems where their decision is needed," as well as "when policy makers need to *find and negotiate certain concrete ad hoc solutions*, so that policy [and governance] becomes part of a complex process of management that is internal to the system itself" (Squazzoni & Boero, 2010, p. 6).

Using the various Web interfaces possible with ABMs allows stakeholders to test alternate solutions for challenging governance design issues. In the next section, we review one governance informatics project to illustrate how this process works.

Application to Transportation Prioritization Project

To illustrate how a governance informatics project proceeds, we highlight a project undertaken in partnership with the Chittenden County Metropolitan Planning Organization (CCMPO) and the Vermont Agency of Transportation (VTrans). This project commenced with the authors' initial interest in studying how regional transportation prioritization networks work and the project eventually morphed into a full-scale implementation of the process described herein (Koliba, Zia, Novak, & Tucker, 2012; Zia & Koliba, 2013).

The goals of the project eventually focused on the desire to simulate state-level transportation project prioritization patterns to gauge the system's effectiveness in achieving system maintenance, economic and environmental objectives, and equity across regions of the state of Vermont (Koliba et al., 2012). To achieve these objectives, a process of stakeholder engagement and participatory model construction was undertaken.

Element 1: Boundaries and Problems

In this example, two initial focus groups were convened, comprised of stakeholders that included regional Federal Highway Administration officials, VTrans officials, congressional staffers, state legislators, metropolitan planning organization staff, CCMPO board members and staff, and town planners. The results of these focus groups were used to refine the boundary conditions and clarify the problem conceptualization. At first, the boundary conditions were set to one county. As VTrans became more invested in the project, it became apparent that the initial boundary conditions of the governance network to be modeled had to be expanded to include regions across the entire state.

The problem conceptualization emerged through the stakeholder focus groups. As details regarding the challenges and opportunities inherent in the existing transportation prioritization process emerged, particular problem domains became clear: VTrans and Federal Department of Transportation officials looked to preserve the existing transportation network, while regional planning and congressional staffers sought to understand how economic and environmental considerations informed transportation prioritization, whereas local government officials and state legislators sought greater equity in the system (e.g., they wanted to know if the current system unduly privileged some localities and regions over others). These problem conceptualizations served as the central features of the problem scenarios to follow.

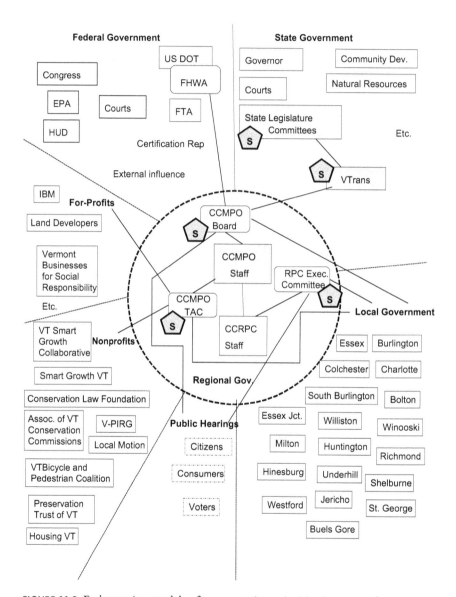

FIGURE 11.2 Early scoping models of transportation prioritization network

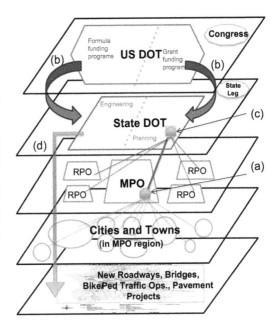

Each town in the metropolitan area is represented on the MPO governing board and technical advisory committee (TAC), and votes on the prioritization of regional projects **(a)**. Regional prioritization accounts for 20% of the statewide prioritization. Federal formula or competitive funding programs provide approximately 80% of funding for most projects **(b)**. The State DOT planning department assimilates the regional prioritization ranking into its own assessments of projects, which accounts for 80% of the statewide ranking **(c)**. The State DOT engineering operations department implements (builds/contracts to build) prioritized roadway, bridge, bike/pedestrian, traffic operations and pavement projects **(d)**.

FIGURE 11.2 (Continued)

Element 2: Scoping Models

Two different iterations of the early scoping models for this transportation prioritization network were developed. The earlier model separated network actors into sectors (public, private, and nonprofit) and further divided the public sector by level of jurisdiction (local, regional, state, and national). The later version of this scoping model placed these actors on plains distinguished by their levels of jurisdiction, while connecting them through the functional morphology observed in this particular network. Figure 11.2 demonstrates the two ways that the scoping models had evolved for this project.

In both versions of the scoping model, action arenas were located in one of two jurisdictional layers: the state level and the regional level and designated as pentagons in both figures. Following extensive data collection and discussion with stakeholders, it became apparent that transportation planning was driven primarily by dynamics set in place by the state department of transportation (VTrans), and are further impacted by obligations to the Federal Highway Administration. These federal-state relations are heavily mediated by federal rules and funding criteria. The processes that unfolded within these action arenas provided the basic architecture of the ABM described in the next section.

Element 3: Computer Simulation Model

A statechart describes the network of events and states, and several were created to formulate a computer simulation model of the transportation project prioritization process (Zia et al., 2012; Zia & Koliba, 2013). Statecharts illustrate the discrete path along which project selection and financial resource allocation takes place, but only shows the transition functions for the "project" classes of agents (see Figure 11.3 for the statechart for this model). Different variables representing project characteristics (e.g., project duration, project cost) were estimated from the "capital book" project database published annually by VTrans. Project characteristics, such as project location, project scores in asset management programs, and regional rankings, were used to calibrate the model to those patterns observed between 1998 and 2010. The focus group data was used to draw the transition functions of the "project" class of agents. A complex interaction of variables (e.g., high project scores and funding availability in a given federal program) resulted in the projects showing a higher ranking for successive phases. At each point, the project moves to the next stage of the process until it is constructed, kept in the running log, or taken off of the books.

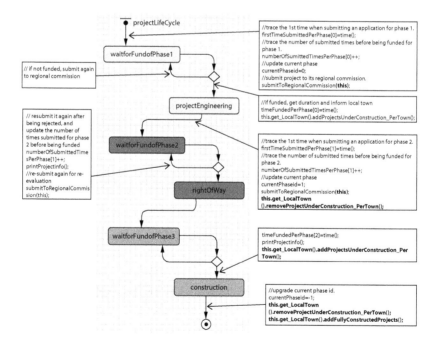

FIGURE 11.3 Statechart for transportation project prioritization agent-based model

The ABM is built as a nested hierarchy of agents. A state agent (i.e., VTrans) contains 10 nested RPCs/MPOs and 600 local towns that are nested within the RPCs/MPOs. Further, transportation projects that arise out of local towns are modeled at the bottom-most layer of this nested hierarchy. During focus group conversations, one experienced participant described the VTrans decision heuristic for funding transportation projects as a "funneling approach" that is captured in the ABM model. VTrans keeps an updated list of problems identified annually and selects a sub-sample on which to undertake feasibility studies, and, subsequently, right-of-way (ROW) approval and, finally, provided funding is available from relevant federal programs and if state and local government matching funds are available, the project is approved for the construction phase. Overall, the ABM module of the computer simulation model for roadway and other project classes captures this "funneling" effect of a sub-sample of initially identified transportation projects that go through three project phases: feasibility, ROW, and construction. A comprehensive overview of this model may be found in Zia & Koliba (2013).

The output of this model is a distribution pattern of funds to state regions and towns. The output can be represented as either a graph, with time on the horizontal axis and funds allocated to each region on the vertical axis, or as a spatial grid, with towns grouped in regions; both represent historically observed patterns extrapolated over 20 additional years into the future.

Element 4: Scenario Development

The problem scenarios initially laid out by stakeholders concerned the extent to which the business-as-usual system adequately maintained the existing system, allowed for environmental and/or economic considerations, or allowed for an equitable distribution of transportation projects across different regions of the state. Thus, the initial problem scenarios were predicated on the efficacy of the existing system.

The range of governance scenarios envisioned at this phase concerned possible alterations to the regional- or state-level weighting scheme for project prioritization, manifested in simulation using sliders in the model's computer interface.

In later phases of this project, an additional set of problem scenarios surfaced as Tropical Storm Irene ravaged the state's transportation infrastructure, leading to extensive damage and a reconsideration of the resiliency of the state's transportation infrastructure. These scenarios were developed collaboratively during a workshop convened by the modelers. A series of problem scenarios related to the anticipated rise in the severity and frequency of extreme flooding events were created.

A follow-up series of meetings with state, regional, and local transportation officials is envisioned. Critical questions concerning the capacity of the existing transportation project prioritization process will be brought forward. The implications for the development of new governance scenarios will be explored. Refining the project evaluation criteria based on the increasing frequency and size of flooding events may be integrated into the existing governance structure. Or, an entirely new way of prioritizing transportation projects may eventually emerge. In either event, the models in place will be of assistance in developing greater infrastructure resiliency.

Element 5: Scenario Runs and Decision Support

By developing ABMs using the institutional actors as agents, we were able to render a close approximation of the funding patterns across the 10 regions of the state for the years 1998–2011. These observed patterns in turn can help anticipate future trends over an additional 20 years.

The ABM calculates annualized flow of financial resources from the state government to regional and town jurisdictions that is contingent upon project prioritization decision-making undertaken by the intergovernmental network of regional and state agents. Approval of projects for different regions on an annualized basis and their associated costs can be used to calibrate the computer simulation model against the observed data for Vermont and other similar states. Users (e.g., policy makers, managers, and other stakeholders) can run business-as-usual and alternate institutional design scenarios by defining the parameters for different scenario runs.

The ABM described here represents current governance configuration of the prioritization process and the computer interface sliders allow for subtle changes to the weighting scheme to consider incremental changes to the governance network. Either the asset management systems or other, more subjective determinants, such as project momentum, can be altered. The number of projects in the pipeline can also be manipulated.

During the course of this project, problem configuration evolved from one grounded in a set of "pre-resiliency" norms to one grounded in the new normal evoked in the wake of catastrophic flooding. As a result, the value of this governance ABM has become even greater, as it is slowly dawning on state and regional transportation planning officials that the current system may not accommodate the kind of alternative futures that may be in store for this region as a result of a changing climate. What this governance informatics project provides for them is a solid basis on which to take what is known about the current system and engage in active experimentation with different configurations in a simulated environment.

Looking Forward: Realizing the Promise of Governance Informatics

The deep uncertainty that persists within the models being devised in these and other projects is still very real (Bankes, 2002). Ongoing considerations to reduce the uncertainty of these models are worth noting.

Determination of Boundary Conditions and Problem Conceptualization

Public problems, policy goals, and even policy solutions are mediated through socially constructed frames of perception (Stone, 2002). One person's problem frame may be another person's solution frame. Determining the functional characteristics of the governance network to be modeled may therefore be contested. Funding sources and the prior dispositions of modelers will likely shape how initial boundary conditions are set. Deciding which parameters are endogenous to the models and which are exogenous may be subjective or predicated on the constraints placed on data availability and theoretical exhortation.

Recruitment of Willing Stakeholders to Participate in the Project

Admittedly, we faced some resistance from stakeholders in the transportation project prioritization network. These stakeholders initially questioned the efficacy of undertaking the process. Questions surfaced concerning the uncertainty associated with models of this nature. Some stakeholders appeared to possess a limited understanding of why a deepened situational awareness of governance dynamics is desirable. These two perspectives were eventually revised with the advancement of the models. Real power disparities can surface when governance networks are modeled. Some will likely be advantaged by having their situational awareness of the governance dynamics deepened. Others will become disadvantaged and, anticipating this, decline participation.

Availability of Data

By collaborating with stakeholders, we believe we were in a better position to gain access to critical sources of data. With the increase in computational power and the advancement of information technology, increasing amounts of data will be made available, suggesting to some that we have entered the era of "big data." Determining how best to use and select data to inform models becomes an important consideration. Realistically, those engaged in the type of governance informatics projects described here will need to either up- or down-scale their expectations about model deliverables, depending on data availability.

Model Validity

The capacity of these models to adequately predict future outcomes with a level of certainty that is amenable to both stakeholders and members of the academic community is of critical importance. Two different kinds of validity are at work here: face and predictive validity. Face validity pertains to the extent that modeled structures and functions align with the perceived realities held by stakeholder groups. Predictive validity concerns the extent to which the conceptual boundaries of the model are clear, distinguishable, and accurately predict future states.

Face validity is critical for creating a governance informatics project that is legitimate in the eyes of stakeholders. If the models produced do not legitimately represent observed phenomena, stakeholders will be less likely to remain engaged. Recall that this process requires a consistent feathering back and forth between engagement and modeling. Without face validity, the ties binding modelers to stakeholders will be undermined. These ties are fragile and politically sensitive. Social scientists have long understood that perception is socially constructed and therefore prone to gaming tendencies. In other words, it is possible that stakeholders will question the face validity of models because they generate results or expose patterns that undermine the existing power structure. Therefore, it must be clear from the beginning that the face validity of the models being produced will be determined through the triangulation of multiple perspectives resulting from the co-production of inter-subjective realities. Ideally, any differences in perception are ironed out during the scoping model phases.

Predictive validity is critically important to engendering models that accurately reflect observed patterns. Modeling complex social systems of this nature must be moderated against acceptance of a certain level of uncertainty. An in-depth discussion of the matter of model construct validity reveals that:

> The level of systemic error that is possible in computer simulation models [of governance networks] can potentially be quite large. . . . Although efforts can be made to reduce the noise of a model, the propensity for large systemic error virtually assures us that the error rates of simulation models of social systems far exceed levels of statistical significance found in linear regression models. We are mindful of why these error rates may be higher in social systems than they are in the more predicable (but still uncertain) areas of natural and biological systems. Social agents maintain a certain level of autonomy in most social systems. The capacity of individual social agents to exert their own free will inevitably lead to a certain level of unpredictability. Agent-based modelers account for this unpredictability in ascribing probability functions to agent behavior that are, ideally, calibrated to empirical observations. Modelers must still make a wide range of choices

in building their models, as they are boundedly rational as well. They make choices around what elements to incorporate into the model and should be prepared to defend those choices.

(Koliba & Zia, 2012, p. 4)

The consideration of model validity and the relationship between validity, legitimacy, and accuracy must be addressed at all stages of a governance informatics project. Governance informatics projects will not and cannot offer stakeholders or researchers a crystal ball to predict the future. What these models provide are opportunities to learn about how different institutional and governance designs may lead to certain results. First and foremost, governance informatics projects build the capacity of systems to learn. In such a process, questions raised about model validity serve as critical features within the learning process. Debates concerning model validation provide the kind of cognitive dissonance needed to stimulate learning (Wenger, 1998). From the early stages of boundary setting and stakeholder engagement, to the middle stages of developing scoping models and collectively learning from these early conceptual models, and finally to the more advanced stages of running alternative scenarios in simulated environments, we believe there are many points of entry to legitimate the investment of time and resources into projects of this nature. The ambitions that underlie these kinds of governance informatics projects are mediated through instances of quick feedback found in the "ah-ha" moments that surface during meetings between modelers and between modelers and stakeholders.

Conclusion

Several trends shaping the landscape of policy informatics are addressed in other chapters of this book. The rise of computational power, growing availability of big data sets, advancement of ABM and other hybrid modeling approaches, and growing appreciation for "complexity friendly" theories that already exist within the public administration and policy studies fields contribute to our optimism regarding our deepened capacity to model governance networks.

The need for more governance informatics has, arguably, never been greater. The persistence of wicked problems, the challenges associated with the complexity of the problems involving matters of geographic and jurisdictional scale, and the role of individual free will and political dynamics are not going away. The situational awareness that we believe is possible when an attempt is made to understand and eventually harness this complexity is worth pursuing. We hope that the methods and approaches outlined in this chapter are only the tip of a proverbial iceberg, signifying a new phase in computer simulation modeling of social/governance phenomena, and perhaps more importantly, building the capacity to harness these models to serve the public good.

Note

Funding to support this work is from the James Jeffords Policy Research Center at the University of Vermont, the University of Vermont Transportation Research Center, and the National Science Foundation (Grant EPS 1101317).

References

Agranoff, R. (2007). *Managing within networks: Adding value to public organizations.* Washington, DC: Georgetown University Press.

Agranoff, R., & McGuire, M. (2003). *Collaborative public management: New strategies for local governments.* Washington, DC: Georgetown University Press.

Bankes, S. (2002). Tools and techniques for developing policies for complex and uncertain systems. *Proceedings of the National Academy of Sciences, 99*(3), 7263–7266.

Baumgartner, F.R., & Jones, B.D. (1993). *Agendas and instability in American politics.* Chicago: The University of Chicago Press.

Endsley, M.R. (1995). Toward a theory of situation awareness in dynamic systems. *Human Factors, 37*(1), 32–64.

Frederickson, H.G. (1999). The repositioning of American public administration. *Political Science and Politics, 32*(4), 701.

Grimm, V., Revilla, E., Berger, U., Jeltsch, F., Mooij, W.M., Railsback, S.F., . . . & DeAngelis, D.L. (2005). Pattern-oriented modeling of agent-based complex systems: Lessons from ecology. *Science, 310*, 987–991.

Guney, S., & Cresswell, A.M. (2010). *IT governance as organizing: Playing the game.* Paper presented at the 43rd Annual Hawaii International Conference on System Sciences, Kona, HI.

Haynes, P. (2003). *Managing complexity in the public services.* London: Open University Press.

IPCC. (2008). Intergovernmental Panel on Climate Change. Retrieved from http://www.ipcc-data.org/ddc_definitions.html

Kettl, D. (2006). Managing boundaries in American administration: The collaborative imperative. *Public Administration Review, 66*(6), 10–19.

Kickert, W.J.M., Klijn, E., & Koppenjan, J.F.M. (1997). Introduction: A management perspective on policy networks. In J.M. Walter, E. Kickert, & J.F.M. Koppenjan (Eds.), *Managing complex networks.* London: Sage.

Koliba, C., DeMenno, M., Brune, N. & Zia, A. (2014). The salience and complexity of building, regulating and governing the smart grid: Lessons from a statewide public-private partnership. *Energy Policy.* Retrieved from http://dx.doi.org/10.1016/j.enpol.2014.09.013

Koliba, C., Meek, J., & Zia, A. (2010). *Governance networks in public administration and public policy.* Boca Raton, FL: CRC Press/Taylor & Francis.

Koliba, C., Mills, R., & Zia, A. (2011). Accountability in governance networks: Implications drawn from studies of response and recovery efforts following Hurricane Katrina. *Public Administration Review, 71*(2), 210–220.

Koliba, C., Reynolds, A., Zia, A., & Scheinert, S. (accepted for publication). Isomorphic properties of network governance: Comparing two watershed governance initiatives in the Lake Champlain Basin using institutional network analysis. *Complexity, Governance and Networks. 1*(2).

Koliba, C., & Zia, A. (2012). Complex systems modeling in public administration and policy studies: Challenges and opportunities for a meta-theoretical research program. In

L. Gerrits & P.K. Marks (Eds.), *COMPACT I: Public administration in complexity.* Litchfield Park, AZ: Emergent Publications.

Koliba, C., Zia, A., & Lee, B. (2011). Governance informatics: Utilizing computer simulation models to manage complex governance networks. *The Innovation Journal: Innovations for the Public Sector, 16*(1), 3.

Koliba, C., Zia, A., Novak, D., & Tucker, M. (2012). *A systems dynamics framework of intergovernmental network innovation: The use of multicriteria analysis to allocate resources within a regional transportation planning network.* Association of Public Policy and Management Annual Conference, Baltimore, MD.

Levy, S. (1993). *Artificial life: A report from the frontier where computers meet biology.* New York: Random House Inc.

McKenzie, E., & Rosenthal, A. (2012). *Developing scenarios to assess ecosystem service tradeoffs: Guidance and case studies for InVEST users.* Washington, DC: World Wildlife Fund.

Milward, H., & Provan, K. (2006). *A manager's guide to choosing and using collaborative networks.* Washington, DC: IBM Center for the Business of Government.

Mitroff, I. (1972). The mythology of methodology: An essay on the nature of a feeling science. *Theory & Decision, 2,* 274–290.

O'Toole, L.J. (1990). Multiorganizational implementation: Comparative analysis for wastewater treatment. In R.W. Gage & M.P. Mandell (Eds.), *Strategies for managing policies and networks.* New York: Praeger Publishers.

Ostrom, E. (2005). *Understanding institutional diversity.* Princeton, NJ: Princeton University Press.

Radin, B. (2006). *Challenging the performance movement: Accountability, complexity and democratic values.* Washington, DC: Georgetown University Press.

Rhodes, R. (1997). *Understanding governance: Policy networks, governance, reflexivity and accountability.* Buckingham, UK: Open University Press.

Rittel, H. W., & Webber, M. M. (1973). Dilemmas in a general theory of planning. *Policy Sciences, 4*(2), 155–169.

Sabatier, P.A., & Jenkins-Smith, H.C. (1993). *The advocacy coalition framework: An assessment.* Boulder, CO: Westview Press.

Sorensen, E., & Torfing, J. (2005). The democratic anchorage of governance networks. *Scandinavian Political Studies, 28*(3), 195–218.

Squazzoni, F., & Boero, R. (2010). Complexity-friendly policy modeling. In P. Ahrweiler (Ed.), *Innovation in complex systems.* London: Routledge.

Stone, D. (2002). *Policy paradox: The art of political decision making.* New York: Norton.

Tucker, M. (2011). *Intergovernmental dynamics and equity considerations in transportation planning and the allocation of funds.* [Master's thesis.] Burlington, VT: University of Vermont.

Van den Belt, M. (2004). *Mediated modeling: A system dynamics approach to environmental consensus building.* Washington, DC: Island Press.

Wasserman, S., & Faust, K. (1994). *Social network analysis: Methods and applications.* Cambridge, UK: Cambridge University Press.

Wenger, E. (1998). *Communities of practice: Learning, meaning, and identity.* Cambridge, UK: Cambridge University Press.

Zia, A., Kauffman, S., & Niiranen, S. (2012). The prospects and limits of algorithms in simulating creative decision making. *Emergence: Complexity and Organization (E:CO)—An International Transdisciplinary Journal of Complex Social Systems, 14*(3), 89–109.

Zia, A., & Koliba, C. (2011). Accountable climate governance: Dilemmas of performance management across complex governance networks. *Journal of Comparative Policy Analysis: Research and Practice, 13*(5), 479–497.

Zia, A. and Koliba, C. (2013). The emergence of attractors under multi-level institutional designs: Agent-based modeling of intergovernmental decision making for funding transportation projects. *Artificial Intelligence (AI) & Society*.

Zia, A., Koliba, C., & Tian, Y. (2012). Governance network analysis: Experimental simulations of alternate institutional designs for intergovernmental project prioritization processes. In L. Gerrits and P. K. Marks (Eds.), COMPACT I: Public administration in complexity. Litchfield Park, AZ: Emergent Publications.

12

THE ROLE OF INFORMATICS IN EDUCATION RESEARCH AND POLICY

Nora H. Sabelli, William R. Penuel, and Britte H. Cheng

Introduction

The ability of complex social systems to resist policy mandates—to revert to the norm—has gained increased attention in educational research. This chapter presents a perspective on the reasons behind the policy resistance of education systems, based on the following considerations: the complex and distributed nature of the education system governance; its close coupling with forces external to the formal system; that, given the local and distributed nature of education practice, top-down approaches will continue to struggle; the primary focus on static student testing data to gauge progress despite the social complexity of the context where students live and learn; and the lack of approaches for evaluating possible policies before implementation. We highlight that the local nature of education governance raises the need to empower education stakeholders to explore proposed policy implementations before they are put into practice.

Together, these considerations lead us to see the potential of modeling and simulation as tools for local stakeholders to use their own system data to answer "what if" questions before choosing and implementing policies. These tools of policy informatics enable a broader understanding of policy implementation by decreasing the barriers to participation, by making complicated data and scientific language tractable, legitimizing diverse perspectives and competing interests, and by enabling cross-perspective debates that allow shared answers to "what if" questions.

In the first section of the chapter, we discuss the role of informatics in education research and policy, including potential uses of tools and approaches to inform critical educational challenges of policy resistance and systemic reform. In the second section, we present exemplar projects to illustrate the potential role of informatics tools, such as modeling and simulation, in education research. In the third section, we discuss specific issues of policy implementation and system reform that informatics can inform, and then we discuss the implications for use of informatics tools and approaches in considering policy issues.

The Role of Informatics as Part of Educational Research and Policy

For many researchers and policy makers, the use of informatics in education focuses on management and testing student achievement. From these data we indirectly infer the outcomes of policies (see, e.g., as part of a large body of work, Baker & Yacef, 2009; Carr & O'Brien, 2010; Collins & Weiner, 2010; Mandinach, Honey, & Light, 2006). For the most part the data are not available to practitioners in a timely manner to be applied to improve instruction, and their use for governance is limited. *Static snapshots* of student learning, by themselves, even if disaggregated, tell us little about the conditions of the educational system and the practices that led to the measured outcomes, and less about the social and cultural context in which the data were obtained—including what is often called the "opportunity to learn." To be of instructional and systemic use, educational informatics requires *dynamic* data on student progress, plus data on how the system itself is changing and learning, and access to modeling tools that facilitate their analysis and interpretation.

More accurate accounts of policy success and failure could be created if they would attend to aspects of complexity, including organizational learning, feedback loops that extend outside the formal education system, the different time lines involved in taking various actions, sensitivity to localized initial conditions, and variability in those conditions (Lemke, 2000). Accounts will also have to consider the actions of stakeholders like parents and community members, advocates of different kinds of reform initiatives, publishers, and others whose views and efforts to change the system affect what goes on in schools and districts (Renée, Welner, & Oakes, 2009). For example, the focus on mathematics and reading produced by No Child Left Behind (NCLB) initiatives resulted in other academic topics getting pushed out of the school day (Dee, Jacob, & Schwartz, 2013; Marx & Harris, 2006). Implementation success also depended on districts having enough resources for well-qualified teachers in the chosen topics. In practice, NCLB led to teaching to weak tests, focusing on short-term strategies to raise test scores, teachers changing test answers, and other negative impacts on education and its public perception.

Information about local system conditions and modeling tools that can operate on such data are not routinely available to most school systems today, partly because of expense and lack of required analytical expertise, and also because education improvement efforts tend to focus on solutions to individual problems within a system—specifically, solutions that appear to have potential to be used at scale. Without the ability to plan for or interpret improvement efforts responsive to systems with particular characteristics, such efforts will produce little information about the kinds of transformations to teaching and learning that reformers desire.

Data about *how an educational system itself learns* are critical to understanding the variation in implementing policies and programs (Graham et al., 2006; Homer & Hirsh, 2006; Palinkas & Soydan, 2012; Scheirer & Dearing, 2011; Woolf, 2008). In education, particular attention has been paid to the ways that local actors adapt policies to local contexts (Coburn, 2003; Cohen & Ball, 1990; Honig, 2004; McLaughlin, 1987). We argue that policy informatics tools, if available to educational stakeholders, will both support effective adaptations of policies and build capacity for effective future implementations. In other words, because variation in local contexts in which policies or interventions are adopted significantly impacts implementation, promoting universal solutions that can be broadly applied is not an efficient strategy for broad educational improvement. New tools for (local) data-driven insight put into the hands of key education stakeholders will allow them to discover, design, and test strategies appropriate for local conditions and expertise. Because general top-down policy mandates depend on local actors' complementary efforts to "craft coherence" out of conflicting policy demands, ideas for improvement, and curricular innovations that circulate within any given district or school (Honig, 2006; Honig & Hatch, 2004), new kinds of performance support tools could facilitate local actors organizing schools and districts to produce *ongoing* improvements to educational equity and excellence. Given schools' and districts' multifaceted complexity of actors, stakeholders, environments, resources, and capacity for action, the appeal of such tools is the ability to build local capacity for continuous improvement, and break the circle of instituting policies for short periods of time and replacing them with others, instead of improving and refining them through iterative cycles of design, testing, and study.

In addition to the need for different kinds of data, we look to policy informatics' use of dynamical simulations and modeling as tools to empower policy makers themselves to test and improve policy strategies before putting them in practice, and to explore both the unanticipated consequences and unrealized opportunities of different policy proposals (Johnston & Kim, 2011). As part of policy decision-making, "what if" experimental scenarios can serve as tests of plausible theories of action and illustrate successes and failures, producing an experimental and reflective improvement ethos not possible without the tools of informatics (Elmore, 2002). From a planning perspective, "a simulation can serve as an operating theory which we 'test' against our experience and best judgment" (McGinn, 2001, p. 1).

Examples of Education Research Using Informatics Approaches

This section briefly describes ongoing projects by the authors that exemplify the role of informatics in educational research and the need to focus on the inherent social nature of education policy implementation. We present this work as a means of understanding policy resistance in educational systems that include not only schools and other formal venues but also the myriad educational resources and communities external to the formal system that directly impact schools' functioning and thus policy implementation.

A first example of using models as tools for representing existing educational experience and a theory of action (Argyris, 1999) is the *Synergies* project (Falk et al., 2013). The project, a collaborative effort among youth, adult community members, and representatives of formal and informal educational institutions and organizations from the community, is focused on analyzing and improving science, technology, engineering, and mathematics (STEM) learning opportunities. The premise of the project is that a better understanding of how and why people, early adolescents within a poor, under-resourced community in particular, develop (or lose) STEM-related interests should make it possible to create a more synergistic and effective STEM education system that supports STEM learning for all, in and out of school.

Community interventions that promote STEM education are varied. Modeling their effects enables examination of the cumulative effects of learner experiences across settings and time. Most educational research focuses on interventions in a single setting, but there is evidence from a number of qualitative studies that children can learn STEM topics across a range of settings, developing interest and knowledge along the way (Bell, Tzou, Bricker, & Baines, 2012). Moreover, psychological theories of interest development (Hidi & Renninger, 2006) imply there are phase shifts in this development, where individuals change from being largely dependent on social and contextual supports to more actively seeking out and creating opportunities to pursue interests. At present, however, we do not know where those thresholds are, and if they are different for different children. Nor do we know how sensitive these thresholds and dynamics are to initial conditions or opportunities within a community. These unknowns all point to the potential value of data and models derived from a perspective that frames formal and informal learning opportunities as part of a complex system.

A critical component of the *Synergies* project is the co-design and use of agent-based models (ABMs) to represent social science theories of interest development, and community members' own theories of action. We believe that these tools are able to capture and display conditions for learning, including inequities that lead some young people to develop a strong interest in science and others to become turned off. ABMs are of value because they help to make important, dynamic mechanisms that characterize complex systems (e.g., feedback loops) visible and thus more accessible (Wilensky & Reisman, 2006). This increases the pool of experts, creates a new medium of evidence-informed communication between interested stakeholders, and provides a shared language for discussions. Moreover, such tools also are potentially powerful for investigating alternate theories of learning and development, including the outcomes of designed learning environments (Blikstein, Abrahamson, & Wilensky, 2007).

A second exemplar project, in earlier stages of development, is an attempt at including specifically the social networking aspects of human interactions in an expanded engineering education system by considering feedback loops among four main foci: *formal education, community impact, private sector components, and federal and state governance* (Cheng, Sabelli, Richey, & Newton, 2012; see also Richey, Newton, Stephens, Backus, & McPherson, 2010; Stephens & Richey, 2011). The proof-of-concept work is based on a system dynamics study of engineering education developed for the Boeing Corporation by Sandia Laboratories (Kelic & Zagonel, 2008) that points to the role of engineering career attractiveness in the graduation rate of engineering students. The Modeling Social Complexity in Education project (Sabelli, Lemke, Cheng, & Richey, 2012) will work with NSF's Energy Education and Centers (EEC) awardees and their data to model the social network dynamics of student, peer, and faculty experience as a function of different institutional learning environments and student backgrounds. One goal of the project is to work with awardees to help them represent the specific dynamical nature of stakeholder interactions in their own organization. Another goal is to develop a framework that aggregates existing partial data into a draft model that could eventually be used to ground a better understanding of the impacts of research, policy, and funding options.

By framing problems of education from a complexity perspective, the project will explore policy implementations and help accelerate the translation of extant research into effective policy. The project poses the following policy-related research questions: What new predictions across extant research outcomes does a coordinated framework enable? Can "data" based on such a framework inform policy and model development, and guide future decisions? What is the role of informal education in engaging a larger and more diverse group of graduates?

Issues of Policy Implementation: Coherence and Sustainability

Research has repeatedly shown that policy mandates impact educational practice in limited and often contradictory ways (Cohen & Ball, 1990; Cohen, Fuhrman, & Mosher, 2007; McLaughlin, 1987; Penuel, Frank, Sun, Kim, & Singleton, 2013). Education researchers have recently joined colleagues in other disciplines in changing their discourse from *scaling up* and *disseminating* innovations to *translating, implementing, and/or integrating research and practice* (Carlile, 2004; Woolf, 2008). Educational innovations entail many different actions and require skilled clinical judgment to use. Like the practice of medicine, moving educational innovations from clinical trials into the field in ways that promote "evidence-based practice" is a difficult problem. Designs to support implementation are worthy objects of research in education, just as they are in all clinical sciences (Penuel & Fishman, 2012).

Central to movement from scaling up to translation is replacing considerations of *fidelity of implementation* of the innovation to considerations of *design for adaptability* to local conditions—that is, to support implementation (Penuel, Fishman, Cheng, & Sabelli, 2011; Sabelli & Dede, 2013; Spillane, Parise, & Sherer, 2011; Spillane, Reiser, & Reimer, 2002). Implementation failure has been linked to the lack of sustainability of innovations in other contexts (Klein & Knight, 2005; Repenning, 2002). In fact, fidelity is not a positive when it opposes a more effective local adaptation (Means & Penuel, 2005). The adaptability of designs implies the need to build into the design local capacity for sustained and principled adaptation (Sabelli & Dede, 2001) and to provide local actors with the learning opportunities necessary for the principled adaptation of programs (Davis & Krajcik, 2005; Penuel, Gallagher, & Moorthy, 2011).

The complexity of educational systems means that policy implementation is always the product of interactions among policies, people, and places (Honig, 2006). The practical importance of focusing on implementation resides in the many semiautonomous professionals who must cooperate to implement a policy; each individual involved brings to the task a set of beliefs, expectations, and practices that must accommodate each other and required pedagogical changes (Coburn, 2004). No Child Left Behind is a case in point; although implemented as a "top-down" policy intended to impact practice through the right combination of incentives for and pressures on schools, it had the opposite results in some cases, in part due to individual and group differences (Penuel et al., 2013).

Also, we know that significant learning by actors within systems takes time and requires coordinated leadership practices and organizational supports (Spillane, Parise, & Sherer, 2011; Spillane, Reiser, & Gomez, 2006). School and district organizations are subject to myriad, conflicting stakeholder demands that change over time (Rowan, 2002). Consequently, a single innovation must adapt to "survive" within an evolving ecology (Zhao & Frank, 2003).

An educational policy informatics focused on implementation must draw upon and build theory to be of use for the field. Often, a bureaucratic theory of rationality undergirds educational policy; setting policy is seen as providing clear and coherent directives, which actors by virtue of their roles within the system will implement. But neoinstitutional theory (Meyer & Rowan, 1977) suggests that local school actors' responses to policies are often symbolic, and do not necessarily transform the technical core of teaching in intended ways. Local actors can work together to implement reforms, although empirical research within this tradition also suggests that in so doing, policies are inevitably adapted and changed by the actors (Frank, Zhao, & Borman, 2004; Penuel et al., 2013).

Comfort with experimentation, for implementing both policy mandates and pedagogical innovations, is often a benefit of collaborations between practitioners and researchers (Coburn, Penuel, & Geil, 2013). Its lack can be a factor in the failure of innovations to scale to new contexts where such needed intermediaries are not available (Fisher & Vogel, 2008; Sabelli & Dede, 2013; Saywell & Cotton, 1999; Sudsawad, 2007; Wolfe, 2006).

Issues of Systemic Reform: Education System Complexity

We consider explicitly the ways that education can be analyzed as a complex social system, given the institutional and social ecology of schooling. Education actors and organizations must constantly adapt to external and internal changes to survive and maintain their organization (Lemke & Sabelli, 2008). But the system is limited in its capacity for collective reorganization, in part because of highly decentralized governance in which the components of educational practice—materials, pedagogies, assessment, resources, instructional workforce quality—are only loosely and infrequently coordinated.

Many individual schools, however, do form cohesive internally interconnected systems, and their implementation of policies typically evidences properties of complex systems, with parts of the system reacting to one another. Leadership at the school level presents a case in point: As a practice, it is *distributed*—stretched across different people (e.g., principal, coach), practices (e.g., observing classrooms, discussions of achievement data), and tools (e.g., protocols for looking at student work together) (Spillane, 2005). Moreover, like in all organizations, leaders and followers have a relationship of mutual dependency where followers depend on leaders for guidance but leaders' authority depends on followers' following for their own success.

The local, distributed nature of leadership among interacting local actors differs from the state and federal governance system whose mandates impact the local system from the outside. A good balance recognizes that the flexibility in the *implementation* of policies must be under local control. Broader policies that align standards—of curriculum, professional development, assessment, and accountability—can help craft a coherent set of incentives for improvement, and create economies of scale within the educational improvement industry.

Studies of the effects of leadership on implementation show the appearance of emergent properties that could not readily have been predicted from individual inputs. In a study of distributed leadership's effects on reform implementation (Penuel et al., 2013), researchers found that policies focused on encouraging teachers to adopt evidence-based reading instruction had the unintended effect of increasing divergence of instructional practices within subgroups. Teachers in subgroups that already enacted high levels of these practices used them more frequently after the policy was put in place, but those in subgroups that engaged in few such practices used them even less after the policy was put in place. This kind of sensitivity to teachers' initial practice—a hallmark of complex systems—could not have been predicted but undercut the policy's aims.

Need for Appropriate Tools to Address Education Policy Challenges

Presently, neither leaders who implement policies nor those who design policies have many tools to help them model critical dynamic interactions. As in other fields that confront changes in complex organizations, local education policy makers would benefit from appropriate tools; asking "what if" questions helps to better understand how their system will behave under specific proposed policies, often cross-jurisdictional and multi-constituency and whose outcomes are not easy to predict.

An aim of developing such tools should be to help realize the ever-elusive goal of *systemic reform*: to bring into alignment standards for learning, assessments, curriculum, and professional development around a coherent set of aims, many of which are shared among localities (Knapp, 1997; Pea & Collins, 2008). Changes are made to systems with histories, experiences, and other forms of institutional momentum—timing should be a resource for implementing policy: for adaptation, for adjustment, for coordinating with other policies and programs in a district, for iterative refinements. Tools that allow one to model such practice constellations and test changes dynamically can play an important role in making the process more efficient.

Tools must grapple with the agency of local actors to interpret policies differently from the way they are intended to be interpreted, which can lead to implementation failure (Spillane, 2004). Tools are needed that help policy makers and leaders anticipate and integrate the diverse ways that teachers might respond to policies, not presuming a single response, but rather a response shaped by prior knowledge and experience (Coburn, 2004). Organizational processes and practices—leadership and the allocation of human, social, and material capital—shape implementation (Spillane, Gomez, & Mesler, 2009). Policies may be undercut when policymakers do not provide incentives for organizations to allocate resources to support implementation, and again, tools that help policy makers model "what if" scenarios related to organizational responses would be useful to explore the nature of incentives that work.

The tools must also grapple with the fact that the "system" from the student perspective includes more than just classrooms in schools. Systemic reforms have often failed to involve an important part of the system—namely, out-of-school influences on both individuals and organizations. Educational reformers often fail to leverage the part of the system outside formal schooling, including peers, family, community, and informal education activities—despite the impact these networks have on learner goals, motivation, and achievement.

Implications for Modeling the Education System

Three points were discussed earlier: (1) the reasons for policy implementation failure, (2) the organizational and structural complexity of education and its organizations, and (3) the need for integrating the macro world of policy makers with the micro world of individual implementers. A logical suggestion is to empower local actors to ask "what if" questions that reflect the specifics of their own situation—human capacity, internal and external social networks, resources, thought leaders, student characteristics, administrators' practices, and so forth—within the external context. Research on distributed leadership (Spillane, 2005; Spillane, Halverson, & Diamond, 2001; Spillane, Parise, & Sherer, 2011) and group model building

(Andersen, Richardson, & Vennix, 1997; Andersen, Vennix, Richardson, & Rouwette, 2007; Vennix, 1999) moves in this direction, and benefits from considering the dynamics of social networking and technological innovations taking place around the participatory, distributed, and collaborative aspects of the Web environment (Lee & McLoughlin, 2008; Lewis, Pea, & Rosen, 2010).

The size and complexity of the education system, as we describe it, suggest that attempting to understand the "education system" will spin out of control in scope, expanding into an unmanageable endeavor; we do not believe that "a" realistic model of the education system would be useful. There is, though, a "sweet spot" between modeling a *local* system—like a school district—in isolation and a comprehensive but unrealistic effort to actually model everything that impinges on the national STEM workforce-and-education system (see, e.g., Ghaffarzadegan, Lyneis, & Richardson, 2011).

A viable research approach to studying such a complex system is to propose simplifying assumptions, which reduce the validity, credibility, and applicability of the models. Insights from such studies can prove generative (Andersen et al., 1997, 2007; Johnson, 1999) but may not be ultimately convincing to practitioners or policy makers. Adequate models require considerations derived from multiple fields and the eventual use for validation of data from multiple sources. Because our main interest is to understand the implications of different policies to enhance the *sustainable* transformation of practice, we emphasize the use of modeling and simulation to build the capacity of local actors to understand the process and, eventually, use its dynamics in support of continuous improvements to practice. But for our addition of the word only, "we are invited not (only) to explore possibilities for understanding the world but to see how the world really works" (Medd, 2001, p. 2).

Building models is dependent on problem definition, particularly on formulations about possible policy actions. "What if" tools enable expanding who can participate in the policy conversation as an approach toward effective alignment, focusing iteratively on what stakeholders see as issues of concern (see, e.g., Roderick, Easton, & Sebring, 2009). Problem definition requires the engagement of teachers who will enact policies and programs in their classroom, but should also productively include parents and community members.

In "research alliances" like the University of Chicago Consortium on Chicago School Research (CCSR) (http://ccsr.uchicago.edu/content/index.php), researchers work alongside district and school officials and community advocates to decide on a multiyear research agenda that identifies key areas of concern to the district. The data that alliances like the CCSR collect are precisely the kinds of system inputs that would permit finer-grained analyses of system dynamics. In addition to data on student outcomes, including achievement and retention data, the Consortium collects data from teachers on the social context of schools, and data from students on their level of engagement. The Consortium makes these data available to schools and district leaders in easy-to-use, static reports via the Web. In addition, CCSR members work with leaders to interpret patterns in data. The delicate balance between goals, and the primacy that trust plays in that balance when researchers work closely with school administrations in interpreting school data, is well described in their publications (see Roderick et al., 2009) and merits further consideration in formulating the role of policy informatics research in education.

What if these static reports, though, became more dynamic? What if there were tools that allowed leaders to simulate the effects of an increase in student engagement on the level of teachers' sense of collective efficacy? Or what if the data could inform the identification of key leverage points for improvement—that is, tipping points in certain indicators of school difficulties that make it nearly impossible for a school to improve?

Existing systems dynamics (SD) models and agent-based models (ABM) of large subsystems have shown promising results as guides for policy-making. These two modeling approaches, starting as they are from either a top-down (SD) or bottom-up (ABM) perspective, can be more powerful if combined to both manage the large-scale interactions of and within institutional-level systems and link these to the local-scale specificity of learning needs and circumstances critical to implementation success (Macal, 2010; Qudrat-Ullah, 2005; Sabatier, 1986). Given the social and cultural differences among actors across locations, and even within a particular location, ABM methods allow researchers to think about data in terms of not only averages (the general) but also the variability (the local) in the data that goes into the model—for example, how career choices correlate with community values and individual expectations.

A case that responds closely to our view of the use of modeling as a policy tool—including the participation of stakeholders outside the formal system—is MinSim (Brauer, 2004.) A group of school administrators, teachers, researchers, and others in Minnesota built a systems dynamics model of a school system in an attempt to understand why some initiatives to improve student achievement succeed and others do not. The work entailed building a model responsive to their policy questions and their own system characteristics. The model allowed schools to input their own data and try out various resource allocation scenarios.

The premise was that education can be seen as a supply-and-demand equation: How much demand do the students bring, and how much in terms of resources does the system have to meet these demands? Two key decisions made during an early phase identified resources and demand with time as the common currency. Each teacher has a certain amount of time (resource) to give each student, with experience and talent governing how efficiently he or she uses it. Starting from the idea that each student requires different amounts of time to learn, the group distinguished academic demand (based on previous assessments) from behavioral demand. Researchers used statistical data for all the teachers in Minnesota to calculate how much demand teachers with different levels of experience can handle. The model was tested in various school districts' beta sites using historical data; older district data were used to initialize the model. The model provided the ability to do a sophisticated cost-benefit analysis of each.

District teams consisted of up to 24 people, including the superintendent, key administrators, teachers, staff, and community members. The teams ran the model using a zero-sum game, where they had to improve student performance by reallocating resources among four achievement levels, then moved to actually testing the goals using data from their own school districts, and used multiple scenarios with budget cuts and increases to see the possible impact.

An ambitious study of the US Education System (Rodrigues & Martis, 2004; Wells, Sanchez, & Attridge, 2007) and a proposal for a similar study of public health (Homer & Hirsch, 2006) focus on systems dynamics methods and discuss both the whole system and the simplifications needed for the study to become doable. Both these studies represent a high-level, top-down approach to SD, whereas the MinSim project (Brauer, 2004) illustrates a localized systems dynamics instantiation, which provides a top-down perspective but with bottom-up input.

In another work that provides a different perspective on the potential usefulness of modeling educational policies, Maroulis et al. (2010) conducted an ABM study of school choice in Chicago. This study represents the bottom-up approach to policy informatics, while still dealing with a general policy context.

Categorizing existing education research as either "mechanism-based" or "effects-based," the authors used ABM methods and network analysis to integrate "micro-level" mechanisms and "macro-level" effects, providing knowledge across two levels of data. Mechanism-based studies provide insight into the motivation and cognition of actors and into social phenomena inside schools. Effects-based research treats factors contributing to academic performance as inputs to yield a particular level of student achievement.

The study modeled allowing households to choose among all public schools in a district, allowing schools with a greater "value-added" (ability to increase student test scores) to enter the district and varying the manner in which students ranked schools. When students valued a school's test scores much more than geographic proximity, new schools of higher, but initially unknown, value-added had difficulty surviving. *"Consequently, a micro-level rule that one might surmise should aid district improvement (placing a relatively high value on school achievement) can also limit district-level performance in certain conditions."* (emphasis added) (Maroulis, et al., 2010, p. 39). The authors conclude, *"By providing tools to characterize and quantify relationships between individuals and to investigate how individual actions aggregate into macro-level outcomes, a complex systems approach can help integrate insights from different types of research and better inform educational policy."* (emphasis added) (Maroulis, et al., 2010, p. 39).

Summary

The formal education system is often considered a closed system rather than what it is, an open system in ongoing interactions with its environment—including peers, family, community and informal education, extracurricular activities, and even the media—that impact learner goals, motivation, achievement, and stability. By focusing only on the components of the formal system we leave out of consideration the dynamic feedback of *all* the components on one another. It is not surprising then that schools, subject to many conflicting pressures, appear to resist change, whether imposed internally or externally.

The systemic reform of education has long been a policy goal that has proven frustrating for many reasons—political, economic, sociocultural, behavioral, demographic, and more. Public health has been leading the field in emphasizing translational and implementation research in ways that have moved us closer to understanding the barriers to achieving practice goals (see Rohrbach, Grana, Sussman, & Valente, 2006, for a review), but the lack of a direct cause and effect relation between policy mandates and effective policy implementation constrains further practical advances in both fields. The policy informatics modeling and simulation tools in use by public health and other social sciences can be used to rethink designing and implementing effective educational policies. It can do this in at least two ways: by helping local education stakeholders model implementation strategies, and by helping researchers and policy makers understand education as a complex system, rather than as a series of loosely coupled separate components.

Modeling is an exploratory tool, rather than a predictive one; "useful answers" vary according to participants' perceptions, and stakeholder tacit knowledge, constraints, biases, and expectations become known through action (McGinn, 2001). Used in this way modeling can simultaneously lead to shared understandings and build the capacity of local actors for sustaining successful strategies.

References

Andersen, D.F., Richardson, G.P., & Vennix, J.A.M. (1997). Group model building: Adding more science to the craft. *System Dynamics Review, 13*(2), 187–201.

Andersen, D.F., Vennix, J.A.M., Richardson, G.P., & Rouwette, E.A.J.A. (2007). Group model building: Problem structuring, policy simulation and decision support. *Journal of the Operational Research Society, 58*, 691–694.

Argyris, C. (1999). *On organizational learning* (2nd ed.). London: Blackwell.

Baker, Ryan S.J.D., & Yacef, K. (2009). The state of educational data mining in 2009: A review and future visions. *Journal of Educational Data Mining, 1*(1), 3–17.

Bell, P., Tzou, C., Bricker, L.A., & Baines, A. D. (2012). Learning in diversities of structures of social practice: Accounting for how, why, and where people learn science. *Human Development, 55*, 269–284.

Blikstein, P., Abrahamson, D., & Wilensky, U. (2007, April). *Multi-agent simulation as a tool for investigating cognitive-developmental theory.* Paper presented at the Annual Meeting of the American Educational Research Association, Chicago, IL.

Brauer, R. J. (2004, October). Testing "what if?" scenarios. *School Administrator.* Retrieved from http://www.aasa.org/SchoolAdministratorArticle.aspx?id=9520

Carlile, P. R. (2004). Transferring, translating, and transforming: An integrative framework for managing knowledge across boundaries. *Organizational Science, 15*(5), 555–568.

Carr, J. A., & O'Brien, N. P (2010). Policy implications of education informatics. *Teachers College Record, 112*(10), 2703–2710.

Cheng, B.H., Sabelli, N.H., Richey, M.J., & Newton, P. (2012). Modeling social complexity in education. Retrieved from http://msce.sri.com/

Coburn, C.E. (2003). Rethinking scale: Moving beyond numbers to deep and lasting change. *Educational Researcher, 32*, 3–12.

Coburn, C.E. (2004). Beyond decoupling: Rethinking the relationship between the institutional environment and the classroom. *Sociology of Education, 77*(3), 211–244.

Coburn, C.E., Penuel, W.R., & Geil, K. (2013). *Research-practice partnerships at the district level: A new strategy for leveraging research for educational improvement.* Berkeley: University of California and University of Colorado.

Cohen, D. K., & Ball, D. L. (1990). Policy and practice: An overview. *Educational Evaluation and Policy Analysis, 12*(3), 233–239.

Cohen, D.K., Fuhrman, S.H., & Mosher, F. (2007). Conclusion: A review of policy and research in education. In S.H. Furhrman, D.K. Cohen, & F. Mosher (Eds.), *The state of education policy research* (pp. 349–382). New York: Routledge.

Collins, J.W., & Weiner, S. (2010). Proposal for the creation of a subdiscipline: Education informatics. *Teachers College Record, 112*(10), 2523–2536.

Davis, E.A., & Krajcik, J.S. (2005). Designing educative curriculum materials to promote teacher learning. *Educational Researcher, 34*(3), 3–14.

Dee, T.S., Jacob, B.A., & Schwartz, N.L. (2013). The effects of NCLB on school resources and practices. *Educational Evaluation and Policy Analysis, 35*(2), 252–279.

Elmore, R. (2002). The limits of "change." Retrieved on April 2, 2008, from http://www.edletter.org/past/issues/2002-jf/limitsofchange.html

Falk, J.H., Dierking, L.D., Staus, N., Haun-Frank, J., Penuel, W.R., Wyld, J., & Bailey, D. (2013, April). *Viewing STEM learning through a community-wide lens: The Synergies Project.* Paper presented at the NARST Annual International Conference, Rio Grande, Puerto Rico.

Fisher, C., & Vogel, I. (2008). Locating the power of in-between: How research brokers and intermediaries support evidence-based pro-poor policy and practice. Retrieved from www.ids.ac.uk/files/dmfile/intconfpaper28Novwebsiteedit.pdf

Frank, K.A., Zhao, Y., & Borman, K. (2004). Social capital and the diffusion of innovations within organizations: Application to the implementation of computer technology in schools. *Sociology of Education, 77*(2), 148–171.

Ghaffarzadegan, N., Lyneis, J., & Richardson, G.P. (2011). How small system dynamics models can help the public policy process. *System Dynamics Review, 27*(1), 22–44.

Graham, I.D., Logan, J., Harrison, M.B., Straus, S.E., Tetroe, J., Caswell, W., & Robinson, N. (2006). Lost in knowledge translation: Time for a map? *Journal of Continuing Education in the Health Professions, 26*, 13–24.

Hidi, S., & Renninger, K.A. (2006). The four-phase model of interest development. *Educational Psychologist, 41*(2), 111–127.

Homer, J. B., & Hirsch, G. B. (2006). System dynamics modeling for public health: Background and opportunities. *American Journal of Public Health, 96*(3), 452–458.

Honig, M.I. (2004). The new middle management: Intermediary organizations in education policy implementation. *Educational Evaluation and Policy Analysis, 26*(1), 65–87.

Honig, M.I. (2006). Complexity and policy implementation: Challenges and opportunities for the field. In M.I. Honig (Ed.), *New directions in education policy implementation: Confronting complexity* (pp. 1–23). Albany: SUNY Press.

Honig, M.I., & Hatch, T.C. (2004). Crafting coherence: How schools strategically manage multiple, external demands. *Educational Researcher, 33*(8), 16–30.

Johnson, B.L., Jr. (1999). The politics of research-information use in the education policy arena. *Educational Policy, 13*(1), 23–36.

Johnston, E., & Kim, Y. (2011). Introduction to the special issue on policy informatics. *Innovation Journal: The Public Sector Innovation Journal, 16*(1). Retrieved from http://www.innovation.cc/volumes-issues/intro_policy_infomatics_v16i1a1.pdf

Kelic, A., & Zagonel, A. (2008, December). Science, technology, engineering, and mathematics (STEM) career attractiveness system dynamics modeling. Sandia Report SAND2008-8049.

Klein, K.J., & Knight, A.P. (2005). Innovation implementation: Overcoming the challenge. *Current Directions in Psychological Science, 14*(5), 243–246.

Knapp, M. (1997). Between systemic reforms and mathematics and science classrooms: The dynamics of innovation, implementation, and professional learning. *Review of Educational Research, 67*(2), 227–266.

Lee, M.J.W., & McLoughlin, C. (2008). Harnessing the affordances of Web 2.0 and social software tools: Can we finally make "student-centered" learning a reality?. In J. Luca & E. Weippl (Eds.), *Proceedings of World Conference on Educational Multimedia, Hypermedia and Telecommunications 2008* (pp. 3825–3834). Chesapeake, VA: AACE.

Lemke, J.L. (2000). The long and the short of it: Comments on multiple timescale studies of human activity. *Journal of the Learning Sciences, 10*(1–2), 193–202.

Lemke, J.L., & Sabelli, N. (2008). Complex systems and educational change: Towards a new research agenda. *Educational Philosophy and Theory, 40*(1), 118–129.

Lewis, S., Pea, R., & Rosen, J. (2010). Beyond participation to co-creation of meaning: Mobile social media in generative learning communities. *Social Science Information, 49*(3), 351–369.

Macal, C. M. (2010). To agent-based simulation from system dynamics. In B. Johansson, S. Jain, J. Montoya-Torres, J. Hugan, & E. Yücesan (Eds.). *Proceedings of the 2010 Winter Simulation Conference (WSC)* (pp. 371-382). Baltimore, MD: IEEE.

Mandinach, E.B., Honey, M., & Light, D. (2006). *A theoretical framework for data-driven decision making.* Paper presented at the annual meeting of the American Educational Research Association, San Francisco.

Maroulis, S., Guimerà, R., Petry, H., Stringer, M.J., Gomez, L.M., Amaral, L.A.N., & Wilensky, U. (2010, October 1). Complex systems view of educational policy research. *Science*, 38–39.

Marx, R.W., & Harris, C.J. (2006). *No Child Left Behind* and science education: Opportunities, challenges, and risks. *Elementary School Journal, 106*(5), 467–477.

McGinn, N.F. (2001, January/February). Computer simulations and policy analysis. *TechKnowLogia*. Retrieved from www.TechKnowLogia.org

McLaughlin, M. W. (1987). Learning from experience: Lessons from policy implementation. *Educational Evaluation and Policy Analysis, 9*(2), 171–178.

Means, B., & Penuel, W.R. (2005). Research to support scaling up technology-based educational innovations. In C. Dede, J.P. Honan, & L.C. Peters (Eds.), *Scaling up success: Lessons from technology-based educational improvement* (pp. 176–197). San Francisco, CA: Jossey-Bass.

Medd, W. (2001). Critical emergence: Complexity science and social policy. Special issue of *Social Issues, 1*(2), 1–6. Retrieved from http://www.whb.co.uk/socialissues/2ed.htm

Meyer, J.W., & Rowan, B. (1977). Institutionalized organizations: Formal structure as myth and ceremony. *American Journal of Sociology, 83*(2), 340–363.

Palinkas, L.A., & Soydan, H. (2012). *Translation and implementation of evidence-based practice.* New York: Oxford University Press.

Pea, R.D., & Collins, A. (2008). Learning how to do science education: Four waves of reform. In Y. Kali, M.C. Linn, & J.E. Roseman (Eds.), *Designing coherent science education: Implications for curriculum, instruction, and policy* (pp. 3-12). New York: Teachers College Press.

Penuel, W.R., & Fishman, B.J. (2012). Large-scale intervention research we can use. *Journal of Research in Science Teaching, 49*(3), 281–304.

Penuel, W.R., Fishman, B.J., Cheng, B., & Sabelli, N. (2011). Organizing research and development at the intersection of learning, implementation, and design. *Educational Researcher, 40*(7), 331–337.

Penuel, W.R., Frank, K.A., Sun, M., Kim, C., & Singleton, C. (2013). The organization as a filter of institutional diffusion. *Teachers College Record, 115*(1), 1–33.

Penuel, W.R., Gallagher, L.P., & Moorthy, S. (2011). Preparing teachers to design sequences of instruction in Earth science: A comparison of three professional development programs. *American Educational Research Journal, 48*(4), 996–1025.

Qudrat-Ullah, H. (2005). Structural validation of system dynamics and agent-based simulation models. In Y. Merkuryev, R. Zobel, & E. Kerckhoffs (Eds.), *Proceedings of the 19th European Conference on Modelling and Simulation* (pp. 481–485). Riga, Latvia: ECMS.

Renée, M., Welner, K., & Oakes, J. (2009). Social movement organizing and equity-focused educational change: Shifting the zone of mediation. In A. Hargreaves, A. Lieberman, M. Fullan, & D. Hopkins (Eds.), *Second international handbook of educational change* (pp. 158–163). London: Kluwer.

Repenning, N. P. (2002). A simulation-based approach to understanding the dynamics of innovation implementation. *Organization Science, 13*(2), 109–127.

Richey, M., Newton, P., Stephens, R., Backus, G., & McPherson, K. (2008, June). *The S&T eco-system: Pressures from kindergarten to globalization.* Paper presented at the Annual Conference of the American Society for Engineering Education, Pittsburgh, Pennsylvania; AC 2008–1063, Volume 11, pp. 7127–7146. Washington, D.C.: American Society for Engineering Education.

Roderick, M., Easton, J.Q., & Sebring, P.B. (2009). *The Consortium on Chicago School Research: A new model for the role of research in supporting urban school reform.* Chicago, IL: Consortium on Chicago School Research at the University of Chicago.

Rodrigues, L.L.R., & Martis, M.S. (2004). System dynamics of human resource and knowledge management in engineering education. *Journal of Knowledge Management Practice, 5.* Retrieved from http://www.tlainc.com/articl77.htm

Rohrbach, L.A., Grana, R., Sussman, S., & Valente, T.W. (2006). TYPE II translation transporting prevention interventions from research to real-world settings. *Evaluation & the Health Professions, 29*(3), 302–333.

Rowan, B. (2002). The ecology of school improvement: Notes on the school improvement industry in the United States. *Journal of Educational Change, 3*(3–4), 283–314.

Sabatier, P.A. (1986). Top-down and bottom-up approaches to implementation research: A critical analysis and suggested synthesis. *Journal of Public Policy 6*(1), 21–48.

Sabelli, N., & Dede, C. (2001). *Integrating educational research and practice: Reconceptualizing goals and policies.* Menlo Park, CA: SRI International. Retrieved from http://ctl.sri.com/publications/downloads/policy.pdf

Sabelli, N., & Dede, C. (2013). Empowering design-based implementation research: The need for infrastructure. In B.J. Fishman, W.R. Penuel, A.-R. Allen, & B.H. Cheng (Eds.), *Design-based implementation research. National Society for the Study of Education Yearbook, 112*(1), 464–480.

Sabelli, N., Lemke, J.L., Cheng, B.H., & Richey, M. (2012). *Education as a complex adaptive system: Report of progress.* Menlo Park, CA: SRI International.

Saywell, D., & Cotton, A. (1999). *Spreading the word: Practical guidelines for locating the power of in-between conference.* Background Paper 17. Research Dissemination Strategies. Loughborough: Water, Engineering and Development Centre.

Scheirer, M.A., & Dearing, J.W. (2011). An agenda for research on sustainability of public health programs. *American Journal of Public Health, 101*(11), 2059–2067.

Spillane, J.P. (2004). *Standards deviation: How schools misunderstand education policy*. Cambridge, MA: Harvard University Press.

Spillane, J.P. (2005). Distributed leadership. *Educational Forum, 69*, 143–150.

Spillane, J.P., Gomez, L.M., & Mesler, L. (2009). Notes on reframing the role of organizations in policy implementation. In G. Sykes, B. Schneider, & D.N. Plank (Eds.), *Handbook of education policy research* (pp. 409–425). Washington, DC: American Educational Research Association.

Spillane, J.P., Halverson, R.R., & Diamond, J.B. (2001). Investigating school leadership practice: A distributed perspective. *Educational Researcher, 30*(3), 23–27.

Spillane, J.P., Parise, L.M., & Sherer, J.Z. (2011). Organizational routines as coupling mechanisms: Policy, school administration, and the technical core. *American Educational Research Journal, 48*(3), 586–620.

Spillane, J.P., Reiser, B.J., & Gomez, L.M. (2006). Policy implementation and cognition. In M. I. Honig (Ed.), *New directions in educational policy implementation* (pp. 47–64). Albany, New York: State University of New York Press.

Spillane, J.P., Reiser, B.J., & Reimer, T. (2002). Policy implementation and cognition: Reframing and refocusing implementation research. *Review of Educational Research, 72*(3), 387–431.

Stephens, R., & Richey, M (2011, July). Accelerating STEM capacity: A complex adaptive system perspective. *Journal of Engineering Education, 100*(3), 417–423.

Sudsawad, P. (2007). Knowledge translation: Introduction to models, strategies, and measures. National Center for the Dissemination of Disability Research at the Southwest Educational Development Laboratory. Retrieved from http://www.ncddr.org/kt/products/ktintro/

Vennix, J.A.M. (1999). Group model-building: Tackling messy problems. *System Dynamics Review 15*(4), 379–401.

Wells, B.H., Sanchez, A., & Attridge, J.M. (2007). *Systems engineering the U.S. education system*. Waltham, MA: Raytheon.

Wilensky, U., & Reisman, K. (2006). Thinking like a wolf, a sheep, or a firefly: Learning biology through constructing and testing computational theories. *Cognition and Instruction, 24*(2), 171–209.

Wolfe, R (2006). *Changing conceptions of intermediaries in development processes: Challenging the modernist view of knowledge, communication and social change*. Brighton, UK: Institute of Development Studies.

Woolf, S.H. (2008). The meaning of translational research and why it matters. *Journal of the American Medical Association, 299*(2), 211–213.

Zhao, Y., & Frank, K. (2003). Factors affecting technology uses in schools: An ecological perspective. *American Educational Research Journal, 40*(4), 807–840.

13

POLICY MODELING OF LARGE-SCALE SOCIAL SYSTEMS

Lessons from the SKIN Model of Innovation

Petra Ahrweiler, Andreas Pyka, and Nigel Gilbert

Introduction

How can modeling and simulation contribute to policy making in complex environments like research and innovation? Policy makers regularly complain about difficulties promoting innovation in the knowledge economy. Every year, governments worldwide spend billions promoting innovation. Innovation policy makers, business managers, and the public often expect that investments in research and development (R&D), higher education institutions, and science-industry networks will immediately produce a flow of products and processes with high commercial returns. When expected returns and positive impacts on labor markets do not materialize, there is considerable disappointment. Legitimation problems reveal the limits of conventional steering, control, and policy functions. Innovation managers mention frustration with the messy, complicated aspects of the innovation process; to them, it simply "does not compute." However, policy informatics offers new approaches for understanding innovation and decision-making about policies to enhance innovation.

Socio-economic systems are confronted with great complexity in the development of new knowledge, its diffusion, and commercial application in innovation. In the case of innovation, agents are confronted with *true uncertainty*, making forecasts impossible. Any analytical approach that offers guidance and support to political decision-makers must acknowledge this intermingling of rich complexity and uncertainty. In the past, policy consultation relied on a linear sequential innovation process, with the invention and innovation phases strictly separated. Consequently, the complex interactions of heterogeneous actors were de-emphasized under the assumption that the actors were endowed with an Olympic rationality, allowing them to apply calculus optimization in their decision-making. However, since the 1980s, the weaknesses of this approach have become very obvious: Different systems are related, the actors within those systems are not endowed with homogenous knowledge, and simplifying assumptions necessary to maintain predictability of the analytic model also stripped it of its usefulness in providing genuine, novel insights into innovation policy.

New approaches enable the modeling of innovation policy initiatives to encompass a greater number of parameters and perform simulations that forecast the potential impact of proposed innovation policy. A conceptual framework that combines the application of empirical research, computational network analysis, and agent-based modeling (ABM) allows for a more integrated and comprehensive understanding of innovation policy making than has hitherto been achieved. The approach may create methodological foundations for the sustainable transformation of research and innovation.

A computational systems approach like ABM can be used to identify and understand the effects of innovation policy strategies and associated knowledge dynamics. In contrast to conventional methods of social research, this method is capable of dealing with the high complexity and non-linearity of innovation processes. ABM users can address questions that involve different levels of the innovation ecosystem: from start-up firms (micro level), to academic-industry partnerships (meso level), and to whole sectors or regions (macro level). ABM decision makers can be informed by large empirical data sets. These factors motivated the model that we have developed and applied over the last decade. Simulating Knowledge Dynamics in Innovation Networks (SKIN) is a multi-agent-based model used to understand innovation policy initiatives that contain heterogeneous agents that act and interact in a large-scale, complex, and changing social environment.

This chapter first describes the details of the SKIN model and explores the development processes associated with modeling large-scale social systems. We then explore four policy-relevant applications of the model. To reveal the value of this approach, each case is broken down into its intended purpose, unique features, how the experiment or activities were conducted, and the specific type of outcomes produced by the application. The chapter concludes with reflections on how the model has been received by the policy-making community over a decade of development and use, including successes, limitations, and contingencies.

What Is the SKIN Model?

The SKIN model is an ABM used to understand innovation policy initiatives that contain heterogeneous agents that act and interact in a large-scale, complex, and changing social environment. The agents represent innovative actors who try to sell their innovations to other agents and end users but must also buy raw materials or more sophisticated inputs from other agents (material suppliers) to produce their outputs. This basic model of a market is extended with a representation of the knowledge dynamics in and between agents. Each agent attempts to improve their innovation performance and sales by improving their knowledge base through adaptation to user needs, incremental or radical learning, and cooperation and networking with other agents.

SKIN is grounded in empirical research and theoretical frameworks from innovation economics and economic sociology. It is the cumulative result of a number of projects carried out over a 12-year period that combined empirical research in innovation networks with agent-based simulation. The work began with the European Union (EU) project, Simulating Self-organising Innovation Networks (SEIN). This project combined four empirical case studies in different sectors of technological innovation that were carried out in different EU member states with agent-based simulation of the case studies (Gilbert, Pyka, & Ahrweiler, 2003; Pyka & Kueppers, 2003). SEIN was used by the European Commission for scenario modeling of current and future innovation policy strategies (Ahrweiler, de Jong, & Windrum, 2003).

The current SKIN model builds on procedures initially designed to fit the biotechnology-based pharmaceutical industry in Europe because it is an excellent example of a knowledge-intensive industry. The key components and capacities of the model are highlighted below.

The Agents

Knowledge is a core concept of an innovation network. Accordingly, one of the first tasks is to find an appropriate knowledge representation. The approach to knowledge representation used in our model is similar to Toulmin's (1967) evolutionary model of knowledge production, which identified concepts, beliefs, and interpretations as the "genes" of scientific/technological development that evolve over time through a process of selection, variation, and retention. Ackermann (1970) interpreted the works of Kuhn and Popper according to this perspective, allowing for different selection systems. In the SKIN model, an analogical concept, the *kene*, is used to represent the aggregate knowledge of an organization (Gilbert, 1997).

The individual knowledge base of a SKIN agent, its kene, contains several units of knowledge. Each unit in a kene is represented as a triple consisting of a firm's capability (C) in a scientific, technological, or business domain, its ability (A) to perform a certain application in this field, and the expertise level (E) the firm has achieved with respect to this ability. In the artificial space of a model, kenes could consist of arbitrary bit sequences of indefinite length. However, the knowledge of the artificial agents must be portrayed in a way that allows comparison with empirical actors. The composition of kenes is inspired by the International Patent Classification (IPC). The IPC provides a hierarchical system of language-independent symbols for the classification of patents according to different areas of technology. IPC codes allow the assignment of technological fields and competences using concordance tables (Schmoch, Laville, Patel, & Frietsch,

2003) to identify industrial sectors. In this sense, the IPC codes can be considered coordinates of an empirical knowledge space and correspond approximately to the units of knowledge comprising the kenes.[1]

Other Agents that Form the Market

Because actors in empirical innovation networks of knowledge-intensive industries interact on both the knowledge and market levels (Garcia, Dunn, & Smith, 2012b, p. 2), a representation of market dynamics in the SKIN model is required.

Each firm agent when set up has a stock of initial capital. It needs this capital to produce its goods for the market and to finance its R&D expenditures. It can increase its capital by selling products. The amount of capital owned by a firm is used as a measure of its size and it influences the amount of knowledge (measured by the number of units in its kene) that it can maintain. In many knowledge-intensive industries, we find both large and small actors, e.g., large pharmaceutical firms, biotech start-ups, former national monopolists, and high technology specialists in the ICT industries (Pyka & Saviotti, 2005). We assume that the large diversified firms are characterized by a larger knowledge base compared to smaller specialized companies (Brusoni, Prencipe, & Pavitt, 2001). To represent this in the model, firms are initially allocated the same starting capital; only a few firms are allocated additional capital.

Firms apply their knowledge to create innovative products with the potential of being successful in the market. The underlying idea for an innovation, the *innovation hypothesis*, is the source an agent uses for its attempts to make a profit in the market and determines the required inputs and the nature of the firm's product.

To engage in production, all inputs need to be obtainable from the market, i.e., provided by other firms or available as raw materials. If the inputs are not available, the firm is not able to produce and must give up the attempt to innovate. If there is more than one supplier for a certain input, the agent will choose the one offering the cheapest price; if there are several similar offers, the one chosen will offer the highest quality. If the firm can go into production, it must determine a price for its product, accounting for its input costs and a possible profit margin. Using a price adjustment mechanism, agents are able to adapt their prices to reflect demand and thereby learn from feedback. An agent applies the knowledge in its innovation hypothesis to make its product and this increases its expertise. This reflects how learning by doing/using is modeled.

The expertise levels of the units in the innovation hypothesis are increased and levels of other units are decreased. Expertise in unused units in the kene is eventually lost and the units are then deleted from the kene; the corresponding abilities are "forgotten" or "dismissed" (Hedberg, 1981). Attempting to be successful in the market means firms are dependent on their innovation hypothesis, i.e., their kene. If there is no demand for a product, the firm must adapt its knowledge to produce something for which there is demand in order to survive (Duncan, 1974). A firm may improve its performance either alone or in cooperation in either an incremental or more radical fashion.

Learning and Cooperation: Improving Innovation Performance

Inspired by literature on organizational learning, SKIN adopts ideas from the framework proposed by Argyris and Schön (1996) to examine the assumption that the greatest competitive advantage for any firm is its ability to learn (de Geus, 1997; Gilbert, Ahrweiler, & Pyka, 2007). Firm agents can:

- use their capabilities (learning by doing/using) to learn to estimate their success via feedback from markets and clients (learning by feedback), and/or
- when feedback is not satisfactory, improve their own knowledge incrementally to adapt to changing technological and/or economic standards (adaptation learning, incremental learning).

If a firm's previous innovation has been successful (i.e., it has found buyers), the firm will continue selling the same product in the next round, possibly at a different price depending on demand. However, if there were no sales, it considers whether change is required. If the firm has sufficient capital, it will carry out incremental research (R&D in the firm's labs). Incremental research (Cohen & Levinthal, 1989) is when a firm attempts to improve its product by altering one of the abilities chosen from the units in its innovation hypothesis, while sticking to its focal capabilities. Alternatively, if a firm is experiencing losses and is under pressure, radical changes in its capabilities may be undertaken to meet completely different client requirements (e.g., innovative learning, radical learning).

A firm may also consider partnerships to exploit external knowledge sources. Deciding whether and with whom to cooperate is based on the mutual observations of each firm, which estimate the chances of success given the requirements of competitors, possible and past partners, and clients. Bolton, Katoka, and Ockenfels (2005), writing from a theoretical viewpoint, and Mitchelet (1992), using empirical evidence, show that when firms know their partner's history of cooperation, greater mutual information improves the conditions for cooperation.

The model allows a choice between two partner-search strategies (Powell, White, Koput, & Owen-Smith, 2005), both of which compare firm capabilities to that of the potential partner. Applying the conservative strategy, a firm will be attracted to a partner with similar capabilities; a progressive strategy will be based on differences between the capability sets.

Previous good experience with contacts generally augurs well for renewing a partnership. For example, Garcia, Dunn, and Smith (2012a, p. 2f) found that in the case of interaction patterns between public research centers and industrial firms, "prior formal relationships are a fundamental element for collaboration. . . . Strong ties (past relationships) appear to be more fundamental in building university-industry ties." This is mirrored in the model: to find a partner, the firm will look at previous partners first, then at its suppliers, customers, and finally, at all others. If a potential partner is sufficiently attractive relative to chosen search strategy criteria, a firm will cease searching and offer a partnership.

If the firm's last innovation was successful and a firm has already established partnerships, it can initiate the formation of a network. This can increase profits because the network will work to create innovations as an autonomous agent in addition to those created by its individual members. The network will distribute any rewards in excess of initial capital investment to its members who may simultaneously continue with their own attempts, thus doubling opportunity for profit. If a network fails and becomes bankrupt, it will be dissolved. If a sector is successful, new firms will be attracted, representing Schumpeterian competition by imitation.

Innovation Policy Initiatives

Using SKIN, policy makers can observe and manipulate patterns of network evolution by changing various parameters, analogous to applying different policy options in the real world. The model affords examination of likely real-world effects of different policy options before implementation. Using real-world datasets and responses to questions put forward by stakeholders, four applications were developed. Collectively, they begin to demonstrate how SKIN can meet multiple policy needs: It can provide precise, detailed information on the effects of specific policy instruments, how well research and innovation networks operate, and how to understand and manage the relationship between research funding and policy goals.

Four Innovation Policy-Relevant SKIN Applications

This section discusses four recent applications of the SKIN model: (1) modeling the impact of university links for business innovation networks, (2) scenario modeling of an individual case from Ireland, demonstrating an MNC-directed innovation policy for indigenous industry networks, (3) simulating the effects of different innovation management strategies of firms in the biotechnology-based pharmaceutical industry, and (4) scenario modeling of European funding schemes based on size and geographical diversity of project teams.

Modeling the Impact of University Cooperation for Business Innovation Networks

Purpose

What role do university-industry links play in innovation generation and diffusion? The first SKIN application (Ahrweiler, Pyka, & Gilbert, 2011) focuses on R&D cooperation between universities and industry partners. These partnerships transfer technology from academic institutions to companies and thus spread knowledge throughout the population. Empirical studies have investigated the knowledge flows from universities to firms (Agrawal & Henderson, 2002; Siegel, Waldman, Atwater, & Link, 2003). The success and failure of technology transfer activities have also been studied (Henderson, Jaffe, & Trajtenberg, 1998; Thursby & Kemp, 2002; Chapple, Lockett, Siegel, & Wright, 2005). However, most of these studies report measurement difficulties due to the unobservable nature and non-linear properties of knowledge flows. Although our hypotheses focus on the same unobservable knowledge flow, using an ABM allows observation of simulated knowledge flows.

R&D and industrial policy assume that university-industry links are the fuel of knowledge-based economies. Considerable public spending on university-industry collaboration may not quickly produce the new and innovative products and processes, increased employment, and high commercial returns expected. However, empirical evidence demonstrating direct commercial profitability of the *science connection* is scarce.

Additional Features

For this application, two new agent types, university agents and spin-off companies, were added to the basic SKIN model. Like the firm agents, university agents have an individual knowledge base (kene). Because the university agents have significant technological capabilities, abilities, and expertise, their kenes are twice as long as those of firm agents. University agents conduct research, form R&D partnerships with firms, and participate in larger networks. If a university agent is successful, it breeds a firm (university spin-off) that can buy, sell, and produce. The kene of a spin-off is a clone of the university's innovation hypothesis. If a spin-off is created, a university loses one unit of its kene and receives a random new kene, representing turnover of researchers. All expertise levels of the spin-off are reduced by one unit. Unlike firm agents, university agents in the extended SKIN model are nonprofit organizations (they do not produce or sell to the market) with stable public core funding. Their success is measured by the economic gains of their firm partners.

Experiments

We compared a scenario with university agents to one without to test whether firms that interact with universities develop more products and experience greater commercial success; whether having universities among the cooperating actors raises the knowledge/competence level of the whole population (i.e., greater variety at the knowledge level, increased innovation performance overall, and increased diffusion in terms of quantity and speed); whether firms that interact with universities are more attractive to other firms when new partnerships are considered with respect to commercialization, financing, and R&D collaboration; and whether firms that interact with universities are better able to adapt to changing environmental conditions.

Outcomes

Performing experiments on the simulation provides new insights that potentially contribute to the formation of a more widely accepted perspective on the effects of university-industry relationships. There is indeed no linear and immediate relationship between increased knowledge inputs and economic profits—our experiments reproduced the ambiguous evaluations of empirical studies. Firms interacting with universities cannot adapt to changing environmental conditions any better than their counterparts without university affiliation. The average lifetime of firms interacting with universities is no greater than those that do not. There seems to be a slight trend toward higher innovation performance in the scenario with universities, but it is not statistically significant. Neither does an increased amount of knowledge automatically lead to increased innovation performance and economic success.

Nevertheless, universities increase the knowledge and competence levels of the entire population of actors, the variety of knowledge among firms, and innovation diffusion in terms of quantity and speed. Furthermore, firms interacting with universities are more attractive to other firms when searching for new partnerships. Although no direct relation between university contributions and economic success can be found, the agent-based simulation confirms that university-industry links improve innovation diffusion and collaborative arrangements in innovation networks.

Scenario Modeling of Irish MNC-Directed Innovation Policies

Purpose

In this SKIN application, we investigated the effects of the presence and embeddedness of multi-national corporations (MNCs) in networks of innovation (Ahrweiler, Schilperoord, Gilbert, & Pyka, 2012). New products and processes can result from the ongoing interactions of innovative organizations such as universities, research institutes, firms such as MNCs and SMEs, government agencies, and venture capitalists.

Our review of knowledge flows and capital stocks was to investigate whether the mere presence of MNCs is beneficial to innovation networks and whether an additional advantage exists if these MNCs are engaged in collaborative R&D with other players in the network. We examined the role of MNCs in innovation networks from the perspective of the host countries of the MNC subsidiaries.

The simulation was grounded in the empirical example of Ireland, enabling us to analyze the role of MNCs in the Irish indigenous industry.

Scenario modeling of the role of MNCs in host countries is highly policy-relevant; in Ireland, there has been growth in the high-technology industry sectors but this growth has only been fostered by foreign-owned MNCs. The MNCs are still poorly integrated into Irish networks, clusters, and innovation centers. Accordingly, Ireland practically has a dual economy: one comprised of SMEs with weak R&D and innovation performance and another with MNCs that have cutting-edge technology and make a substantial contribution to the country's wealth. To prevent MNCs from leaving Ireland to relocate in more competitive manufacturing locations and to help the indigenous industry benefit from MNCs, Irish policy makers attempt to convince MNCs to locate their R&D in Ireland and cooperate with Irish organizations (SMEs, universities, etc.) in innovation networks. ABM can contribute to assessing the impacts and effects of these policies by modeling scenarios that correspond to Irish MNC policy options.

Additional Features

In the adapted model for this application, MNCs are normal SKIN agents producers (manufacturers) using inputs from suppliers and selling their products to customers. They can employ all learning and networking activities usually available to SKIN agents. However, MNC agents usually have a larger knowledge base and greater capital stocks, giving them a better chance to enter foreign markets and market new technologies due to their commercialization experience and distribution networks.

Experiments

When conducting the experiments, we operationalized the policy questions in relation to the Irish economy: how important is the knowledge integration function of MNCs as knowledge hubs and financial magnets to regional innovation networks? Does a firm population that contains MNCs perform better in terms of knowledge diffusion and innovation performance than a uniformly sized population of small and medium firms? What effects do MNCs and their activities have on the knowledge level of the firm population?

To deal with these questions, we constructed two different comparisons. First, we compared a model with MNC agents to a model without MNC agents. Second, we compared a model containing MNC agents that conducted R&D in cooperation with the indigenous industry to a model where MNCs follow a go-it-alone innovation strategy, while all other firms in that scenario cooperate with each other. Go-it-alone innovation strategies were described above as incremental and radical learning, which can be performed in the R&D labs of a firm without having to cooperate with external partners. Cooperation includes the aspects of cooperating as partnerships and networks.

The initial configuration of the experiments was set up to mirror the conditions in the Irish economy. Modeling results were validated against empirical findings from Irish data.

Outcomes

Our results strongly confirmed the current Irish MNC policy strategies. Attracting and retaining MNCs does result in increased capital availability and innovation performance for the indigenous industry. Surprisingly, even the mere presence of MNCs in the indigenous economy raises knowledge flows in the host country's industry because firms can more safely engage in R&D and marketing activities. This is intensified when MNCs engage in local learning activities and embed themselves into the R&D network of regional innovation. The agent-based simulation confirmed that MNCs that collaborated with indigenous innovation networks in R&D did improve the knowledge and competence level of the entire industry, along with innovation diffusion and collaborative arrangements in the host country.

Simulating the Effects of Different Learning Strategies

Purpose

This SKIN application (Gilbert, Ahrweiler, & Pyka, 2007, 2010) is about simulating the effects of different learning activities of firms in the biotechnology-based pharmaceutical industry (i.e., go-it-alone strategies, such as incremental and radical learning, and learning through R&D partnerships and innovation networks). According to organizational learning literature, the greatest competitive advantage a firm has is its ability to learn. The SKIN model was applied to simulate the learning competence of firms to improve policies and management decisions for supporting innovation.

The results of this application were closely observed by a large MNC in the UK pharma industry. At the time, the company's management was considering future cooperation and embeddedness strategies for their enterprise. Performance of an agent through a simulation using properties similar to that of the company was tracked. However, different combinations of learning and cooperation strategies were applied and provided interesting insights into policy and management outcomes for navigating through complex innovation networks and improving its position in the network to exploit its resources to the best advantage.

Additional Features

This application did not require the addition of any features to the basic model.

Experiments

Experiments were conducted to test different learning and cooperation strategy combinations against a baseline scenario where firms were not able to learn or cooperate at all.

Outcomes

The simulation experiments underlined the importance of innovation and learning and finding new capabilities external to the firm, either through partnering or radical research. The results also showed that simpler forms of organizational learning, such as learning by doing and learning by feedback, are of limited value alone when operating in the highly dynamic environment of modern knowledge-based market sectors, although they are significant when combined with other forms of learning.

The pharma company that was observed wished to determine whether its preferred go-it-alone strategies were indeed sufficient for satisfactory innovation performance. However, it was found that performance could be improved through partnerships and networks. The simulation also revealed possibilities that would be impossible to carry out in the real world. Clearly, it would not be practicable to constrain actual firms in their learning capacities in order to study the effect on their success and determine any causal consequences, still less to do this for a whole market sector. But as this application has shown, it is possible and revealing to do so using a model.

Scenario Modeling of European Funding Schemes

Purpose

This SKIN application investigated the interaction among political governance, structures, and functions of R&D collaboration networks that are induced by policy programs, particularly with respect to the networks that have emerged in the seven European Framework Programmes. The goal was to identify ways to evaluate efficient network structures for R&D collaboration networks and to develop policy designs for their creation.

Additional Features

SKIN was adapted to the context of EU-funded research networks where the agents are R&D organizations. Instead of triples, kenes consist of quadruples: research direction (RD), capabilities (C), abilities (A), and expertise (E). RD represents the general knowledge orientation, ranging from completely theoretical knowledge (i.e., basic research) to exclusively applied research. These research directions allow for a differentiation between university and firm actors.

C represents different knowledge areas and technological disciplines, e.g., biochemistry, telecommunication, or mechanical engineering. For modeling purposes, the number of capabilities chosen must be sufficiently large to cover all potential research areas in the European Framework Programmes. The market of the basic SKIN model is "switched off."

The behavior of agents follows the empirical understanding of processes of network formation and evolution in the European Framework Programmes:

- **Definition of *incentives and rules* for R&D collaboration:** The EU provides funding for collaborative research. The rules are defined in the European Framework Programmes (e.g., rules for project consortia, research topics, time span).
- **Consortium formation/partner choice:** The actors (research institutes, firms, etc.) form project consortia. Partner choice mechanisms apply.
- **Proposal production:** The actors collaboratively produce a proposal, which is the initial representation of the collaborative knowledge of the consortium partners, and submit the proposal to the Commission.
- **Proposal selection:** The Commission evaluates each proposal according to a template that emphasizes content (program match), quality, and architecture of the consortium (i.e., minimum number of members, industry involvement). The Commission selects projects to fund; availability of funding limits the number of accepted projects.
- **R&D cooperation:** The research and cooperative learning activity projects are initiated. The required deliverables are produced (i.e., publications, patents, reports).
- **Performance evaluation:** The Commission changes its work program, taking into account the success of projects. The process repeats.

The following flowchart (Figure 13.1) summarizes this cycle (Scholz et al., 2010).

Experiments

This application works with a baseline scenario, reproducing the network structures of FP6 obtained by using the calibration data and policy rules of a particular European Framework Programme. The policy experiments were designed as an implementation of new rules in the baseline scenario. To analyze the impact of the policy experiments on the evolution of the research networks, we compared network structures and network performances. For example, one experiment dealt with an increase in the minimum number of teams participating in a research project, another suggested a scenario with large projects and the addition of geographical location as one of the project selection criterion used by the European Commission. For large projects, we implement the policy rule that project consortia must be greater than 2.5 times the size of a normal project and consortia members must be from at least four different geographical areas.

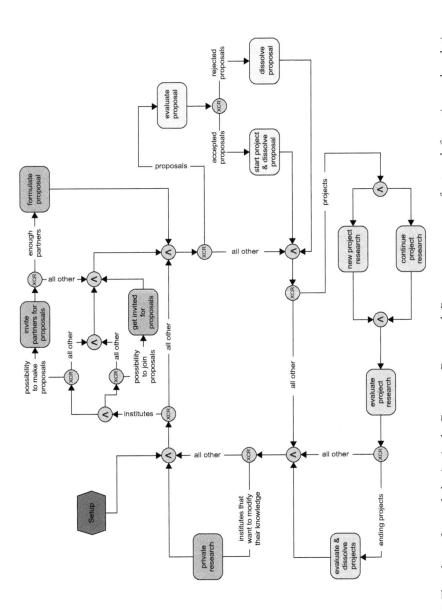

FIGURE 13.1 Flowchart of agent behavior in the European Framework Programmes process of network formation and evolution

Source: Scholz et al., 2010, p. 310. Reprinted with permission.

Outcomes

The simulation showed that larger consortia with greater geographical diversity diminish project output and performance—both in quantity and quality. The model demonstrated that it can act as an *in silico* laboratory for policy instruments. It allows experimentation with a variety of policy parameters to investigate both effectiveness and consequences on the creation and behavior of collaborative networks.

By offering a conceptual framework for the combined application of empirical research, computational network analysis, and ABM, the basic model demonstrates that SKIN is a new, powerful tool that allows a more integrated and comprehensive understanding of innovation policy making than previously possible.

Assessing the Experience with the SKIN Model

In this section, we assess our experience with the model and its applications in terms of its successes, limitations, and contingencies. This relates to managing expectations of stakeholders, general apprehensions around the predictability of future developments, data availability issues, and validation questions.

The Scope of the Model

The SKIN model, like any model, is an abstraction from reality. Not every aspect of innovation networks can be included; if they were, the resulting model would be so complicated that nothing of value could be learned. The art of modeling is to include only the essentials, neither more, nor less. Hence, each example application of SKIN could be criticized for failing to model some relevant characteristics; it is not until the model is validated against the empirical world that we can assess whether it has been over-simplified. For example, the extent of heterogeneity between universities is not well represented in the model of university-industry links. Adding such heterogeneity, for example, in the balance of universities' teaching and research or the extent of their entrepreneurial culture, could add realism and accuracy to the model, but would be at the cost of considerable complication and the requirement for additional data for calibration. Similarly, in the model of Irish MNCs, further development would be required to recognize heterogeneity as integrated players, global innovators, implementers, and local innovators (Gupta & Govindarajan, 1991, 1994).

The Data Base

Applying the SKIN model to specific policy topics as done here requires not only adaption of the model (e.g., attribute agents with the appropriate characteristics and behavior), but also calibration of the model to the situation being simulated.

For example, design and calibration of the model of European funding schemes required both a detailed analysis of FP6 funding policies and data about the partners involved in FP6 project consortia. The adequacy of the simulation depends on the reliability of available data for calibration. Such data can be difficult to obtain and process into a form suitable for analysis. In the case of the model of European funding schemes, significant effort was required to *clean* data because organization names changed from year to year and some companies were merged.

The Policy Questions

Different versions of the SKIN model were developed to provide answers to policy questions. However, determining what the policy questions should be was often not simple. Many issues can arise. Whose questions are to be answered? This requires identifying relevant stakeholders. Which questions do these stakeholders want answered? Our experience suggests that policy makers often do not have clearly articulated policy questions readily formulated and lengthy discussion may be required before general interest in a topic can be honed down to a specific question. Furthermore, some questions may not be answerable using simulation models, or may depend on normative concepts that defy clear definition. For example, one of the objectives of the European funding scheme model was to assess the effect of various policy changes on network performance. But what policy makers meant by *performance* was not well defined and required work to operationalize in a form that policy makers could relate to.

An additional issue was managing expectations of policy makers about what simulation models of complex social phenomena could achieve. The SKIN model contains numerous stochastic elements, thus generating slightly different results for each run. Point predictions are therefore not possible. Of greater concern, because the model is non-linear, small changes to parameters can have significant effects on results. Thus, at best, the model can be used to examine broad scenarios and assess the likely effects of policies in qualitative terms, but cannot provide the certainty and precision policy makers would ideally like.

Conclusion

In this chapter, we have shown that our observations do not necessarily lead to the rejection of a scientific analysis of innovation processes. We have introduced an agent-based model that places at its center the network organization of innovation processes. Network organization means that single actors consciously choose their connections to other agents. Some may be refused and others dissolved due to failure or lack of trust. The ties are a necessary condition for knowledge flow among agents.

We applied the network model to four very different cases to show how this modeling tool can be used to derive policy and strategic conclusions. The first case dealt with university-industry networks, which are central to innovation policies. The connection of basic researchers with applied researchers should guarantee quick transfer of knowledge and technology and consequently improved competitiveness and economic performance. The results of the experiments show there is no simple relationship between basic and applied research. The interaction processes were found to be more subtle, and positive effects generally emerge indirectly from the increased variety of knowledge and improved capabilities of the agents involved.

The second case focused on actors in a sectorial innovation system. Large MNCs interacted with specialized local firms, encouraging economic development. The experiments identify targets and mechanisms through which economic policy can improve the possibility that a country will catch up to global competition.

The third variant of the model was used to disentangle the complexities of learning processes in firms. These are empirically unobservable, but can be measured in the artificial world of the model. With the help of a model designed for the particularities of the biopharmaceutical industry, we were able to improve understanding of innovation and cooperative behavior in this sector. Results are not only relevant to innovation policy, but also for development of strategy.

The final two examples can be considered *policy engineering*. The agent-based simulation model was adapted to the characteristics of the research networks generated in the European Framework Programmes. The variables were calibrated with rich empirical information and the model trained to reproduce the historical path of the emergence of the European research area. The potential effects of new policy designs that modify size, spread of disciplines, and nationality requirements on this European research landscape were then analyzed.

ABMs that focus on mutual knowledge exchange and learning in research and innovation networks facilitate modeling of interactions between heterogeneous agents confronted with uncertainty. The examples in this chapter illustrate the flexibility of this class of models. The basic mechanisms of knowledge exchange and knowledge creation remains similar in each of the cases. However, the agents are modified to reflect different sizes, location, and knowledge. The chapter also demonstrates that agent-based simulation models can be adapted to address the information interests of different target groups (e.g., firms and policy makers).

Agent-based simulation models allow analysis of dynamic and historical processes using so-called *in-silico* experiments. Because they can be calibrated and trained with empirical information, they improve understanding of the mechanisms and dynamics that have shaped past developments. The results of a scenario analysis using such models must be interpreted subject to the *epistemological disclaimer* that point predictions of the future are impossible. On one hand, the results tell us what mechanisms and dynamics modelers think are relevant to explain past developments. On the other hand, extensions to the future can only show that certain development paths might lead to bottlenecks and should be avoided. Additionally, because these are non-linear interactions, certain self-organizing, and therefore surprising, changes are possible. Modelers and policy makers must therefore remember that ABMs do not predict precise developments, although they can identify general trends, including the possibility of surprises. In other words, agent-based simulation models help support instinctive design of policy instruments. Even if the veil of ignorance cannot be lifted, it at least becomes more transparent.

Note

1   The analogy is not exact; kene units can represent items of tacit knowledge that cannot be made explicit like IPC codes and can also represent non-technical knowledge, such as business strategies, marketing techniques, and management competences, none of which are patentable.

References

Ackermann, R. (1970). *The philosophy of science*. New York: Pegasus.

Agrawal, A., & Henderson, R. (2002). Putting patents in context: Exploring knowledge transfer from MIT. *Management Science, 48*, 44–60.

Ahrweiler, P., de Jong, S., & Windrum, P. (2003). Evaluating innovation networks. In A. Pyka & G. Kueppers (Eds.), *Innovation networks: Theory and practice*. Cheltenham, UK: Edward Elgar Publishers.

Ahrweiler, P., Pyka, A., & Gilbert, G.N. (2011). A new model for university-industry links in knowledge-based economies. *Journal of Product Innovation Management, 28*(2), 218–235.

Ahrweiler, P., Schilperoord, M., Gilbert, G.N., & Pyka, A. (2012). Simulating the role of MNCs for knowledge and capital dynamics in networks of innovation. In M. Heidenreich (Ed.), *Innovation and institutional embeddedness of companies* (pp. 141–168). Cheltenham, UK: Edward Elgar.

Argyris, C., & Schon, D.A. (1996). *Organizational learning: A theory of action perspective*. Reading, MA: Addison-Wesley.

Bolton, G. E., Katoka, E., & Ockenfels, A. (2005). Co-operation among strangers with limited information about reputation. *Journal of Public Economics, 89*, 1457–1468.

Brusoni, S., Prencipe, A., & Pavitt, K. (2001). Knowledge specialisation, organizational coupling and the boundaries of the firm: Why do firms know more than they make? *Administrative Science Quarterly, 46*, 597–621.

Chapple, W., Lockett, A., Siegel, D., & Wright, M. (2005). Assessing the relative performance of U.K. university technology transfer offices: Parametric and non-parametric evidence. *Research Policy, 34*, 369–384.

Cohen, W. M., & Levinthal, D. (1989). Innovation and learning: The two faces of RD. *The Economic Journal, 99*, 569–596.

de Geus, A. (1997). *The living company*. London: Brealy.

Duncan, R.B. (1974). Modifications in decision structure in adapting to the environment: Some implications for organizational learning. *Decision Sciences, 5*, 705–725.

Garcia, R., Dunn, D.T., & Smith, L.A. (2012a). Boundary spanners and social networks surrounding research centers. *Partners for Innovation Program*. [Working Paper: NSF 04–556.]

Garcia, R., Dunn, D.T., & Smith, L.A. (2012b). Performance of knowledge interactions between public research centers and industrial firms: An application for Spanish public research centers based on project-level data. [Working Paper.]

Gilbert, G.N. (1997). A simulation of the structure of academic science. *Sociological Research Online, 2.* Retrieved from http://socresonline.org.uk/2/2/3.html

Gilbert, G.N., Ahrweiler, P., & Pyka, A. (2007). Learning in innovation networks: Some simulation experiments. *Physica A, 378,* 100–109.

Gilbert, G.N., Ahrweiler, P., & Pyka, A. (2010). Learning in innovation networks: Some simulation experiments. In P. Ahrweiler (Ed.), *Innovation in complex social systems.* London: Routledge.

Gilbert, G.N., Pyka, A., & Ahrweiler, P. (2003). Simulating innovation networks. In A. Pyka & G. Kueppers (Eds.), *Innovation networks: Theory and practice* (pp. 169–196). Cheltenham, UK: Edward Elgar.

Gupta, A.K., & Govindarajan, V. (1991). Knowledge flows and the structure of control within multinational corporations. *Academy of Management Review, 16*(4), 768–792.

Gupta, A.K., & Govindarajan, V. (1994). Alternative value chain configurations for foreign subsidiaries: Implications for coordination and control within MNCs. In H. Thomas, D. O'Neal, & R. White (Eds.), *Building the strategically responsive organization* (pp. 375–392). Chichester, UK: John Wiley.

Hedberg, B. (1981). How organizations learn and unlearn. In P.C. Nystrom & W.H. Starbuck (Eds.), *Handbook of organizational design.* Oxford: Oxford University Press.

Henderson, R., Jaffe, A., & Trajtenberg, M. (1998). Universities as a source of commercial technology: A detailed analysis of university patenting, 1965–1988. *Review of Economics and Statistics, 80*(1), 119–127.

Michelet, R. (1992). Forming successful strategic marketing alliances in Europe. *Journal of European Business, 4,* 11–15.

Powell, W.W., White, D.R., Koput, K.W., & Owen-Smith, J. (2005). Network dynamics and field evolution: The growth of inter-organizational collaboration in the life sciences. *American Journal of Sociology, 110,* 1132–1205.

Pyka, A., & Kueppers, G. (2003). *Innovation networks: Theory and practice.* Cheltenham, UK: Edward Elgar.

Pyka, A., & Saviotti, P.P. (2005). The evolution of R&D networking in the biotech industries. *International Journal of Entrepreneurship and Innovation Management, 5*(1/2), 49–68.

Schmoch, U., Laville, F., Patel, P., & Frietsch, R. (2003). Linking technological areas to industrial sectors. *Final Report to the European Commission, DG Research.* Karlsruhe, Germany: Fraunhofer Institute for Systems and Innovation Research. Retrieved from http://www.obs-ost.fr/sites/default/files/TechInd_Final_Report_Revision.pdf

Scholz R., Nokkala T., Ahrweiler P., Pyka A., & Gilbert N. (2010). The agent-based NEMO model (SKEIN): Simulating European framework programmes. In P. Ahrweiler (Ed.), *Innovation in complex social systems.* London, UK: Routledge.

Siegel, D.S., Waldman, D., Atwater, L., & Link, A.N. (2003). Commercial knowledge transfers from universities to firms: Improving the effectiveness of university-industry collaboration. *Journal of High Technology Management Research, 14,* 111–133.

Thursby, J., & Kemp, S. (2002). Growth and productive efficiency of university intellectual property licensing. *Research Policy, 31,* 109–124.

Toulmin, S. (1967). *The philosophy of science: An introduction.* London: Hutchinson.

14

LEND ME YOUR EXPERTISE

Citizen Sourcing Advice to Government

William H. Dutton[1]

Introduction: Capturing the Potential of Public Expertise

The diffusion of the Internet and Web has expanded the potential for distributed collaboration (e.g., through sharing documents, contributing comments, and co-creating information), as illustrated by the success of open source software development.[2] A growing number of visionaries see these initiatives as heralding a revolution in how organizations, including governments, function—tapping the wisdom of crowds—the idea that the many are smarter than the few (Surowiecki, 2004; Tapscott & Williams, 2006; Malone, Laubacher, & Dellarocas, 2009). These visions have been defined as *crowdsourcing* and *mass collaboration*. However, the counterintuitive appeal of crowdsourcing can be viewed as highly problematic for officials charged with the responsible management of public bodies in liberal democracies (Coleman & Blumler, 2011).[3] Is this a responsible approach for the public sector?

Based on a variety of case studies and participant-observation of public initiatives, this chapter argues that crowdsourcing, when properly qualified, can be well suited to capturing the value of distributed expertise—often interchangeably referred to as *distributed intelligence*—in support of public policy and regulation. To accomplish this, networks of individuals must be cultivated and managed, not as crowds, but as collaborative network organizations (CNOs). Champions of such initiatives can use a variety of networking platforms and management strategies to develop CNOs that can capture the value of expertise distributed among citizens for citizen sourcing expertise.

This chapter seeks to clarify a workable vision and strategies for tapping distributed public intelligence. Properly understood and managed, distributed expertise presents a significant prospect for the public sector to address critical problems, from bringing individual experts together to manage a crisis, to tapping the *civic intelligence* of the general public (Schuler, 2001).

Approach

This chapter addresses these questions by drawing a number of general reflections from four complementary bodies of evidence. The first is a set of 12 exploratory case studies of distributed problem solving networks in a wide variety of fields, including high-energy physics, biomedical sciences, IT software, and filmed entertainment (Dutton, 2008; Chui, Johnson, & Manyika, 2009). These studies were conducted to provide empirical evidence to ground debate over the performance of CNOs and the merit behind increasing practitioner support of such collaborations. These well-known projects, such as Wikipedia, are associated with peer-produced, distributed problem solving, as well as novel cases that employed different approaches in other areas of application, ranging from scientific collaboration to film production. Although not focused on government initiatives, they illustrate the potential for distributed networks in fields that suggest potential for the public sector.

Second, this research was augmented by a continuing review of case studies of user-driven innovation communities across Europe that were supported by the Danish Research Council, such as a study of community networking initiatives (Zorina & Dutton, 2013). A third set of studies was based on the study of innovations in digital research, which included studies of new forms of collaboration across the disciplines, such as citizen science (Dutton & Jeffreys, 2010; Dutton, 2011). Finally, the chapter was informed by the author's participant-observation in efforts to foster development of networked collaboration in the public sector over several years.

Thus, there is ample evidence that the Internet can provide a viable platform for building CNOs. These networks can perform a variety of functions important to bringing expertise to bear on policy and decision-making. The chapter provides an overview of how networks can be used by governments to harness distributed expertise and outlines lessons learned from early cases, including key opportunities and risks.

The Rise of Collaborative Network Organizations

In academic research, private industry, and civil society, new forms of collaboration have begun to emerge that offer real promise for the public sector. Distributed problem solving networks are composed of a variety of applications that can be classified in numerous ways. They may link individuals within an existing community or organization via Internet-enabled applications that aim to solve particularly complex and novel problems (e.g., addressing *bugs* in software). Or they may be pre-structured by Internet platforms that enable new inter-organizational networks to generate or mine insights gathered from the interaction of distributed actors, such as between medical specialists.[4] Others have classified networks by whether they focus on *information aggregation, prediction model aggregation,* or *problem solving.*[5]

The diffusion of the Internet, Web, and related information and communication technologies (ICTs) has increased the potential to network experts, locally and worldwide. This idea is not new. But recently, it has given rise to the concept of *distributed problem solving networks.*[6]

However, the so-called *wisdom of crowds* is counter-intuitive. On the one hand, one reason is that it has captured the imagination of many—it is interesting. The advantages of tapping distributed intelligence, compared to information systems in formal organizations, include the potential to:

- Improve an individual's judgment by pooling the views of multiple people, provided they have no prejudice and a greater than even likelihood of being correct (de Condorcet, 1994 [1785]);
- Aggregate geographically distributed information and intelligence;
- Enhance diversity, bringing together heterogeneous perspectives and approaches (Page, 2007);
- Enable more rapid diffusion of questions and answers by permitting simultaneous review rather than sequential processing;
- Avoid negative aspects of small group processes, such as *groupthink* (Sunstein, 2004);
- Enable more people to engage in and understand public issues; and
- Support greater independence of and less control by established institutions with rooted interests and ways of thinking (Dutton, 2009).

On the other hand, the wisdom of crowds has also been the subject of much skepticism, particularly if the concept is taken literally. Crowds include multiple people that are often prejudiced, particularly in the political settings of government and public policy, and often have a low probability of being correct—violating basic assumptions behind the value of pooling judgments. However, crowds do not write Wikipedia entries; managed sets of networked individuals contribute to such co-productions (Dutton, 2008). In most case studies of crowdsourcing, a minority of *core participants* is responsible for the majority of activities within the network.

Three Types of Collaborative Network Organizations

Our studies of crowdsourcing began with a focus on distributed problem solving. However, case studies illustrate the wide variety of problems that various networks sought to address. These initiatives often address many different problems simultaneously. This makes it unrealistic to group initiatives or approaches by any specific category of problem, as the choice of network is seldom driven by a simple rational solving of a pre-defined problem. More often, a network such as a wiki becomes a solution space looking for emergent problems to solve.[7]

It was therefore useful to characterize these networks by the activities they supported—categorized by type of CNO—rather than the purposes or problems they addressed, which were many and shifting (Dutton, 2008; Richter, Bray, & Dutton, 2010). Illustrations and examples of each type are provided in Table 14.1.

The first, most common type of CNO focuses on information sharing. Networks were primarily used to transmit and link objects in a distributed network

TABLE 14.1 Three types of collaborative network organizations

Type	Illustrative Application	Example
1.0 Sharing	Using email and shared documents for the design and management of large-scale, distributed projects that share information	The Atlas[8] research project at CERN used email, attachments, and Web-based documents to support collaboration among 1,900 physicists in 37 countries working on a high-energy physics experiment
	Use of shared, viewable databases for coordinating distributed collaboration	The Bugzilla[9] project used a database to track software defects and manage repairs for Firefox and other Mozilla open source software, enabling individual contributions to be allocated
	Broadcast search: networking problem holders and solvers through awards, prizes, and other incentives	InnoCentive[10] uses broadcast search to join solution "seekers" with problem solvers who compete for prizes by generating solutions[11]
	Deep linking and search: enables both documents and data to be linked and searched	Neurocommons[12] enables access to biomedical information through deep searching and natural language processing of open abstracts and datasets (Wilbanks & Abelson, 2010)
2.0 Contributing	Aggregating and prioritizing news content	Digg[13] and other news platforms find, aggregate, rate, and prioritize news
	Sharing insights, information, and opinions among experts in a field	Sermo[14] links licensed physicians in the United States to share information and assist each other and sponsoring organizations
	Predicting outcomes	Information markets[15] are used to aggregate the judgments of individuals to predict public and private events

Type	Illustrative Application	Example
3.0 Co-Creating	Collaboration through massive multiplayer online games (MMOGs)	Seriosity[16] uses MMOG to help prioritize and manage email and information overload
	Open source software development	Firefox developers use the Internet to prioritize key features to produce a more user-friendly version of the Mozilla browser[17]
	Open wiki content creation, allowing users to collaborate—add and edit online content	Creators of Simple Wikipedia use Media Wiki to write and simplify complex text entries in Wikipedia[18]
	Open production of creative artifacts, such as films	A Swarm of Angels[19] uses the Internet for international creator-led collaborative development in making a film

context, thereby reconfiguring access to information. The idea of a linked data web (i.e., the Web) to support more intelligent search, linkage, and retrieval of information further advances the role of such a network in sharing information.

The second CNO focuses on enabling network members to both share and contribute information, employing group and social networking applications of the Web for communication, thereby reshaping who contributed information to the collective group. Efforts to aggregate judgments in information markets or joint rating systems represent typical examples of this type of networking.

The third type of CNO enables members of the network to co-create information. These CNOs facilitate cooperative work toward shared goals, thereby reconfiguring the sequencing, composition, and role of contributors. Wikipedia exemplifies such a network of enabled co-creation of information.

Key features of these types of collaboration networks overlap. For instance, networks enabling user-generated content also exploit the hypertext linkages so valuable in finding and sharing documents. Likewise, cooperative joint collaboration—enabled by Collaboration 3.0—exploits the potential for user-generated content as well as hypertext links, while focusing on the collaborative production of documents or other information products.

The Value of Collaborative Networks in the Public Sector

Can these new forms of collaborative networking be of value to the public sector? This section briefly sketches early efforts to exploit computer-mediated communication for accessing distributed expertise and the ways in which the technical and user context has changed. It also addresses the degree to which electronic communication in the public sector has focused on citizen consultation over citizens as sources of expertise, arguing that such a focus only on consultation could miss an opportunity to use the Internet in new ways.

Electronic Networks of Expertise: A Brief History

Emergence of CNOs is the latest stage in a 40-year thread of initiatives to harness distributed expertise. The RAND Corporation's 1960s development of Delphi techniques (Dalkey, 1967) in forecasting sought to reduce bias created by influential individuals in social dynamics of co-located face-to-face groups of experts. The potential for computer-based communication networks to enable sharing of expertise accelerated the drive toward distributed collaboration in the 1970s with computer conferencing, group decision-support systems, and later initiatives in computer-supported cooperative work.

An early innovation in computer conferencing was driven by the ambition to create a platform to quickly network experts in national emergencies and policy issues. Developed in 1971 at the United States Office of Emergency Preparedness, the system was called Emergency Management Information Systems and Reference Index (EMISARI). Using Teletype terminals linked to a central computer over telephone lines, the system employed many features of contemporary collaboration technologies, including applications for real time chat, polling, and threaded discussion.[20]

The diffusion of personal computers across organizations shifted focus away from computer conferencing to the development of *groupware*, executive boardrooms, and other applications to connect individuals within and across organizations through networks.[21] Various groupware and computer-supported cooperative work projects pursued many of the same objectives tied to the networking of distributed expertise to share and jointly create information products and services.

Today, government convened experts do not need to be available at the same time or in the same time zone to collaborate effectively. The technical platforms to support distributed intelligence are further advanced and tied to a critical mass of users that did not exist in the 1970s, even within academia.

A Transformation in the Communicative Power of Individuals

The Internet enabled the rise of *networked individuals* (Dutton, 2009; Dutton & Eynon, 2009). Through search and social networking tools, individuals can source their own information and build virtual networks within and beyond any given organization. Individuals can decide who should be communicating with whom and mine distributed intelligence within and outside their formal organizations to enhance their performance.

However, truly collaborative organizations can capture the value of distributed intelligence and literally expand and blur the boundaries of the organization, as they mine distributed intelligence. If cultivated within an organization, distributed networking has the potential to get an organization closer to ground level, empowering individuals having real contact with the actual customer, service, or problem to be solved.

Organizations can be more focused on networking their organization than the networking choices and patterns of individual Internet users (Dutton & Eynon, 2009). This focus enforces protection of organizational information systems and traditional institutional networks. It can conflict with new developments, like social networking or crowdsourcing, and lead to some confusion if managers continue to focus only on intra-organizational systems.

Moving Beyond Citizen Consultation to Citizen Sourcing Advice

With some exceptions, governments have historically shied away from fully exploiting distributed intelligence and have tended to focus on engaging networked individuals as citizens, rather than as experts. More recently, attention has shifted to citizen consultation, which has allowed governmental units, like parliamentary committees in the United Kingdom, to obtain public feedback on issues (Coleman, 2004). New Zealand organized an online consultation on the ethical issues of pre-birth testing—a debate that could determine whether children are born.[22] In the United Kingdom, the development of a system for citizens to draft and endorse petitions was an innovative approach aimed at connecting the Prime Minister's Office with citizens.[23]

Citizen consultation is one clear role that the Internet can support in the public sector, but it can be used for more. There has been recent discussion about the potential for crowdsourcing in the public sector through such concepts as *wiki government* or *collaborative democracy*. These imply a move beyond consultation, but may blur distinctions between citizen consultation and expert advice. Noveck has defined collaborative democracy as "using technology to improve outcomes by soliciting expertise (in which expertise is defined broadly to include both scientific knowledge and popular experience) from self-selected peers working together in groups in open networks" (2009, p. 17). This is a useful definition, incorporating both consultation and distributed expertise, but it might be useful to draw a sharper distinction between these two very different roles that networking can play in government.

One role is gauging opinion, which comes closest to collaborative democracy. As an example, it may involve asking citizens to respond to policy options on the basis of their experience. The other role is engaging expertise that may be scientific, technical, or experiential. Citizens are not necessarily experts, but any given citizen may have expertise in a specific area.

The *expert* has long been a critical aspect of governance. Machiavelli's *The Prince*, written in 1513, and present day political scientists have wrestled with the role of experts in governance, particularly in democratic regimes (Benveniste, 1977) where technocrats can be an object of mistrust (Dutton & Kraemer, 1985).

Expertise is embedded in routine practice; governments hire consultants, conduct studies, or build models to advise public officials on particular issues. However, the Internet provides mechanisms to create distributed problem solving networks that can complement and be a substitute for in-house expertise and paid consultants. This tool could garner timely and effective advice in ways that could reduce costs while engaging citizens in new and meaningful ways in the process of governance.

Understanding how to engage and respond to expertise can be as essential as consultation to the vitality of democratic institutions and processes. Supporting consultation is an important role for the Internet. The very legitimacy of decision-making in a liberal democracy depends on government responsiveness to public opinion. Public opinion polls, committee hearings, and consultation exercises are largely geared to understanding the balance of public opinion concerning policies and decisions. New ICTs, such as the Internet, have the potential to enable more direct and frequent patterns of consultation.

The challenge for government will be to avoid airing public issues only to gauge public opinion if citizens are accepted as experts. Another problem is finding relevant experts on the basis of merit and a spirit of voluntarism wherever they live or work. Further, it will be important to facilitate bringing citizen expertise to bear on particular questions in a timely and effective manner.

Using the Internet in Sharing, Contributing, and Co-Creating Public Information

With the Web and subsequent advances, such as linked data (i.e., the Semantic Web), great emphasis has been placed on employing the Internet to better inform the public. There have been initiatives around the world to put public information online and more recent initiatives to open government data for deep linking, search, and reuse.[24]

Wikis in particular have been used for serious purposes, ranging from supporting primary school children who walk to school,[25] to use by military and intelligence communities, including the United States military's Milipedia, which is used to share information across the armed forces. The Office of the Director of National Intelligence, Intelligence Community Enterprise Services, inspired by an essay on wikis and blogs (Andrus, 2004; Thompson, 2006), developed Intellipedia, a wiki comparable to Wikipedia, for sharing information within the United States Intelligence Community. Intellipedia includes information that can be accessed at different levels of classification, from top secret to unclassified.

Additionally, public sector initiatives around the world aim to use the Web, social networking sites, and mobile phones and texting (SMS) to engage citizens in discussion forums, e-consultations, polling, and petition systems (UN, 2010, pp. 83–91). President Barack Obama's Twitter feed has had a large following; it created an opportunity to follow the president day-by-day.

Access to government information and data is important to distributed problem solving. If individuals can obtain government information online and collaborate with others over the Internet, they have the potential to hold government and other institutions more accountable (Dutton, 2009). Even serious online games, e.g., Games with a Purpose (GWAP), are designed to better inform and engage citizens on public policy issues such as the environment or to solve practical problems such as tagging photographs or other images.[26]

Compared to the public sector, development of systems that enable collaboration in everyday life, business, and science, has been far more prominent. These same approaches could bring distributed expertise into governmental processes. Technical advances, such as those embodied in social networking applications, have made this easier by enabling users to rate the comments. Government has not taken quick advantage of these opportunities due to challenges that are different from those faced by business and civil society. The next section provides an overview of the kinds of systems that *are* most prominent within government.

Lessons Learned and Ways Forward

Governments worldwide have been experimenting with ways to use advances of the Internet and Web to engage the public in ways never before feasible, such as through the analysis of big data. There are also inroads being made to bring citizens into government as experts. The Mercyhurst College Institute for Intelligence Studies created its Police Act Review Wiki in 2007 to enable citizens of New Zealand to register and contribute to the revision of this legislation.[27] The United Kingdom Treasury asked the public for suggestions for budget cuts.[28] In this spirit, the United States Patent and Trademark Office (USPTO) initiated Peer-to-Patent and invited members of the public to become community reviewers of patent applications to complement and support the work of patent examiners.[29]

However, there have been numerous failures and dashed expectations—a frequent result of such significant technological, generational, social, and economic change (Tapscott & Williams, 2006). Experiments with collaborative network organizations provide a number of lessons for embarking on this strategy.

Technology: Using Existing eInfrastructures that Support Collaboration

Open source software has contributed a variety of tools for CNOs, such as Media-Wiki, software that can be used to create wiki-based collaborative environments, facilitating a proliferation of such infrastructures. The public uses the Internet as an open platform for distributed collaboration, using wikis and collaborative software, such as Google Docs. The United States intelligence community has an unclassified collaboration space, Intelink-U,[30] a set of Web-based services, tools, and technologies that includes Intellipedia. The current availability of such platforms and tools is one of the major enablers of collaboration in all sectors. An individual or agency does not need to start from scratch.

The Value of Top Management Support

CNOs can develop without top management support. The Internet enhances the resources accessible to networked individuals, allowing them to choose to collaborate with whom they wish to fulfill their own objectives and often enhance their productivity in the workplace without top management support. This will be difficult for managers to stop.

However, if organizations in the public sector wish to foster development of CNOs to solve their problems, top management must create a climate that is supportive of such development, recognize the value of such initiatives, and be visible participants in their use. Networks can be successful until a change of management undermines their support within the organization, such as what occurred with Feet First, a site designed to encourage children to walk to school.

The Importance of Managing Collaboration

The most prominent design feature that emerged from empirical studies of CNOs was reconfigurations to regulate who communicated what to whom and when within the network. This demonstrates that users are seldom viewed simply as "crowds" involved in collaboration, but are controlled in part through architecture and in part through network management via assignment of editing rights and privileges. For example, broadcast search might reach out to millions to find one problem solver, which is the focus of systems like InnoCentive. Likewise, a Wikipedia entry is likely to be written, edited, and updated by a small group of experts in a particular subject area, not a crowd. Most successful CNOs have very strong leaders or champions, able to resolve issues in a timely way and continuously press forward (Cassarino & Geuna, 2008).

A Core Set of Contributors from a Critical Mass of Users

Successful networks are able to build a critical mass of users. This often takes time and work in recruiting members who form a community, feeling a sense of ownership and value within the network. It is possible to bring a problem to an existing community of users. Although many users are necessary, a small minority of core participants often make the majority of contributions to the network (Richter, 2008), conforming to the so-called *power law*, where only a few are most active, with levels of activity quickly trailing off to form the long tail of the distribution; most people make very few contributions. Public initiatives should focus simultaneously on solidifying a similar structure of a core set of users to drive content development and a critical mass of users who follow and occasionally contribute to the network. This can require a cultural change for organizations that do not have a tradition of opening communication to the broader public.

Incentives for Networked Individuals and Networked Institutions

Unlike older management information systems designed top-down to provide management with information about the organization and its parts, CNOs are built by networked individuals to inform and meet the needs of users (Dutton & Eynon, 2009). It is in the process of users sharing, contributing to, and co-creating information and services of value that a system of value is created for the organization. William Heath refers to this as a control shift (Ctrl-Shift).[31] Users innovate for themselves and, in this sense, democratize the information system (von Heppel, 2005, p. 1). Top managers are no longer in as much control, but are in a position to cultivate or kill the development of CNOs (Richter et al., 2010) and should be more open and supportive of giving up control over content and features to empower users and provide them with a genuine sense of ownership.

Motivations behind the Modularization of Tasks

Successful systems often need to be extremely modular in their allocation of work or tasks. For instance, one of the earliest and still successful uses of email has been for broadcast search. A typical query is: "Does anyone know someone knowledgeable about a particular topic?" Questions like "Who knows?" can be ignored by those who do not know the answer and quickly dealt with by those who do know. When a reader of Wikipedia reads an entry and sees a mistake or the need for an update, it is possible to edit the entry and complete the task in a matter of minutes. It is the cumulative contributions of many editors making small contributions that have resulted in the unpredictably successful growth and quality of this online encyclopedia. By keeping tasks modular and easy to complete, it is possible for a multiplicity of individuals with a plethora of diverse motivations to drive collaboration. When the task becomes more comprehensive and difficult, or when you are asking very hard questions, more structured incentives are required, such as offering a prize to the problem solver as with InnoCentive, or remuneration as to a paid consultant.

The Public Sector Meets Collaborative Network Organizations

The Transportation Safety Authority is credited with creating the first United States government blog in 2008 (Noveck, 2009, p. 15), illustrating the significant lag behind other sectors. Rationale for the laggard nature of adopting new communication trends in the public sector include:

1. *Risk Aversion.* Fear of exposure to highly visible failure if few participants engage is likely due to inexperience with platforms.
2. *Concerns over Low Levels of Participation.* Efforts to citizen-source expertise can be successful especially with small numbers of experts.
3. *A Focus on Evidence-Based Policy.* Public policy can be undermined by flaws in evidence supporting decisions, but collaborative networks can provide sufficiently good information to move ahead until better information is available (Nielson, 2011).
4. *Gaming of Outcomes in a Political Arena.* The idea that the many are wiser than the few assumes individual participants have no prejudice and there is a greater than even probability of being correct.[32] Gaming by nominating or encouraging experts that are known to favor a particular solution (Dutton & Kraemer, 1985, pp. 1–20) in a transparent setting can hold actors more accountable.
5. *Control over Communication.* Unauthorized leaks to the press are often costly and potentially set back progress on plans or decision-making (e.g., WikiLeaks). An option is to start small and get managers involved and then grow into other more complex platforms.
6. *Concerns over Civility and a Lack of Expertise.* Several effective incentive-based rules can counter contributors who do not have the expertise valued by a collaborative network, e.g., level of participation allowed (i.e., word count limits) and user rating schemes (Bray, Croxson, Dutton, & Konsynski, 2008).
7. *Committing Politicians and Officials.* Public agencies avoid public consultation, particularly direct polls, to avoid suggesting a public mandate. Tapping advice from distributed experts will result in a more diverse range of expert views and be less threatening than reliance on a single advisory report.

A Way Forward for Government and the Public Sector

There are many uncertainties raised by efforts to capture the potential of distributed expertise. However, there are strong counter arguments to these uncertainties and the potential remains great and relatively untapped. The benefits of distributed intelligence need to be demonstrated through early projects that strategically exploit the lessons learned by other initiatives around distributed intelligence.

First, engaging networked institutions such as governments to support efforts to use distributed intelligence is key. Governmental actors cannot simply legislate or pronounce the existence of a collaborative network organization, but they can do a great deal to frustrate development when they restrict departments or individuals from using tools like blogging in official capacities. Managers and executives have an array of strategies for blocking activities, but they have less capacity to initiate CNOs. Individuals inside and outside of government are instrumental.

They not only choose to participate in distributed problem solving networks, but encourage participation in them throughout the organization.

The motivations can be manifold. Experts might participate to demonstrate their expertise, to gain a reputation, to pursue their passion for a subject, or simply for the joy of contributing. There is no single motive. CNOs need to engage communities of users from which they can draw on the expertise of key members, while keeping costs of participation low. Platforms such as Intellipedia and Wikipedia foster the development of communities of networked individuals. Networked individuals usually participate in several different user communities that cross boundaries of single platforms and organizations. Prominent politicians and government officials have the ability to appeal to these communities for help.

It is also necessary to manage the tensions between these two actors. Networked institutions often value control and secrecy, whereas networked individuals thrive on autonomy and openness. Reconciling these tensions can be approached by establishing a set of policies, principles, and guidelines that are viewed as reasonable to both actors and enable the bottom-up development of initiatives.

It is critical to set policies and rules of order that apply for government CNOs, just as *Robert's Rules of Order* might be applied in public meetings, to ensure that procedures are fair and permit views to be heard (Dutton, 1996). A useful starting point can be found in the policies and guidelines adopted by Wikipedia, including its *five pillars*, which state that participants should "interact in a respectful and civil manner."[33] Ideally, basic guidelines for participation in citizen sourcing activities that are applicable across a wide variety of communities and policy areas should be adopted, even if some communities must tailor them to their particular needs. Potential issues include whether to moderate discussions, who can participate, and whether anonymous contributors can rate the contributions of others.

These policies and guidelines must be negotiated to reassure managers and executives, the networked institutional guardians, while assuring experts they will have sufficient autonomy and visibility to make meaningful contributions. The exact nature of the guidelines is less important than transparency and that they are mutually agreed to from the start. Changing guidelines by the imposition of moderating a forum that was initially not moderated is less desirable than having the right level of moderation in the first instance. That said, all communities have rules that evolve over time.

Summary: Distributed Public Intelligence

Expertise is distributed geographically, institutionally, and socially. It has become a cliché, but no less correct, that not every expert in any given field works for your government or any other single organization. In a multitude of cases across the public sector, expertise is often located closer to a local problem or across the globe—beyond the reach of government officials when and where advice is most needed. This chapter has argued that government can creatively harness the

Internet to tap the wisdom of distributed public expertise and points to a set of challenges, guidelines, and strategies for realizing this potential for networking with citizens, not only as constituents, but also as advisors and experts.

There are many reasons that public officials will cite for not experimenting with innovations in distributed collaboration, but these concerns can be addressed and countered by a strong set of valid reasons for moving forward on initiatives. A range of small but visible projects for tapping the wisdom of distributed civic intelligence could be an incremental step for radically transforming how governments connect with citizens as experts. To get these started, champions need to emerge, individuals who understand that their agency or department is supportive of their use of networking and establishes a basic set of tools, policies, procedures, and guidelines that can be built upon and not reinvented by each initiative. Developing these policies and guidelines, as well as following a few general strategies, such as documenting existing success stories, provides a place to start in citizen sourcing advice to government.

Notes

This chapter originated as a policy brief (Dutton, 2010) prepared for the Occasional Paper Series in Science & Technology, Science and Technology Policy Institute (STPI) in Washington, DC.

1 The author is grateful to David Bray, Michael Chui, Paul David, Jane Fountain, Brad Johnson, James Manyika, Yorick Wilks, anonymous reviewers at STPI, and Kevin Desouza, for their comments on this chapter. The research underpinning this chapter was supported by several projects, including the "Performance of Distributed Problem solving Networks," supported by McKinsey, the Fifth Estate Project at the OII, and a grant from the Danish Council for Strategic Research for the Research Consortium on "The Governance and Design of Collaborative User-driven Innovation Platforms" at the Technical University of Denmark.
2 Weber (2004) documents the success of open source software.
3 In this sense, I would not agree with the thrust of Malone et al. (2009). There is not a strict dichotomy between hierarchy and crowdsourcing, as CNOs are managed networks, albeit not strictly hierarchical in all activities.
4 A definition and rationale for this early classification is provided by David (2007).
5 S. E. Page, "Diversity in Distributed Problem solving Networks." Text of lecture at an OII Forum, Saïd Business School, Oxford, 31 January 2008.
6 This label is derived from terminology of InnoCentive, adapted by Paul A. David, that defined a project at the Oxford Internet Institute, entitled "The Performance of Distributed Problem solving Networks." http://www.oii.ox.ac.uk/research/?id=45
7 This is similar to the "garbage can" model of organizational decision-making in which people in organizations have solutions looking for problems to which they can be applied, such as outsourcing a problem (Cohen, March, & Olsen, 1972).
8 http://atlas.ch/
9 http://www.mozilla.org/bugs/
10 http://www2.innocentive.com/
11 See Lakhani, Jeppesen, Lohse, and Panetta (2007).
12 http://sciencecommons.org/projects/data/background-briefing/
13 http://digg.com/about/
14 http://www.sermo.com/

15 Croxson and Bray (2008)
16 http://www.seriosity.com/
17 http://www.mozilla.com
18 den Besten and Loubser (2008)
19 http://aswarmofangels.com/
20 An overview of EMISARI is provided by OEP (1973) and is a core feature of an innovative book on computer conferencing (Hiltz & Turoff, 1978).
21 Johansen (1988) provides an overview of groupware and other early collaborative tools.
22 http://goodpracticeparticipate.govt.nz.customer.modicagroup.com/levels-of-partici pation/collaborative-processes-and-partnerships/bioethicscasestudy.html
23 http://epetitions.direct.gov.uk/
24 http://opengovernmentdata.org/. Also see Wilbanks and Abelson (2010).
25 http://www.feetfirst.govt.nz/about
26 http://en.wikipedia.org/wiki/Game_with_a_purpose
27 This was part of the Mercyhurst Innovative Use of Wikis Project of the Institute for Intelligence Studies. http://wikispacesintel.wikispaces.com/
28 http://webarchive.nationalarchives.gov.uk/20130405170223/http:/www.hm-treasury.gov.uk/spend_spendingchallenge.htm
29 http://www.peertopatent.org/
30 http://ra.intelink.gov/ provides remote access to Intelink-U, which is content hosted on the network DNI-U, a secure United States Government network maintained by the Intelligence Community Enterprise Services Office (ICES). Intelink-U and DNI-U were formerly an unclassified network, called the Open Source Information System (OSIS).
31 https://www.ctrl-shift.co.uk
32 This is called the 'Jury Theorem' (Condorcet, 1994 [1785]).
33 http://en.wikipedia.org/wiki/Wikipedia:Five_pillars

References

Andrus, D.C. (2004). *Toward a complex adaptive intelligence community: The wiki and blog.* Retrieved from https://www.cia.gov/library/center-for-the-study-of-intelligence/csi-publications/csi-studies/studies/vol49no3/html_files/Wik_and_%20Blog_7.htm

Benveniste, G. (1977). *The politics of expertise* (2nd ed.). San Francisco: Boyd & Fraser.

Bray, D., Croxson, K., Dutton, W.H., & Konsynski, B. (2008). Sermo: An authenticated, community-based, knowledge ecosystem. [Working Paper.] Oxford: Oxford Internet Institute.

Cassarino, I., & A. Geuna, A. (2008). Distributed film production: Artistic experimentation or feasible alternative? The case of a Swarm of Angels. [Working Paper.] Oxford: Oxford Internet Institute.

Chui, M., Johnson, B., & Manyika, J. (2009). *Distributed Problem-Solving Networks: An Introduction and Overview* [OII DPSN Working Paper No. 18]. Oxford: University of Oxford, Oxford Internet Institute. Available at SSRN: http://ssrn.com/abstract=1411739

Cohen, M.D., March, J.G., & Olsen, J.P. (1972). A garbage can model of organizational choice. *Administrative Science Quarterly, 17*(1), 1–25.

Coleman, S. (2004). Connecting parliament to the public via the Internet. *Information, Communication and Society, 7*(1), 1–22.

Coleman, S., & Blumler, J.G. (2011). The wisdom of which crowd? On the pathology of a listening government. *The Political Quarterly, 82*(3), 355–364.

Croxson, K., & Bray, D. (2008). Information markets: Feasibility and performance. [Working Paper.] Oxford: Oxford Internet Institute.

Dalkey, N. C. (1967). *Delphi*. Santa Monica, CA: RAND. Retrieved from http://www.rand. org/pubs/papers/P3704.html

David, P. A. (2007). Toward an analytical framework for the study of distributed problem solving networks. [Working Paper.] Oxford: Oxford Internet Institute.

de Condorcet, M. (1994 [1785]). *Essai sur l'application de l'analyse a la probabilité des decisions rendues a la pluralité des voix*. Paris: l'Imprimerie Royale.

den Besten, M., & Loubser, M. (2008). Resolving simple and contested entries on Wikipedia. [Working Paper.] Oxford: Oxford Internet Institute.

Dutton, W. H. (1996). Network rules of order: Regulating speech in public electronic fora. *Media, Culture, and Society, 18*(2), 269–290.

Dutton, W. H. (2008). The wisdom of collaborative network organizations: Capturing the value of networked individuals. *Prometheus, 26*(3), 211–230.

Dutton, W. H. (2009). The fifth estate emerging through the network of networks. *Prometheus, 27*(1), 1–15.

Dutton, W. H. (2010). Networking distributed public expertise: Strategies for citizen sourcing advice to government. *Science and Technology Policy*. Washington, DC: Science and Technology Policy Institute, Institute for Defense Analyses.

Dutton, W. H. (2011). The politics of next generation research: Democratizing research-centered computational networks. *Journal of Information Technology, 26*, 109–119.

Dutton, W. H., & Eynon, R. (2009). Networked individuals and institutions: A cross-sector comparative perspective on patterns and strategies in government and research. *The Information Society, 25*(3), 1–11.

Dutton, W. H., & Jeffreys, P. (Eds.). (2010). *World wide research: Reshaping the sciences and humanities*. Cambridge, MA: The MIT Press.

Dutton, W. H., & Kraemer, K. L. (1985). *Modeling as negotiating: The political dynamics of computer models in the policy process*. Norwood, NJ: Ablex Publishing Corporation.

Hiltz, R., & Turoff, M. (1978). *The network nation: Human communication via computer*. Reading, MA: Addison-Wesley.

Johansen, R. (1988). *Groupware: Computer support for business teams*. New York: The Free Press.

Lakhani, K. R., Jeppesen, L. B., Lohse, P. A., & Panetta, J. A. (2007). The value of openness in scientific problem solving. [Working Paper No. 07–050.] Cambridge, MA: Harvard Business School.

Malone, T. W., Laubacher, R., & Dellarocas, C. (2009). Harnessing crowds: Mapping the genome of collective intelligence. [MIT Sloan Research Paper No. 4732–09.] Cambridge, MA: Center for Collective Intelligence, Sloan School of Management, MIT.

Nielson, M. (2011). *Reinventing discovery: How online tools are transforming science*. Princeton: Princeton University Press.

Noveck, B. S. (2009). *Wiki government: How technology can make government better, democracy stronger, and citizens more powerful*. Washington, DC: Brookings Institution Press.

OEP. (1973). EMISARI: A Management Information System Designed to Aid and Involve People. Paper presented at the 4th International Symposium on Computers and Information Science. Miami Beach, Florida: OEP.

Page, S. E. (2007). *The difference: How the power of diversity creates better groups, firms, schools, and societies*. Princeton: Princeton University Press.

Richter, W. (2008). Intellectual property law and the performance of distributed problem solving networks (DPSNs). [Working Paper.] Oxford: Oxford Internet Institute.

Richter, W. R., Bray, D. A., & Dutton, W. H. (2010). Cultivating the value of networked individuals. In J. Foster (Ed.), *Collaborative information behavior: User engagement and communication sharing*. Hershey, PA: IGI Global.

Schuler, D. (2001). Cultivating society's civic intelligence: Patterns for a new 'world brain.' *Information, Communication and Society, 4*(2), 157–181.

Sunstein, C.R. (2004). *Infotopia: How many minds produce knowledge*. New York: Oxford University Press.

Surowiecki, J. (2004). *The wisdom of crowds: Why the many are smarter than the few and how collective wisdom shapes business, economies, societies and nations*. New York: Doubleday.

Tapscott, D., & Williams, A. D. (2006). *Wikinomics: How mass collaboration changes everything*. New York: Penguin.

Thompson, C. (2006, December 3). Open-source spying. *The New York Times Magazine*. Retrieved from http://www.nytimes.com/2006/12/03/magazine/03intelligence.html?pagewanted=print

UN. (2010). *United Nations e-government survey 2010*. New York: Department of Economic and Social Affairs, United Nations.

Von Heppel, E. (2005). *Democratizing innovation*. Cambridge, MA: Cambridge University Press.

Weber, S. (2004). *The success of open source*. Cambridge, MA: Harvard University Press.

Wilbanks, J., & Abelson, H. (2010). 'Open access' versus 'open viewing' for a web of science. In W. H. Dutton & P. Jeffreys (Eds.), *World wide research: Reshaping the sciences and humanities* (pp. 322–324). Cambridge, MA: The MIT Press.

Zorina, A., & Dutton, W.H. (2013). Building broadband infrastructure from the grassroots: The case of home LANs in Belarus. *Journal of Community Informatics, 10*(2).

PART V

Governance Infrastructure

15

SYNTHETIC INFORMATION ENVIRONMENTS FOR POLICY INFORMATICS

A Distributed Cognition Perspective

Christopher Barrett, Stephen Eubank, Achla Marathe, Madhav Marathe, and Samarth Swarup

Introduction

Policy making is a distributed cognition problem.[1] It almost always involves multiple social, economic, and infrastructure systems, as well as multiple stakeholders with different views of the problem and different goals. In addition, the systems involved change rapidly and are influenced by the collective actions of large numbers of people. Thus, any policy action can have cascading consequences that uniquely affect different stakeholders.

Despite all of this, a policy problem can be viewed sufficiently abstractly to conceive of it as a single problem. Consider the problem of responding to and containing an epidemic. We shall take this as a running example in this chapter. When an infectious disease outbreak occurs, society must respond to contain or eliminate it. This single statement encapsulates what needs to be a complex and distributed, yet coordinated and adaptive, response from many individual and institutional actors. It also enables us to see it as a single cognitive problem. It is a cognitive problem because the policy makers are situated in a perception-action feedback loop, which consists of continuously monitoring the status of the epidemic and adaptive implementation of policies to minimize the outbreak. However, it is a distributed cognition problem because no single policy maker or other stakeholder has access to all of the information, nor can they single-handedly take action. Policy making and implementation is carried out through the collective reasoning and action of a large network of people who are spatially distributed and may have limited communication with each other.

Comfort (2007) frames crisis management as a complex adaptive system and posits that a cognitive frame is necessary to understand the interaction between communication, coordination, and control. She notes that "without cognition, the other components of emergency management remain static or disconnected and, as shown by the record of operations during Hurricane Katrina, often lead to cumulative failure." (p. 189)

It is commonly said that to process the large amount of data and information involved in policy making, an informatic approach is required. We argue here that this is not sufficient. The approach must be one that enables and facilitates distributed cognition. The goal is not simply to get stakeholders talking to each other. They already do that. The goal, therefore, is to make their decision-making processes interdependent in the right way.

We do this by building a synthetic information environment (SIE), as described in Box 15.1. This is a simulation environment that dynamically combines data from multiple sources and, when used for decision-making, creates a coherent set of views for different stakeholders, thereby enabling coordination and distributed cognition.

An SIE can anonymize individual elements (people or otherwise) so that they are not identifiable, if so desired. This is important for anonymity, privacy, and security. In addition to nominative data, individual elements as well as appropriate sub-systems are assigned numerical, declarative, and procedural information, resulting in a realistic and complete representation of the social information

---

BOX 15.1

Synthetic Information Environment (SIE) Defined

A Synthetic Information Environment (SIE) is a collection of software systems, data sets, and protocols that provides policy informatics with:

- Adaptability through support for multiple views and optimization criteria for multiple stakeholders;
- Extensibility through the ability to incorporate multiple data sources;
- Scalability through the ability to model very large, interacting networked systems; and
- Flexibility by allowing evaluation of a large class of possible interventions and policies.

An SIE of a given socio-technical system is built by combining data from multiple sources of information, including the Census, infrastructure elements, time-use data, and geo-spatial data, to create a disaggregated model of the population for a region of interest. This "synthetic" information preserves the structural and causal relationships between interacting elements that constitute the underlying socio-technical system.

environment. SIEs have been used to study topics ranging from the spread of epidemics, to human behavior in the aftermath of a nuclear explosion.

Viewing the problem as one of distributed cognition changes the focus of analysis from individual policy makers to a system in which policy makers are integrated. It brings to the fore the problem of developing and maintaining a coherent and consistent picture of the ongoing epidemic (or other policy problem) among a team of decision-makers, even if they are separated in space and time. If an SIE is to be a tool for creating distributed cognition, it must be integrated into the cognitive loop of policy makers. When multiple policy makers are part of the same SIE, it effectively induces an overlap in their cognitive systems and provides a structured and consistent state. In this chapter, we elaborate on this perspective.

In this chapter, we describe the complexity and coupling between behavior and infrastructure during an epidemic and argue that policy making in this context can be abstracted as a cognitive problem. We describe its distributed nature and the constraints induced by this perspective. Then we describe our attempts to address this problem using synthetic information technology and two case studies in which our system was used: in the first instance for pandemic response training and in the second for actual pandemic response during the H1N1 outbreak of 2009.

An Epidemic as a Complex System

An epidemic, as with most policy problems, is a temporally extended event and often, by the time the problem is recognized, it is too late to prevent it entirely. For example, in the H1N1 "swine flu" pandemic of 2009, although there is evidence that it had been spreading several months before, the first case was not diagnosed until March 17, 2009, in Mexico. From that point on, the virus spread very rapidly. In the United States, the first case was reported in California toward the end of March. In April, authorities in Mexico City began implementing control measures, closing down schools and public places, as the severity of the outbreak was recognized. The same strategy was followed in Texas, although data about transmissibility and mortality rates were still unavailable. It was early May when the first epidemiological evaluation of the outbreak was published (Fraser et al., 2009). By then, it was too late to attempt to prevent the rapid spread of H1N1 around the world. By June, the size of the outbreak was large enough that the World Health Organization (WHO) raised their pandemic alert to phase six ("global pandemic").

The question of how to most effectively respond to pandemics such as H1N1 is complicated, involving public health systems, regional and urban population dynamics, economic effects, critical infrastructure availability, and public policy. An integrated decision-support system in which all of these factors can be studied simultaneously is needed to identify effective response strategies.

It is well understood that planners must take individual behavior into account when preparing for crises. However, it is not as well appreciated that social responses to public policy can significantly impact the efficacy of public policy and disaster response (Sterman, 2006). In the context of a large-scale crisis, when reactions may produce very unusual social contact patterns, it is exceedingly complicated to predict and prepare an optimal response (Comfort, 2007). Human response, public policies, and specific crisis situations are intricately intertwined with one another, making it impossible to obtain a clean, simple formal model and solution.

Uncertain consequences and conflicting motivations between micro and macro levels for individuals, federal, state, and local authorities are at the heart of issues such as non-compliance with public policy and, more generally, breakdown of the rule of law in society. The examples, although complex, are amenable to analysis.

How can mathematical models help in formulating and evaluating policies for such complex problems? Developing models for such complex systems that aim to make a point prediction (e.g., exactly how many people will be sick on some day) is of very little use. On the other hand, the use of models for decision support and policy making as an aid in planning, analyzing counter-factual experiments, promoting compliance, and targeting response strategies with limited resources are much more reasonable goals. Over the last 100 years, researchers have developed sophisticated mathematical and computational models aimed at supporting many of these objectives (e.g., Hethcote, 2000; Eubank et al., 2004).

Models must be capable of taking information about the state of the epidemic as input and analyzing the effect of one or more policy decisions on the course of the epidemic. If the models are able to provide a real-time or better response, they can be embedded in the decision cycle of policy makers. We refer to this decision cycle as the measure-project-intervene cycle, but it is not specific to our approach. It is an abstraction of the process followed in all policy making. There are many advantages, however, to using explicit computational and/or mathematical models (Epstein, 2008). Primarily, they make assumptions explicit, can be much more complex than intuition, and allow rigorous testing.

In the measure-project-intervene cycle, the measure step consists of performing surveillance to estimate the current state of the epidemic. The project step consists of using models to estimate the possible future evolution of the epidemic. The intervene step consists of implementing policy to control the epidemic. The effectiveness of this approach depends on the models used in the projection step.

To complete the loop, feedback from the measurement step should be incorporated into models to create a projection for the next step. A policy or intervention, therefore, must be an adaptive response to the state of the problem. In the case of the H1N1 pandemic, for example, the initial policy to close schools temporarily was relaxed during the later phase of the pandemic due to a better understanding of H1N1 and feedback from the first round of school closures.

Policy Making as Distributed Cognition

To understand how computational tools can enable more efficacious responses to complex policy problems, we develop a distributed cognition perspective. There are three essential points we wish to make here: (a) mental models are susceptible to systematic errors when faced with complex problems; (b) if properly designed, computational tools can function as cognitive augmentation to address these problems; and (c) our approach, in particular, facilitates interaction between multiple policy makers in a consistent manner by unifying them into a single, distributed cognitive system.

Mental Models and Cognitive Errors

It is well accepted that cognition involves making mental models of the environment and the self (Neisser, 1967; Picton & Stuss, 1994; Grush, 2004). These include models of dynamic processes (Gentner & Stevens, 1983) and are based on a combination of experience, intuition, and formal instruction. For instance, Vosniadou and Brewer (1992) showed that children's mental models of the shape of the earth are based on their intuition and experience, modified by what they learn in the classroom. They found that children often start out with a rectangular-earth or disc-earth model. When told that the earth is round, they modify this naive model in various ways, such as by assuming that the earth is a hollow sphere and we live on a flat surface deep inside.

Thus, new information is incorporated into an existing mental model instead of replacing it. This may lead to subtle errors in reasoning and planning because the inconsistencies in the mental models only come to light upon detailed examination. In the above experiment, the children themselves were unable to completely verbalize their mental model. On being asked about the shape of the earth, most would claim it to be round or spherical. However, when asked detailed questions like what would happen if we were to keep walking in a straight line, some responded that we would reach the end of the earth, but would not fall off because we are inside the earth, thus implying a mental model of a hollow earth.

Similarly, people make systematic errors in their mental models of dynamic processes. For example, Sterman (1989) studied decision-making in a simulated "beer distribution game," which involved inventory management in the face of uncertain demand. He found that in the face of complex dynamics, subjects tended to adopt an "open-loop" mental model, which can hinder learning and efficient decision-making. These and other cognitive and neurological studies have been formalized into mathematical models based on predictor-corrector filters (such as Kalman filters) (Rao, 1999; Grush, 2004).

This model assumes that the brain maintains an internal estimate of the state of the world and is constantly generating predictions about observations through a forward model. These predictions are compared with actual observations and are used to "correct" the estimate of the state and generate new actions. Effective performance depends on the quality of the internal forward model and how feedback is incorporated into it. This highlights the importance of updating the forward model and the controller based on feedback.

However, as we see from the discussion above, relying solely on mental models can be hazardous, especially when dealing with complex socially coupled systems. People either fail to utilize feedback when faced with complex dynamics or try to modify pre-existing erroneous models to incorporate new information. The latter is especially difficult to determine because such conceptual errors are not immediately obvious.

This is where computational and mathematical models become very important. If built correctly, they can incorporate nonlinearities, delayed effects of actions, and complex dynamics, and can be rigorously tested to analyze their phase space, their sensitivities to parameters and initial conditions, and the variance due to stochasticity. Thus, these models can extend the cognitive abilities of policy makers and enable them to adapt more effectively to changing circumstances.

Extended Cognition

It is important to distinguish this extended cognition from simple, temporary tool use. Clark and Chalmers (1998) discuss at length the conditions that must apply for a cognizer + artifact to constitute an extended cognitive system. They rely in part on the notion of epistemic actions (Kirsh & Maglio, 1994), which are actions performed to aid cognitive processes as opposed to actions performed to change the environment to a preferred state. Kirsh and Maglio (1994) studied subjects playing the computer game Tetris and showed that the subjects' physical rotation of blocks on the computer screen were faster than they could perform mental rotations of the same blocks and that, while playing the game, the subjects use these physical rotations instead of mental rotations.

Thus, some cognitive activity (rotation of blocks) has been offloaded to the physical system. These actions are desirable for the information they generate (where the blocks will fit) and not for any intrinsic benefit in the Tetris game to rotating the blocks while they are falling. Therefore, these are epistemic actions.

Under certain conditions, Clark and Chalmers (1998) argue that it is natural to see the cognitive system as extending to encompass the physical artifacts that are being used to perform epistemic actions. They discuss an example of an Alzheimer's patient using a notebook to keep track of information. They argue that this person + notebook should be considered a single, extended cognitive system because it satisfies four conditions, which we paraphrase below and summarize in box 15.2.

First, the artifact (notebook) is a constant for the particular cognitive activity (remembering). Second, the information provided by the artifact is readily available. Third, the information obtained from the artifact is automatically endorsed

---

BOX 15.2

Extended Cognition and Distributed Cognition

Extended Cognition (Clark & Chalmers, 1998): An individual and an artifact (e.g., an Alzheimer's patient and the notebook he uses to keep track of information) are said to constitute an extended cognitive system when:

- The artifact is a constant in the individual's life for a particular activity.
- The artifact makes information readily available to the individual.
- The individual automatically endorses information obtained from the artifact.
- Information was endorsed when it was put into the artifact.

Distributed Cognition: Multiple individuals and artifacts constitute a distributed cognitive system when they collectively satisfy the four conditions noted above.

---

on retrieval. Fourth, the information obtained from the artifact must have been endorsed in the past and put there as a consequence. If these conditions are all met, the notebook occupies exactly the same position in the life of the Alzheimer's patient as (internal) memory does for a person not afflicted with Alzheimer's.

If we accept their thesis, it has multiple implications for the design and use of a tool for policy informatics. First, the tool must be designed in a way that makes the relevant information *readily available*. This involves both good software and interface design and good training of users. Second, users must trust the output of the tool so that they *automatically endorse* its results. Third, the tool must be integrated into the cognitive loop of the policy makers, so that it is *a constant* for the activity of policy making. Fourth, they must have some understanding of the assumptions in the models and the technology behind the tool, i.e., *endorsement in the past*.

In reality, the picture is much more complicated because multiple stakeholders are involved in both policy planning and policy implementation. The cognitive abstraction is helpful in developing a picture of the overall process. We can now unpack the effect of multiple actors by taking a distributed cognition view.

Distributed Cognition

In epidemic policy, as in other policy spheres, decision-making is divided between federal and local levels. Although decisions such as how much vaccine to manufacture and when, where, and to whom to make it available are centralized, decisions such as when and for how long schools should be closed are made at the local level (based on federal recommendations).

At both levels, decisions may be affected by factors other than the epidemic itself. For example, decisions about vaccine manufacture may be affected both by cost and distribution logistics. Decisions about school closure may be affected by how many children depend on school-provided lunches and how many class hours would be lost.

These interdependencies among the epidemic, infrastructure, and socio-economic systems not only make decision-making complex and distributed, they result in an implementation of policy that is different from what is intended. In addition, policy implementation requires interpretation of directives within the local context. This can often mean that even if there is complete "buy-in" at the local level about federal policy recommendations, implementation can still turn out other than intended as local authorities discuss, interpret, and apply directives. This has been analyzed from a distributed cognition perspective in the realm of education policy by Spillane, Reiser, and Gomez (2006). They found that policy implementation requires extended "sense-making" activities that are carried out through discussions and are influenced by actual classroom experiences in an iterative fashion. Thus, the distributed cognition perspective leads to new insight by changing the focus of analysis.

In this view, individual cognition is not constitutive of distributed cognition; rather, it is constituted by it. The unit of analysis is the system of practice, not the individual (Hutchins 1995, 2001). To make sense of cognitive activity that is distributed across multiple individuals and their artifacts, it is necessary to understand how information is represented, communicated, and used in this system.

In contrast to Hutchins (2001), we take a narrower view based on the extended cognition perspective discussed above. In our view, a system of multiple individuals and artifacts constitute a distributed cognitive system precisely if it satisfies the same four conditions, but with information distributed across all individuals and artifacts instead of just one individual and one artifact (Box 15.2).

Essentially, it is the difference between a group of people and a team. A group of people can interact with each other, but do not have a clear division of cognitive activity, anticipate the actions of each other, or work in a coordinated manner toward some goal.

Optimizing team performance has been the subject of research for many decades and the role and importance of shared mental models in team performance is recognized (Cannon-Bowers, Salas, & Converse, 1993; Artman & Garbis, 1998). However, attention has focused on social and interpersonal interactions between team members and effects on performance.

The role of distributed representation on problem-solving performance has received relatively less attention, although Zhang (1998) shows that when decision-makers only have partial views of a system, there can be a net performance loss because they must expend time and effort on communication, cross-checking, and constraint propagation. In such a situation, a team can actually perform worse than an individual expert.

In the case of epidemic policy, different actors (and different institutions) have access to different sources of data and make decisions under different constraints. For example, policy makers at the United States Department of Health and Human Services (DHHS) may not have access to all data available to epidemiologists at the CDC, and may not need it. The epidemic system, in any case, is too complex for a single individual to be an expert in all aspects of the problem. Thus, we are always in a situation where multiple policy makers have only partial views of the problem. Therefore, different groups must be able to coordinate their decision-making so they do not work at cross-purposes.

An informatic approach to this policy problem must have the ability to integrate information from multiple sources and present each group with a relevant view. Such a system would provide a substrate for distributed cognition, allowing individual actors to be separate in space and time, but still tie them together as a team.

Use of an explicit computer model as the forward model allows multiple stakeholders to access a common representation and expected outcomes of interventions along the dimensions most relevant to them.

For several years now, we have been developing SIEs to accomplish this objective. These tools implement complex network science-based models of epidemics on high-performance computing architectures, but keep these details hidden from users, presenting only a clean, easy-to-use interface that allows epidemiologists to construct experiments to evaluate various interventions based on the most current data available. They reduce or eliminate the necessity for explicit cross-checking and constraint propagation between different stakeholders by carrying out these processes automatically and, if necessary, opaquely. When multiple policy makers use an SIE as part of their extended cognitive system, it creates an overlap in their cognitive states, effectively integrating them into a single, distributed cognitive system. Our approach aims to facilitate distributed cognition by providing better than real-time evaluations so that policy makers can use them as part of their decision-making loop. The models used are capable of representing interventions flexibly, which can account for implementation differences on a location-by-location basis. We next describe how we construct these synthetic information environments, and how they have been used for policy making during epidemics and other situations.

An Approach Rooted in Synthetic Information, Interaction Modeling, and Multi-Perspective Decision-Making

In this section, we describe an approach developed by our group over the past 15 years that is actively used to support public and defense policies as they pertain to large-scale crises and is embodied in the Comprehensive National Incident Management System (CNIMS).

As a first step, we develop synthetic multi-theory multi-layered (MTML) networks. The coupled multi-networks are composed of social contact networks that serve as the substrate for disease transmission, social friendship networks, and information and economic networks that facilitate information dissemination and economic activity as it pertains to epidemics. This MTML network synthesis is achieved using a first principles approach that combines diverse sources of data and information with well-accepted social and behavioral theories to synthesize MTML networks (Monge & Contractor, 2003; Barrett, Eubank, & Marathe, 2006; Barrett et al., 2009; Barrett, Bisset, Leidig, Marathe, & Marathe, 2010; Barrett et al., 2011).

Available observations and knowledge are normally not structured specifically for a particular question. We overcome this data problem by using available, sometimes imperfect, information in the form of data and procedures to synthesize an integrated representation of what is known in the context of the decision to be made. Additionally, the synthetic data created (e.g., population, contact networks, activities) by our approach have the following features: (1) they are statistically equivalent to real data; (2) they are anonymous, which helps overcome issues related to human subjects; (3) they are comprehensive and provide justification for certain kinds of policy decisions; (4) their components can be replaced with real data as available; and (5) they represent integrated interaction-related information from multiple sources. It is important to note that these networks are not available explicitly—we assert that it is simply impossible to construct representations of such networks based solely on measurements.

In the second step, we develop high-performance computing enabled models to study dynamical processes on the MTML networks synthesized in the first step. Traditionally, researchers have focused on epidemic reaction diffusion processes. But, as we discussed in the introduction, planning for and controlling epidemics requires understanding not only disease propagation, but public policies and intricate behavioral responses as well. We have developed a system called Simdemics to support this. It contains models of various reaction diffusion processes pertinent to supporting public health epidemiology.[2] The final step involves overlaying a decision support environment that allows decision-makers to carry out what-if experiments. The pervasive computing-based environment allows decision-makers to design experiments and provides analysts with automatic methods for ranking interventions and studying interaction effects, contrasts, and main effects. The SIE system is schematically illustrated in Figure 15.1, constructed through a multi-step process that combines information from multiple unstructured sources, driven by a query and context. The resulting information structure supports multiple stakeholder (policy maker) views and maintains constraints between them.

Different stakeholders or policy makers can have restricted and partially overlapping views of the system as desired. The SIE system maintains coherence between their views and updates structures as new information becomes available. It acts as cognitive augmentation by allowing users to offload the cognitive activity of forecasting onto the system. Because the same system is used by all the

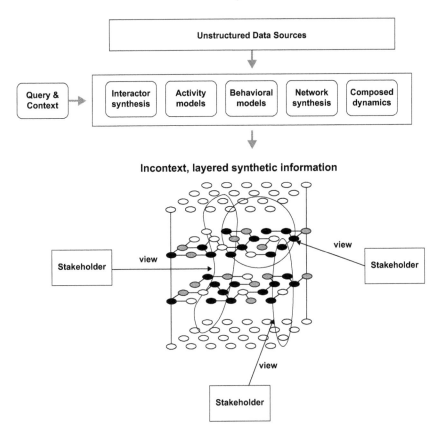

FIGURE 15.1 A schematic illustration of a Synthetic Information Environment

stakeholders, it draws them into a single, distributed cognitive system without the need for explicit communication, constraint propagation, or negotiation.

Applications

In our final section, we outline two recent epidemiological studies/exercises that were conducted using the synthetic information technology discussed above. Each of these studies highlights the role of realistic modeling environments in developing and analyzing public and military policy. Policies to control epidemics and maintain force readiness were evaluated in these dynamically evolving scenarios.

Unified Combatant Command Pandemic Study

This work was conducted as part of an exercise hosted by a Unified Combatant Command to prepare defense decision-makers for the information and response environment likely to be encountered in future influenza pandemics (Barrett,

---

BOX 15.3

How SIEs Enable Distributed Cognition

Synthetic Information Environments (SIEs) enable distributed cognition in the following ways:

- They provide rigorous models that can incorporate nonlinearities, delayed effects of actions, and complex dynamics. They can be rigorously tested to analyze their phase space, their sensitivities to parameters and initial conditions, and the variance due to stochasticity. Thus, an SIE extends the cognitive capabilities of a policy maker by allowing him to partially offload the cognitive activity of reasoning about a complex system.
- If well designed, they can be incorporated into the mental life of policy makers to extend their cognitive abilities. This requires designing a flexible and intuitive user interface, appropriate user training, and embedding the SIE into the cognitive loop of policy makers.
- When the same SIE is used by multiple policy makers, it provides a consistent (although not necessarily identical) state for all of them, even if they are concerned with different aspects of a problem. This means that the cognitive systems of multiple policy makers can be brought into concordance, facilitating distributed cognition.

Beckman, et al., 2011). This tabletop exercise sought to provide participants with a realistic course of events, information flows, and stakeholder interests that would be involved during a nationwide spread of influenza. Specifically, we constructed and ran a multi-scale simulation of the spread of influenza, which served as the *ground truth* of the event, guided the exercise scenario, and informed the exercise white team. The simulation was run for a total of 300 days and illustrated how influenza spreads from the west coast to eastern parts of the United States. Our high-resolution models enabled enhanced situational assessment and leveraged surveillance information available to local public health authorities.

Using synthetic surveillance system techniques, an estimate of what surveillance signals the epidemic would create was calculated. These results were interpreted from several different stakeholder perspectives. Additionally, the nation-wide simulation was used to run simulations on a more localized and highly detailed scale, enabling more realistic estimations of the epidemic's impact on realistic areas of potential interest.

The actual state of the epidemic, surveillance reports, detailed assessment of the impacts on the areas of interest, the interpreted actions of public health and United States Department of Defense (DoD) officials, and simulation support events were all organized on a single timeline. This timeline was used to drive the tabletop exercise. Its interactive display concisely conveys multiple levels of information tailored to the particular interests and influence of different stakeholder communities, while enabling a vision of the overall progress of the exercise.

Requirements defined during support of this exercise led to several novel uses of existing analytical tools and techniques and to significant advances in very large scale, highly detailed social epidemiological simulation technology. This marks the first time that knowledge of existing surveillance systems was applied to simulation results to provide a realistically obscured situational awareness. This enhanced the realism for participants. The exercise required integration of simulations on different scales (nation-wide and regional). Technologically, this had not been done before and led to the development of a framework to enable this linkage. The use of the timeline to organize data from different sources and actions of different stakeholders, while presenting a unified view of the events, was important for the success of the exercise.

Federal Interagency Support during the Emergence of H1N1 Influenza in April and May 2009

This effort was in direct response to the initial reports of the emergence of the H1N1 influenza virus that eventually caused a global pandemic. At first, infections were confirmed in Mexico, California, and Texas and, a few days later, New York. The rapid spread combined with initial overestimates of its mortality rate raised serious concerns of a repeat of the 1918 influenza pandemic. Initial reports about the disease characteristics were unreliable, with wide variations placed on important parameters like proportion of symptomatic individuals and duration of infectious period. Having developed DIDACTIC, a Web-based front-end to our SIE for just this purpose, we were able to quickly run a series of studies to learn the likely impact of parameter variation on the United States population. A quick report was drafted about the impact of disease characteristics on the size and shape of the expected epidemic curve. Several variants of disease models were added to the DIDACTIC tool.

As H1N1 influenza continued to spread in the United States, DHHS teamed up with the Defense Threat Reduction Agency to place the DIDACTIC tool in the hands of government analysts to provide day-to-day modeling results. At a daily interagency meeting at DHHS led by the Office of the Assistant Secretary for Preparedness and Response (ASPR) and the Biomedical Advanced Research and Development Authority (BARDA), analysts and policy makers discussed overnight projections of the pandemic and possible interventions. These projections were generated using DIDACTIC as one of two available tools that were capable of rapidly providing detailed forecasts and analysis.

This integration inside the 24-hour decision cycle running the federal government's response to this emerging crisis would not have been possible without the development of highly optimized modeling software and the Web-enabled interface. The analysts at DHHS were able to perform course-of-action analysis to estimate the impact of closing schools and workplaces.

DIDACTIC, in conjunction with other modeling tools, was used to study a number of important policy questions, including: (a) the role of anti-virals as prophylactic drugs to reduce the severity of the pandemic and delay its peak, (b) the effect of school closures on slowing the spread, and (c) the role of travel restrictions, especially international travel restrictions.

Officials were able to use modeling tools such as DIDACTIC to evaluate these scenarios. This experience demonstrates the importance and feasibility of placing sophisticated modeling tools in the hands of public health decision-makers and highlights the role that highly detailed modeling can play in the process of responding to an emerging crisis. A distributed cognition viewpoint is important here.

We continued to interact with BARDA officials as a part of the National Institute of Health (NIH) MIDAS project to support pandemic response as the pandemic continued to unfold. During the later stages of the pandemic, a rapid-turnaround study was undertaken to understand what conditions would need to occur to create a "third wave" of the pandemic, as had been experienced in previous pandemics (1918, 1958). BARDA was charged with procuring the vaccines and ensuring appropriate distribution, making the prospect of a third wave particularly relevant to their activities.

The study (Eubank & Lewis, 2009) showed that with significant increases in pathogen transmissibility and an underestimation of the current number of infected individuals (or immunized through vaccination), a large third wave could be created. Alternatively, if the pathogen were to significantly change its antigenicity, thus weakening the level of immunity people had achieved through infection or vaccination, and increased in transmissibility as well, a large third wave could occur. It did not seem likely that any one of these changes alone would create a third wave of significant size. This study and others helped alleviate concerns of a third wave and allowed DHHS to refocus their efforts.

Together, these studies demonstrated the utility of SIEs and associated high-performance computing-oriented pervasive modeling environments for situation assessment and public health logistic problems during an ongoing world-wide crisis.

Conclusion

We have described a synthetic information-based approach for policy making and analysis as it pertains to public health and military preparedness.

The creation and implementation of policy is a complex, authorized, meta-decision-making process. Policies, institutions, individuals, and associated common resources (e.g., use of societal infrastructures) interact as composed processes and constraints. The dynamics are complex throughout the composed system and of themselves produce new properties and constraints. Socially coupled policies for such areas as public health influence and are influenced by the collective behavior of millions of individuals who participate, interact, and respond to policy plans and interventions. Due to their complexity and interdependencies, these policies can have cascading, unexpected consequences that are often beyond the unaided cognitive capacity of an individual to understand.

Advances in computational and communications technology, network and algorithm science, complex system dynamics, and cognitive science are creating a revolutionary change in these sorts of distributed, socially coupled system dynamics. In such settings, the ideas of an objective observer with access to a representational "world in a bottle," a single "decision-maker" with purview over that world, and a compliant population become ever more obsolete. Our approach thus involves improving understanding of and producing decision support environments for complex, distributed, socially coupled real world decision systems. Here we have developed a distributed cognition perspective to explain our methodology.

There are multiple aspects to the policy making problem that are addressed by our approach. First, public health as a policy domain is very complex due to the scale of the system, its nonlinearities, feedback loops, and social coupling. Mental models are unable to deal with the scope of such domains and can often contain subtle, unspoken assumptions and errors. SIEs externalize the models and can account for their complexity and stochastic variability and can act as cognitive augmentation by allowing users to offload the cognitive activity of projecting the consequences of interventions in such a complex system.

Second, by virtue of the clean and intuitive interface design of DIDACTIC and its technologically enabled quick response time, the SIE can be integrated into the decision cycle of policy makers, as demonstrated by ASPR and BARDA analysts during the H1N1 pandemic response.

Third, an SIE is capable of maintaining different views for different stakeholders and internally ensuring coherence between these views. This reduces the need for explicit communication and constraint propagation between stakeholders, primary factors affecting team performance when representations are distributed. Thus, an SIE draws multiple stakeholders into a single, distributed cognitive system by inducing an overlap between their cognitive states. This helps make decision-making processes interdependent and reduces the likelihood of stakeholders working at cross-purposes.

Our systems have been used by multiple government agencies, both for training and response. This process has taught us that policy informatics is not simply a matter of organizing information; it must also take into account the cognitive processes of policy makers and stakeholders. The perspective presented here motivates the need for developing pervasive synthetic information architectures and provides a natural way to understand distributed decision and policy making.

Acknowledgments

We thank our external collaborators and members of the Network Dynamics and Simulation Science Laboratory (NDSSL) for their suggestions and comments. This work has been partially supported by NSF HSD Grant SES-0729441, NSF PetaApps Grant OCI-0904844, NSF NETS Grant CNS-0831633, NSF CAREER Grant CNS-0845700, NSF NetSE Grant CNS-1011769, NSF SDCI Grant OCI-1032677, DTRA RD Grant HDTRA1-09-1-0017, DTRA Grant HDTRA1-11-1-0016, DTRA CNIMS Contract HDTRA1-11-D-0016-0001, NIH MIDAS Grant 2U01GM070694-09, NIH MIDAS Grant 3U01FM070694-09S1, NIH Grant 1R01GM109718, and the George Michael Scholarship from Lawrence Livermore National Laboratory.

Notes

1 Informally, distributed cognition refers to the coordinated cognitive activity of multiple individuals, and analysis is focused on the system of practice, not just the individuals (Hutchins, 1995, 2001).
2 See *Bisset, Chen, Feng, Vullikanti, and Marathe (2009)* and *Barrett et al. (2010)* for details on model description. See *Chen, Marathe, and Marathe (2010)* on how such a system can be used.

References

Artman, H., & Garbis, C. (1998). Situation awareness as distributed cognition. In T. Green, L. Bannon, C. Warren, & J. Buckley (Eds.), *Proceedings from the 9th Conference of Cognitive Ergonomics* (pp. 151–156). Limerick, Ireland. Retrieved from http://www.diva-portal. org/smash/record.jsf?pid=diva2:479674

Barrett, C., Beckman, R., Bisset, K., Chen, J., Eubank, S., Feng, A., ... Swarup, S. (2011). Modeling to support a tabletop exercise for USNORTHCOM. [Tech. Rep. No. 11-141.], NDSSL. Blacksburg, VA: Virginia Bioinformatics Institute, Virginia Tech.

Barrett, C., Beckman, R.J., Khan, M., Kumar, V.S.A., Marathe, M.V., Stretz, P.E., ... Lewis B. (2009). Generation and analysis of large synthetic social contact networks. In *Winter Simulation Conference* (pp. 1003–1014). Austin, TX: Water Simulation Conference Foundation. Retrieved from http://www.informs-sim.org/wsc09papers/095.pdf

Barrett, C.L., Bisset, K.R., Leidig, J., Marathe, A., & Marathe, M.V. (2010). An integrated modeling environment to study the co-evolution of networks, individual behavior and epidemics. *AI Magazine, 31*(1), 75–87.

Barrett, C.L., Eubank, S., & Marathe, M.V. (2006). An interaction based approach to modeling and simulation of large biological, information and socio-technical systems. In D. Goldin, S.A. Smolka, & P. Wegner (Eds.), *Interactive computation: The new paradigm* (pp. 353–392). New York: Springer.

Barrett, C., Eubank, S., Marathe, A., Marathe, M., Pan, Z., & Swarup, S. (2011). Information integration to support policy informatics. *The Innovation Journal, 16*(1), 2.

Bisset, K., Chen, J., Feng, X., Vullikanti, A., & Marathe, M. (2009). EpiFast: A fast algorithm for large-scale realistic epidemic simulations on distributed memory systems. In *Proceedings from 23rd ACM International Conference on Supercomputing (ICS)*. New York, NY:

Association for Computing Machinery. Retrieved from http://dl.acm.org/citation. cfm?id=1542336

Cannon-Bowers, J.A., Salas, E., & Converse, S. (1993). Shared mental models in expert team decision making. In N.J. Castellan Jr. (Ed.), *Individual and group decision making: Current issues* (pp. 221–246). Hillsdale, NJ: Lawrence Erlbaum.

Chen, J., Marathe, A., & Marathe, M.V. (2010). Coevolution of epidemics, social networks, and individual behavior: A case study. In *Advances in social computing, third international conference on social computing, behavioral modeling, and prediction* (Vol. 6007, pp. 218–227). New York: Springer.

Clark, A., & Chalmers, D. (1998). The extended mind. *Analysis, 58*(1), 7–19.

Comfort, L.K. (2007, Dec). Crisis management in hindsight: Cognition, communication, coordination, and control. *Public Administration Review, Special Issue: Administrative Failure in the Wake of Katrina*, S188–S196.

Epstein, J.M. (2008). Why model? *Journal of Artificial Societies and Social Simulation, 11*(4).

Eubank, S., Guclu, H., Kumar, V.S.A., Marathe, M., Srinivasan, A., Toroczkai, Z., & Wang, N. (2004, May 13). Modelling disease outbreaks in realistic urban social networks. *Nature, 429*, 180–184.

Eubank, S., & Lewis, B. (2009). *Conditions for a 3rd wave of pH1N1 (slides).* [Tech. Rep. No. 09–507.], NDSSL. Blacksburg, VA: Virginia Bioinformatics Institute, Virginia Tech.

Fraser, C., Donnelly, C.A., Cauchemez, S., Hanage, W.P., Van Kerkhove, M.D., Hollingsworth, T.D., . . . WHO Rapid Pandemic Assessment Collaboration. (2009). Pandemic potential of a strain of influenza A (H1N1): Early findings. *Science, 324*(5934), 1557–1561.

Gentner, D., & Stevens, A.L. (Eds.). (1983). *Mental models.* Mahwah, NJ: Lawrence Erlbaum.

Grush, R. (2004). The emulation theory of representation: Motor control, imagery, and perception. *Behavioral and Brain Sciences, 27*, 377–442.

Hethcote, H.W. (2000). The mathematics of infectious diseases. *SIAM Review, 42*(4), 599–653.

Hutchins, E. (1995). How a cockpit remembers its speeds. *Cognitive Science, 19*(3), 265–288.

Hutchins, E. (2001). Distributed cognition. In N.J. Smelser & P.B. Baltes (Eds.), *International encyclopedia of the social & behavioral sciences* (pp. 2068–2072). Oxford: Pergamon.

Kirsh, D., & Maglio, P. (1994). On distinguishing epistemic from pragmatic action. *Cognitive Science, 18*, 513–549.

Monge, P.R., & Contractor, N.S. (2003). *Theories of communication networks.* Oxford: Oxford University Press.

Neisser, U. (1967). *Cognitive psychology.* New York: Appleton-Century-Crofts.

Picton, T.W., & Stuss, D.T. (1994). Neurobiology of conscious experience. *Current Opinion in Neurobiology, 4*, 256–265.

Rao, R.P.N. (1999). An optimal estimation approach to visual perception and learning. *Vision Research, 39*, 1963–1989.

Spillane, J.P., Reiser, B.J., & Gomez, L.M. (2006). Policy implementation and cognition: The role of human, social, and distributed cognition in framing policy implementation. In M.I. Honig (Ed.), *New directions in education policy implementation: Confronting complexity* (pp. 47–64). Albany, NY: SUNY Press.

Sterman, J.D. (1989). Modeling managerial behavior: Misperceptions of feedback in a dynamic decision making experiment. *Management Science, 35*(3), 321–339.

Sterman, J.D. (2006). Learning from evidence in a complex world. *American Journal Public Health, 96*(3), 505–514.

Vosniadou, S., & Brewer, W.F. (1992). Mental models of the earth: A study of conceptual change in childhood. *Cognitive Psychology, 24*, 535–585.

Zhang, J. (1998). A distributed representation approach to group problem solving. *Journal of the American Society of Information Science, 49*(9), 801–809.

16

PARTICIPATORY SIMULATION AS A TOOL OF POLICY INFORMATICS

Definitions, Literature Review, and Research Directions

Gerard P. Learmonth, Sr. and Jeffrey Plank

Introduction

As our environment comes under increasing stress, the need for effective natural resource governance is clearly urgent. Not only are the challenges daunting, but traditional methods of analysis and inquiry, as well as existing institutions, are seen to be largely inadequate. The physical sciences provide insight and understanding of natural processes, and the social sciences provide insight into individual and collective human behavior. Both are rooted in the positivist tradition of the scientific method—observation and experimentation. The analysis of *complex* coupled natural and human systems, however, requires a new approach; one that can support the development of new, more adequate institutions.

For example, policy makers neither need to know nor understand the minute details of individual natural processes. Often such processes are described in terms of deep mathematical models, the fruits of which are not readily accessible to the untrained observer. More intricate models that attempt to integrate multiple co-evolving natural processes—ecological models—are even more challenging, as the relationships of the interacting processes are elusive to all but those in the scientific community. The social sciences likewise have their own specialized methods for examining individual and group behavior, where natural processes provide the stage for that behavior.

Historically, there has been a separation between the natural and social sciences, each with its own history, theory, and methods. More than ever, this separation is outmoded and counterproductive. The new field of *socio-environmental systems* (SES) explicitly acknowledges the need to integrate human activity and behavior with the natural sciences (Janssen & Ostrom, 2006). Although SES provides a broad foundation for the study of complex adaptive natural and human systems, *policy informatics* provides a toolset for actionable science.

An important and useful method in this toolset is *participatory simulation*. Here, an underlying simulation model of natural processes is combined with human role-playing in a game-like platform. Participatory simulation combines tools and methods such as agent-based models (ABMs) and role-playing games (RPG) into *serious games* as an innovative method for engaging stakeholders and policy makers in the joint exploration of the structure and dynamics of coupled natural and human systems.

This chapter provides an overview of participatory simulation and some of its unresolved challenges, highlights relevant literature, describes an exemplary prototype, and proposes a research agenda.

Simulation Modeling of Coupled Natural and Human Systems

Complex, Coupled Natural and Human Systems

Coupled natural and human systems are inherently *complex systems*. A complex system is characterized as consisting of a number of autonomous elements that (1) interact in a *non-linear* fashion, (2) remain in a *non-equilibrium* state, (3) have *no central coordinating authority*, and (4) display macro behavior that is *not predictable* in the long run. One method for understanding the structure and behavior of a complex system is computer-based *simulation*.

The traditional physical and social sciences acknowledge the use of abstract simulation in problem situations where the underlying problem is deemed intractable using conventional deductive methods of experimental science. But, there is an implicit assumption that normative science remains at the core. Complexity science, with simulation as a principal tool, is the alternative method we propose be used to better understand coupled natural and human systems. Bradbury (2006) makes this case succinctly:

> What we would move on to is what I call *deep complexity*. It promises to be a science that accepts the contingent complexity of the world, adapts its ideas about explanation, prediction, and control as goals of science, and learns a new concept of understanding. . . . It will be complex not simple, drawing its strength more from the coming ubiquity and transcendent power of computing rather the analytical power of mathematics.

Computer-based simulation has evolved from an adjunct tool of the normative sciences to a respectable approach to scientific inquiry in its own right. Although the traditional sciences have yet to fully accept *complexity science* and complex system simulation as an authentic scientific method, we propose this approach (in particular, participatory simulations) as a valuable tool in policy informatics.

Traditional Simulation Modeling

Computer-based modeling and simulation is a mature technology, being among the first applications developed for use on digital computers. Simulation modeling as a tool is deemed appropriate by traditional science when analytic closed-form methods for describing the behavior of a complicated system are unavailable or computationally intractable. A validated and verified simulation model can be developed to characterize the salient structure and behavior of a system of interest. It can provide an opportunity to gain a deeper understanding of the nature of a system and how it might respond under hypothesized scenarios.

Traditional simulation models are typically expressed in the form of *equation-based models*. These relate a set of input parameters and assumptions through a collection of equations that produce system states and outputs. Moving through a series of discrete time steps, the equations are evaluated and the system's states and performance criteria are updated. Often, a given simulation model is executed systematically over a range of inputs and assumptions called a *parameter sweep*. The benefit of such execution is to understand the sensitivity of the system (model) to changes in assumptions individually and in combination. This approach can be computationally challenging, but it allows for experimentation that otherwise is impractical.

For aspects of a system that are not known precisely, the equations of an equation-based simulation model are based on received theory or, if not, are estimated based on statistical analysis of empirical data. Some model parameters and/or assumptions may be represented *stochastically (probabilistically)*. Regardless, *discrete-event (stochastic) simulation* models are *validated* against theoretical expectations or expert opinion (e.g., subject matter experts) and the simulation model's computer output is *verified* against known results where available (Law, 2006). However, human behavior in traditional simulation models is also represented in equation form. Such representation is far from realistic as it lacks dynamic choice and novelty, basic characteristics of human behavior.

Jay Forrester (1961) introduced a new approach to modeling complicated and complex systems—*system dynamics* modeling. Forrester's observation was that traditional equation-based modeling failed to adequately capture the inherent feedback mechanisms characteristic of system behavior. Originally presented as a set of *differential equations*, system dynamics modeling was challenging for those without a mathematical background, yet developed a devoted following. The ideas and concepts embodied in system dynamics modeling have now become more widely accepted with the introduction of visual software tools for model development and experimentation together with accessible popular literature.

Original tools represented the natural environment with *stock-and-flow* models, that is, with an appropriate collection of *stocks* (aggregations of undifferentiated units of some object) and the *flows* between and among them. Importantly, stock-and-flow models incorporate *feedback loops* among the stocks and flows to capture the inherent dynamics of the system being modeled. However, when limited to stocks and flows of undifferentiated units, human behavior is only captured in the aggregate. Regardless, system dynamics models have been widely used in ecology, natural resource management, and climate change modeling. To popularize *systems thinking* and system dynamics modeling and simulation, authors (Meadows, 2008; Senge, 1990) have done much to advance wider acceptance of this approach.

Both equation-based and system dynamics simulation represent top-down modeling methodologies. They epitomize a *deductive* approach consistent with the traditional normative scientific method. In each, macro-level system components and behavior are modeled and micro-level outcomes are observed.

In contrast to the deductive approach of discrete-event simulation modeling, there are *inductive* methods and tools that are well suited for simulating *complex* natural and human systems. In a complex system, macro-level behavior cannot be inferred from micro-level behavior. Thus, the simulation of a complex system through non-linear interaction of its micro-level components often yields insight into unanticipated macro-level behavior and performance—so-called *emergent* outcomes.

Von Neumann (1966) proposed a method to simulate systems made up of a collection of components interacting in a non-linear fashion without a central coordinating authority—hallmarks of a complex system. Von Neumann's *cellular automata* consist of simple elements, *cells*, on a lattice structure. In their simplest form, cell states are characterized by discrete state values. Each cell exists in a *neighborhood* made up of its four or eight adjoining cells. At each time step, the cell displays simple decision-making through a *rule* by which it observes its current state and those of its neighbors and updates its state based upon the rule. For example, a cell's rule might state that if a majority of the cell's neighbors are in one particular state, then this cell adopts that state too—"follow the crowd," as it were. Over time, all the cells on a lattice might "freeze" into a non-changing pattern, continuously change in a chaotic manner, or display changing, yet recognizable patterns. As simple as this modeling method is, it has been used extensively to successfully capture the dynamics of human systems.[1]

Wolfram systematically examined the behavior of simple one-dimensional cellular automata and showed that certain cell update rules lead to quite regular behavior over discrete time steps, in some instances "freezing" in uninteresting singular patterns. Some rules produce recognizable repeating patterns while others yield continuously changing, unpredictable patterns. With a simple one-dimensional model and its simple update rules, the ability of cellular automaton to produce unpredictable behavior offers a counterpoint to normative science. Many in the scientific community vigorously challenged Wolfram's advocacy of this new science of complexity and complex systems.

The notion of individual, discrete units—cells, in this case—equipped with local information (the state of its neighborhood) and simple decision-making rules proved appealing to the budding computer science field of *artificial intelligence* (AI). Individual units equipped with the property of *agency* could interact with other agents, producing interesting, unpredictable, system-wide outcomes.

A logical extension of the cellular automaton/AI approach is to generalize cells/agents to admit a wider range of states for each cell, allow more sophisticated rules and decision-making, and to relax the neighborhood concept, *individual-based modeling* (IBM), or *agent-based modeling (*ABM). IBM is commonly used in ecological systems (Grimm & Railsback, 2012), whereas the term ABM is used in more general modeling contexts.

Agent-based modeling provides an opportunity to model natural system processes with required fidelity and directly represent human behavior in the form of decision-making actors—software *agents*. Agents of various types are equipped with decision rules and also have access to certain (typically local) information about the state of the system in time. Because agent choices and behaviors can also be influenced (constrained) by external policies, ABMs can serve as useful tools for policy experimentation.

An example of this use in the exploration of policy alternatives for restoring and sustaining the health of the Chesapeake Bay Watershed is the University of Virginia's Computing for Sustainable Water project. This very large-scale ABM was run on the IBM World Community Grid and examines the effect of a large number of proposed policy interventions. Preliminary analysis of the results of this simulation reveals that certain well-accepted policies for nutrient reduction (Best Management Practices) in the Chesapeake Bay are ineffective under plausible assumptions about human behavior.

Although valuable insights can be gained from agent-based modeling, modeled human behavior remains limited to the skill of the developer in representing the scope of agent decision-making and behavior, and the rules by which software agents act upon the information they receive throughout model execution.

Participatory Simulation

ABMs take a significant step toward integrating human behavior into a simulation model of natural processes using artificial software agents to represent human behavior. These do not fully capture the richness and variety in human behavior, especially that which is non-rational (i.e., emotional, speculative, or simply ill informed).

To address this limitation, ABMs have been incorporated with RPGs, where live human players engage actively with the simulation model. RPGs have an extensive history in the education and entertainment domains. From simple board games to visually rich video games (e.g., SimCity and massive multiplayer online games, MMOGs, such as Farmville and Second Life), the art of integrating live players and their decision-making into simulated environments has become quite sophisticated. The combination of a simulation model with a role-playing game yields a *participatory simulation*.

As described by Guyot and Honiden (2006), the purpose of a participatory simulation may be threefold: (1) training and education, (2) participant observation, and/or (3) action to solve a particular common resource problem. The authors emphasize that these "kind[s] of games are not designed to be fun but are based on the idea that participants behave in the game as they do in real life and that they will consequently use the same strategies."

For our purposes, we define participatory simulation as the combination of an underlying computer-based simulation model with human game players interacting directly with that simulation model. This combination allows the representation of a complex system (e.g., a socio-environmental system) in a simulation model with the advantage of representing human behavior realistically through role-playing. The important characteristic is the ability of human players to receive information from the simulation model about the current state of the environment and to input decisions based on that information. The system, then, is a simulation model of natural processes coupled with actual human behavior.

The underlying computer-based simulation model may be as simple as a spreadsheet or as sophisticated as a complete agent-based model with virtual software agents complementing the live player agents. Various approaches have been offered for the type of simulation model (Barreteau, Bousquet, & Attonaty, 2001; Barreteau, Antona, et al., 2003; Bousquet et al., 2002; Guyot, Drogoul, & Lemaître, 2005).[2]

In the UVA Bay Game, we use a hybrid approach wherein a system dynamics (aggregated) simulation model is coupled with live agent players. The behavior of the larger aggregations of undifferentiated roles (e.g., crop farmers) is influenced by the actions of the relatively small number of live decision-making agents. This approach captures the varied and changing behavior of individual players while mapping that behavior to the larger socio-environmental scale of the entire Chesapeake Bay Watershed.

The number of human players and their respective roles need not be a complete representation of the system being modeled. An additional feature of participatory simulations is that the process of designing and building it may itself be a participatory process. Engaging stakeholders and subject matter experts in the design process assures that the model and the representation of player-agents conforms to the desired environment. Early recognition of this process produced a charter adopted and agreed to by leading researchers in the field (Barreteau, Le Page, & D'Aquino, 2003). The Companion Modeling, or ComMod, approach is especially focused on participatory modeling and participatory simulation in the domain of natural resource management. Companion modeling implies an iterative process of design and development wherein stakeholders engage with modelers to design, build, validate, and verify the resultant participatory simulation.

Is the participatory modeling of an ABM of a complex system an incremental step in the series we have described? Or is it a discontinuous leap? We believe it is a fair question, and one with very significant implications for modelers and model users. In the participatory modeling process, stakeholders from different sectors express their interests for representation, agree to a shared level of abstraction, contribute to an iterative process of calibrating and synchronizing those interests, and validate the representation of stakeholder relationships. This process is a social process fundamental to sustainable cultures. It is important to realize that the extension of modeling procedures, from mathematical models to ABMs, may call into question the authority of traditional modeling conventions, particularly in relation to other conceptual tools for understanding, facilitating, and expressing the relation of human behavior to natural processes.

Regardless, designing and building a participatory simulation is not without its challenges. Among these is finding the appropriate scales—spatial, temporal, and social. In natural resource governance, a spatial and temporal scale determines the underlying simulation model's ability to accurately capture the essential dynamics of the natural environment, especially as a proper response to player decision-making (Guyot, Drogoul, & Lemaître, 2005). In equation-based simulation models of natural processes, there is a tendency to achieve ever-finer levels of spatial and temporal description. For purely scientific research, this may be justified. In a participatory simulation, appropriate spatial and temporal scales are often higher-level, yet are sufficient to reproduce an authentic environment in which the model plausibly responds to game players' decision-making choices.

Another potential disadvantage is that each participatory simulation gameplay is unique and thus not reproducible because the same players usually make different decisions in subsequent gameplay based on their prior gameplay experience. An advantage of this same observation is that repeated gameplay by the same players can lead to "learning." In the classroom, we see accelerated learning about complex systems, especially about the relation of information and individual and collective decision-making. Outside the classroom, participatory simulations can provide real-world stakeholders with the opportunity to better understand and perhaps change their own roles and behavior. Competitors for limited water resources in rural and urban areas may use participatory simulations to determine how to cooperate as suppliers and consumers of ecosystem services. To this end, a participatory simulation offers a powerful tool for carefully designed social learning and policy experimentation.

Participatory Simulation in Natural Resource Policy

The history of participatory simulation as a tool for policy informatics is deeply rooted in the study of approaches to renewable resource management and is attributed Olivier Barreteau (Guyot & Honiden, 2006), who strongly emphasized the need to involve local stakeholders in the actual process of designing the participatory simulation—participatory modeling. These principles were codified in the ComMod Charter (Barreteau, Antona, et al., 2003).

Research Directions

Participatory Simulation as a Ritual of Social Integration

There are opportunities, challenges, and barriers in using participatory simulations for education and for natural resource management. We initially designed the UVA Bay Game to enable university students to see how the part they study in one class fits into a complex socio-ecological system—and to participate in a complex system whose important properties emerge in relation to the actions of independent agents. Students learn that information for decision-making is always incomplete and that innovative data correlations can add disproportionate value to incomplete data; that their decisions indeed alter their world; and that complex systems thinking can enable not only the intellectual but also the social conventions required for adaptive management. This sense of personal and collective agency is an important lesson, and it is achieved through a game experience that converts player expectations and a lively gameplay to serious ends.

Outside the classroom, participatory simulations can support multi-sector stakeholder engagement and capacity building for community-based natural resource governance. Industrial water users, such as Coca-Cola, and NGOs, such as The Nature Conservancy and World Resources Institute, acknowledge that increasing competition for water requires better stewardship, initially through reductions and efficiencies in water use for manufacturing and production and ultimately through multi-sector cooperation. The complexities require participants to understand that information for decision-making is always incomplete, innovative data correlations can add disproportionate value to incomplete data, decisions do alter their world, and complex systems thinking can not only enable the intellectual, but also the social conventions required for adaptive management.

We distinguish participatory (i.e., live agent-based) simulations from strictly deterministic models that function without live agents and aim for, but necessarily fail to deliver, predictive fidelity. In contrast to models that attempt predictive fidelity, participatory simulations yield results that, although inherently unpredictable and unrepeatable, yield richer insights into the interaction between the human and natural dimensions. Because participatory simulations depend on live players for activation, they have the capacity to involve stakeholders in model design, generation of hypotheses about system drivers, and testing of innovative policies or business practices. Participatory simulations offer the modeling community an alternative to deterministic modeling that puts scientific, economic, and policy information in the hands of real stakeholders; one that requires close collaboration with those stakeholders. Participatory modeling, i.e., engagement of multi-sector stakeholders in the construction and application of watershed simulations, thus can help develop complements or alternatives to environmental policy as government regulation or new institutions for natural resource management. According to the

National Research Council (2011), adaptive management institutions are the highest national priority for climate change adaptation.

As an educational tool and a shared research laboratory, participatory simulations such as the UVA Bay Game provide distinctive opportunities for complex systems learning and for investigating the relation among complex systems thinking, data correlation, communication, decision-making, social behavior, and behavior change. These simulations can also support the representation or modeling of innovative policies or institutions, such as ecosystem services markets. In university classrooms, the conventions of commercial digital games also accelerate faculty and student acceptance of participatory simulations. Widespread familiarity among the university community along with large-scale online game conventions make the mastery of large amounts of data possible, which, in academic contexts, are associated with science, technology, engineering, and mathematics disciplines. Higher education institutions, however, are especially hospitable to innovation and our capacity to extend participatory simulations for education and research are limited only by our capacity for multi-disciplinary collaboration.

Beyond higher education, the institutional situation for participatory modeling and participatory simulations is more problematic. For a special journal issue on 'Modeling with Stakeholders,' Bousquet and Voinov (2010) write:

> While we are getting better at the process of building models, our track record in actually using models for making timely and appropriate decisions is quite poor. Despite our improvement, it is still difficult to convince decision and policy makers to use them, especially when they are facing unpopular and controversial issues.

Their aim is to:

> explore how we can better undertake our modeling with stakeholders. Our goal was to focus on the stakeholder part of the modeling activity . . . We invited papers that showed links to planning and policy making, how stakeholders could be involved and how they played a role in both identifying the model goals and more generally in providing guidance throughout the process.
>
> *(Bousquet and Voinov, 2010)*

For Bousquet and Voinov (2010), adopting a new tool for natural resource governance—a tool by which community members represent their relationships with each other and with an ecosystem over time—is a problem of social interaction: Increasing the interactions of modelers with the agents they model can improve models and increase opportunities for their institutionalization.

Each paper written on the topic thus makes assumptions about different kinds of change—individual, collective, and institutional—and about the kind of process "modeling with stakeholders" is. That these assumptions are not examined or made explicit is not surprising: The aim is to improve the modeling process, that is, to present multiple instances of participatory modeling to inform best modeling practices. If assumptions about theories of change are not explicit, the degree of change associated with adopting participatory modeling is obscured. And not making explicit assumptions about integrative social processes obscures the similarities of participatory modeling to other integrative social rituals, which anthropologists observe in all cultures. If participatory modeling is a contemporary ritual of social integration, what can modelers learn from anthropologists about their interactions with stakeholders?

For anthropologists, there is extensive literature that deals with participatory modeling and simulation models in which the metaphorical term *model* refers to physical artifacts and rituals of social integration. Lansing and Kremer (1993) argue that "simulation models are uniquely appropriate for addressing the issues of adaptation and determinism in the development of complex social systems like the water temples of Bali." Lansing and Kremer characterize Balinese water temples as physical artifacts of socio-ecological systems constructed and maintained by human agents to distribute water to optimize rice harvests. The concept of a complex system thus enables Lansing and Kremer (1993) to test this characterization: They derive a simulation model from the physical artifact; populate it with historical data for annual rainfall, irrigation flow, crop yields, water stress, and pest damage; generate multiple scenarios based on different assumptions about human agency and social coordination; and observe results, including emergent properties. The historical data for rice harvests suggest that Balinese water temple networks constitute complex adaptive systems at optimal scales of social coordination in which farmer associations, or *subaks*, and water temple congregations adapt their behavior in relation to ecosystem variability through rituals of interdependence.

Our point in citing anthropological research on Balinese water temples and the corporate architecture of the Central Andes is to call attention to the rich scholarly context for research on the process and social function of participatory modeling and participatory simulations (see, for example: Stark, 1999; Plog, 2003; Carballo, 2007; Schmidt, 2001). For anthropologists, Lansing and Kremer conclude, the concept of a complex adaptive system of human agents and natural processes and its associated analytical techniques has very significant explanatory power. The quantification associated with complex adaptive system simulations provides additional rigor for more adequate anthropological explanations of physical artifacts of social integration than do models that discount human agency. As Lansing and Kremer point out, the adequacy of our models of "living" artifacts, such as Balinese water temples, has very significant real-world consequences.

Their argument thus has important implications for modelers interested in "using models for timely and appropriate decision-making," particularly their quite explicit exposition of the relation of qualitative and quantitative models of social integrations and their concluding remark on the need for comparative modeling.

Just as the metaphorical character of the term "model" makes accessible conceptual tools across disciplines, so the methodological issues associated with explaining physical artifacts as models of social integration identified by anthropologists provide conventions and standards for assembling and disseminating tools for social action. For Stanish and Haley, for example, these artifacts enable the scaling up of cooperation among politically egalitarian groups in simple "chiefdom" societies and mark the beginning of more complex forms of sociopolitical organization. To explain the development of "complex architecture" in the Central Andes in the third millennium bc, that is, "corporate architecture . . . that is built and designed to be used and seen by the community as a whole," Stanish and Haley fashion a methodology drawn from economic anthropology, evolutionary game theory, comparative behavioral psychology, and evolutionary psychology. Indeed, the language of their thesis statement indicates the importance of these disciplines to their argument:

> We argue that large-scale public architecture and related ritual practices represented cultural innovations that did in fact effectively tap into human cooperative psychology by allowing the efficient broadcasting of information about participants' behavior and by allowing greater opportunities to demonstrate to a large audience of potentially willing cooperators (conditional cooperators) that a reciprocity system was operating according to local norms of fairness and free-rider punishment, thus making possible large and productive systems of reciprocity and labor organization.
>
> *(Stanish and Haley, 2004)*

Interestingly, they use the concept of "agent-based models" to explain how a community shifts from one artifact of social integration to another:

> A model that incorporates human decision-making is agent-based, reducing causality to the level of individual choices based upon perceived costs and benefits. We contend that such an agent-based approach not only is not inconsistent with an evolutionary model but in fact is the theoretical underpinning for a new generation of cultural evolutionary theory.
>
> *(Stanish and Haley, 2004).*

Their explanation thus includes treatment of changes in "corporate architecture" in the Early Horizon period, a discussion that points to important questions about the spatial and temporal scales for models of social integration.

What is remarkable about this essay in anthropological explanation is that it very explicitly describes the strengths and limits of conventional academic methods for dealing with physical artifacts that function as models of social integration. The new anthropological explanation they need draws on methods from economics, game theory, and social and behavioral psychology, and combining these methods leads to historical explanation as well, particularly of cultural innovation and methodological change.

As participatory modeling and participatory simulations become subjects for scholarly research in the academy, we anticipate that practitioners will benefit from greater familiarity with the conventions of cultural studies. By understanding that cultural traditions extend conventions for social integration that are both constitutive and descriptive, practitioners can access more conceptual tools, find new collaborators, and prioritize goals. For practitioners, as Bousquet and Voinov suggest, the institutionalization of participatory modeling and simulations may present the greatest challenges and the greatest opportunities. Paradoxically, if the institutionalization of participatory modeling and simulations requires measurable results, then data about costs and outcomes, especially over longer temporal scales, will be essential; that data, along with the benefits of social integration and community capacity-building, can come only through institutionalization.

Four Questions for Modelers and Stakeholders

For the continued development and application of participatory simulations of coupled human and natural systems, we pose four questions that suggest directions for further multi-disciplinary research.

Do We Have Institutions for Multi-sector Engagement at the Ecosystem Scale?

In our Chesapeake Bay watershed, governmental jurisdictions are not synchronized with the ecosystem: Six states, in addition to the District of Columbia, with more than 170 counties, make independent land-use decisions; some state policies for marine resources, land development, and agriculture are made in isolation. Ecosystem services markets are institutions that stimulate multi-sector engagement. But market forces constrain stakeholder participation. If multi-sector groups typically are convened in crises, at moments of institutional failure, then these convenings likely provide weak conditions for healthy multi-sector stakeholder engagement. A crisis rivets stakeholder attention, but the crisis context precludes the development of multi-sector trust, the exploration of problem definition that can build on long timescales, and collective risk taking. In the absence of crisis convenings, can participatory modeling constitute an institution for on-going multi-sector collaboration?

Do We Have the Data We Need to Model Complex Systems for Multi-sector Stakeholder Engagement?

At best, we now have spatial and temporal data about natural processes for regulation but not for multi-sector collaborative decision-making. Much of the data we have is interpolated. For natural systems, the measuring of seasonal pulses and local variations is essential. Without monitoring, how can the results of behavior change be measured? There are very significant open questions about the spatial and temporal scale at which participatory simulations need to work to provide multi-sector social or policy benefits. For example, ecosystem services markets function at the spatial scale that is least understood by economists because business management conventions and contemporary tools for economic development typically work at the local, national, or global scale, not at regional or cultural scales (Fujita, Krugman, & Venables, 1999).

Do We Know How to Model Social or Community Processes?

There is a growing body of information, typically from practitioner case studies, but few comparative studies, about social or community processes, such as how different sectors and stakeholders exchange information, develop trust, and in some iterative way begin to act collectively. For example, recent case studies in community-based collaboration suggest that local knowledge of ecosystems carries with it a stewardship ethos that is strong enough to sustain multi-sector negotiation (Dukes, Firehock, & Birkhoff, 2011). Top-down policy that homogenizes and discounts local difference is doomed because it eliminates these essential resources. But what spatial scale is local? What are the community process timescales? How do we quantify real-world community processes? How do we link human behavior and natural processes across spatial and temporal scales?

What Is the Real Relation of Mathematical Models to Policy Institutions?

Policy makers are used to regulation so they are comfortable with mathematical models that promise, but never achieve, predictive fidelity. So the relation of modeling to policy making is troubled: mathematical models in fact retard policy making. But we know that mathematical models cannot represent the distinctive nonlinear and emergent properties of complex systems that are produced by live agents—and that live agent models yield unpredictable results. So here is a fascinating relationship: regulation fits with mathematical models and adaptive management with participatory simulations of complex systems. Put another way, as we earlier suggested, the process of building, validating, and applying participatory simulations for policy making will involve changing policy institutions. Institutional design is no small undertaking, especially for adaptive management, for how will executive and legislative branches of government cede their authority for decision-making that changes in relation to unpredictable or uncertain system change?

Let us conclude by noting that policy informatics can use participatory modeling and participatory simulation to guide the reorganization and application of knowledge about the relation of human behavior to natural processes to support the development of more adequate public and private sector institutions. For institutional design, there may be no better partner than the major research university because no knowledge-based institution is as intellectually diverse, has such a long view, and maintains such a large inventory of models of social and socio-environmental integration. However, as long as its knowledge is distributed among isolated units, the special resources of the research university may not be accessible to other sectors. For multi-sector partnering, however, research universities will need to integrate knowledge that is distributed through different academic disciplines. In our experience with the UVA Bay Game, the construction and application of a participatory simulation of a complex socio-ecological system supports and sustains multi-disciplinary collaboration (Plank, Feldon, Sherman, & Elliot, 2011).

The concept of a complex system of human behavior and natural processes and the "serious game" format of the participatory simulation are ideal for multi-sector stakeholder engagement. They hold the promise of releasing hard policy problems from ideological stalemate. For multi-sector communities, the anthropological tradition should be liberating because it exhibits the variety of solutions, the inevitability of change, and the recurring need for the design of institutions adapted to our environments. Here, the "policy" in policy informatics takes on a broader significance as collective human action. With conceptual tools borrowed from anthropology and other humanities and social science disciplines, we can create virtual and dynamic simulations for—and artifacts of—social integration, without which we cannot imagine a sustainable culture or a sustainable environment.

Notes

1 See, for example, Thomas Schelling's (1971) seminal work on modeling segregation in communities. For more extensive applications in urban dynamics, see the work of Michael Batty (2005). Von Neumann's ideas have been extensively examined by Stephen Wolfram in his magnum opus, *A New Kind of Science* (2002).
2 A good discussion of approaches to building the underlying simulation modeling is found in Le Page, Becu, Bommel, and Bousquet (2012).

References

Barreteau, O., Antona, M., d'Aquino, P., Aubert, S., Boissau, S., Bousquet, F., . . . Weber, J. (2003). Our companion modelling approach. *Journal of Artificial Societies and Social Simulation, 6*(2). Retrieved from http://jasss.soc.surrey.ac.uk/6/2/1.html

Barreteau, O., Bousquet, F., & Attonaty, J. M. (2001). Role-playing games for opening the black box of multi-agent systems: Method and lessons of its application to Senegal River Valley irrigated systems. *The Journal of Artificial Societies and Social Simulation, 4*(2), 5. Retrieved from http://www.vcharite.univ-mrs.fr/PP/rouchier/cours Master/3iemeCours/BarreteauRiver.pdf

Barreteau, O., Le Page, C., & D'Aquino, P. (2003). Role-playing games, models and negotiation processes. *The Journal of Artificial Societies and Social Simulation, 6*(2), 10. Retrieved from http://jasss.soc.surrey.ac.uk/6/2/10.html

Batty, M. (2005). *Cities and complexity: Understanding cities with cellular automata, agent-based models, and fractals.* Cambridge, MA: MIT Press.

Bousquet, F., Barreteau, O., d'Aquino, P., Étienne, M., Boissau, S., Aubert, S., & Castella, J. (2002). Multi-agent systems and role games: Collective learning processes for ecosystem management. In M.A. Janssen (Ed.), *Complexity and ecosystem management: The theory and practice of multi-agent systems.* Cheltenham, UK: Edward Elgar Publishing.

Bousquet, F., & Voinov, A. (2010). Preface. In *Environmental Modelling and Software, 25*(11), 1267.

Bradbury, R. (2006). Towards a new ontology of complexity science. In P. Perez & D. Batten (Eds.), *Complex science for a complex world: Exploring human ecosystems with agents* (pp. 21–26). Canberra: Australian National University Press.

Dukes, E., Firehock, K., & Birkhoff, J. (Eds.). (2011). *Community-based collaboration: Bridging socio-ecological research and practice.* Charlottesville, NC: University of Virginia Press.

Forrester, J.W. (1961). *Industrial dynamics.* Waltham, MA: Pegasus Communications.

Fujita, M., Krugman, P., & Venables, A. J. (1999). The spatial economy. Cambridge, MA: MIT Press.

Grimm, V., & Railsback, S. (2012). *Agent-based and individual-based modeling: A practical introduction.* Princeton, NJ: Princeton University Press.

Guyot, P., Drogoul, A., & Lemaître, C. (2005). Using emergence in participatory simulations to design multi-agent systems. In *Proceedings of the fourth international joint conference on Autonomous agents and multiagent systems* (pp. 199–203). New York, NY: Association for Computing Machinery.

Guyot, P., & Honiden, S. (2006). Agent-based participatory simulations: Merging multi-agent systems and role-playing games. *Journal of Artificial Societies and Social Simulation, 9*(4), 8.

Janssen, M.A., & Ostrom, E. (2006). Empirically based, agent-based models. *Ecology and Society, 11*(2), 37. Retrieved from http://www.ecologyandsociety.org/vol11/iss2/art37

Lansing, S., & Kremer, J. (1993). Emergent properties of Balinese Water Temple networks: Coadaptation on a rugged fitness landscape. *American Anthropologist, 95*, 97–114.

Law, A. M. (2006). *Simulation modeling and analysis* (4th ed.). New York: McGraw-Hill.

Le Page, C., Becu, N., Bommel, P., & Bousquet, F. (2012). Participatory agent-based simulation for renewable resource management: The role of the Cormas simulation platform to nurture a community of practice. *Journal of Artificial Societies and Social Simulation, 15*(1), 10.

Meadows, D.H. (2008). *Thinking in systems: A primer.* White River Junction, VT: Chelsea Green Publishing Company.

National Research Council. (2011). *America's climate choices.* Washington, DC: National Academies Press.

Plank J., Feldon, F., Sherman, W., & Elliot, E. (2011). Complex systems, interdisciplinary collaboration, and institutional renewal. *Change: The Magazine of Higher Learning, 43*(3), 35–43.

Schelling, T. (1971). Dynamic models of segregation. *Journal of Mathematical Sociology, 1*, 143–186.

Senge, P. M. (1990). *The fifth discipline.* New York: Doubleday/Currency.

Stanish, C., & Haley, K. (2004). Power, fairness, and architecture: Modeling early chiefdom development in the Central Andes. *Archaeological Papers of the American Anthropological Association, 14.*

von Neumann, J. (1966). *Theory of self-reproduction automata.* Urbana: University of Illinois Press.

Wolfram, S. (2002). *A new kind of science.* Champlain, IL: Wolfram Publishing.

17

ACTION BROKERING FOR COMMUNITY ENGAGEMENT

A Case Study of ACTion Alexandria

Jes A. Koepfler, Derek L. Hansen, Paul T. Jaeger, John C. Bertot, and Tracy Viselli

Introduction

In early rural American life, the barn raising tradition arose out of necessity; small efforts and contributions from large numbers of people were required to construct a building that would benefit the larger community as a whole. Brought together by the harsh realities of frontier living, settlers responded to the needs of their community and peers. Whether motivated by altruism, an expectation of a returned favor, or devotion to a shared value system, the resulting collective action helped build strong social ties as well as physical structures within communities.

Today, communities still face challenges as monumental as barn raising that require just as many hands to construct, such as ensuring community food pantries are stocked to feed the hungry and rebuilding homes after a natural disaster strikes. Social technologies like community listservs, wikis, and other communication and idea generation tools enable individuals to contribute in small ways to issues they care about. However, there is an increasing need to move from individual actions to collaborative or community efforts to address social challenges. People use existing social media sites like Twitter and Facebook to help promote collective action, and increasingly systems are specifically designed to meet the unique needs of collective civic participation. These systems often include technical features and functions that enable crowd-sourcing through social networks, but action does not occur simply because a system enables it. Although crowds do form on their own to solve social problems, there are many civic challenges that may go unnoticed by local citizens.

One major challenge of enabling civic participation is pairing those who are willing to take action with those who have legitimate needs. In the increasingly compartmentalized and busy lives of everyday citizens, individuals are often unaware of the most pressing needs in their local communities. When community problems are known, it is not always clear how one can contribute or collaborate in an unstructured, volunteer-based environment. These challenges emphasize the role of intermediaries who can help match potential actors (i.e., volunteers) with those in need and with the nonprofits and government agencies that serve them. We call this intermediation *action brokering*.

Action brokering through websites and social media (Hansen et al., 2014) can be seen as part of a broader set of public engagement goals to promote what has been called *collaborative governance* or *participatory governance*—the inclusion of public agencies, nonprofit civic organizations, and individuals in addressing community issues (Ansell & Gash, 2007; Callahan, 2007; O'Leary, Gerard, & Bingham, 2006; Page, 2010). Such initiatives "blur traditional boundaries between organizations, sectors, and policy design and implementation" and their success depends on meaningful involvement from all stakeholder groups in a community (Page, 2010, p. 246).

In this paper, we discuss ACTion Alexandria,[1] a platform designed to promote local collective action in the Alexandria, Virginia, community. The ACTion Alexandria platform uses social structures and online tools to support the critical community function of action brokering. Here, we identify the social practices and technical features that can be used to implement action brokering and assess their impact, highlighting key factors that contribute to successful collective action, the challenges, and policy implications of platforms that promote action brokering. We end with a recap of best practices learned.

Action Brokering

Defining Action Brokering

This section introduces action brokering, its evolution, and the differences between it and other strategies for online community engagement. Action brokering is best defined by describing its core elements.

Action. The word "action" connotes "something that is done or performed, a deed, an act" (Action, n.d.). In the context of civic participation, actions include acts of service (e.g., labor), provision of resources (e.g., items, money, access), sharing of expertise, and the active promotion of a cause. Actions are the primary goal of civic participation initiatives that aim to promote pro-social behaviors. The needs and interests of the specific community determine the nature and scope of the actions that are undertaken and the behaviors that will result.

Brokering. The word "brokering" connotes an intermediary who negotiates an exchange between two parties (Broker, n.d.). The primary responsibility of a broker in a traditional financial sense is to bring together buyers and sellers. In the context of civic action, action brokers help address social issues by matching individuals interested in performing actions, *action seekers* (e.g., community members as potential volunteers), with those who help organize and offer actions, *action providers* (e.g., nonprofits or government agencies that work on behalf of individuals in need). Brokering may also occur between organizations that share a common mission, but would not otherwise know about one another or be able to work together effectively without the intermediary.

There are several benefits of action brokering. As in the financial world, the advantage of using a broker for community action is that the broker has a comprehensive view of the market and has established relationships with different parties. Maintaining a comprehensive network of connections takes considerable time and effort. By delegating social network maintenance to a broker, individuals and organizations can devote their scarce resources, primarily time, to their own domain, yet remain connected to potential partnerships via the broker. In addition, brokers can effectively organize collective efforts that would not be feasible for disconnected groups to perform on their own.

Action brokering takes place informally across many types of action seekers and action providers in a community. Action brokering occurs when a mother invites her friend to serve on the PTA (Parent-Teacher Association), when a religious congregation encourages its members to serve at the local soup kitchen, and when the local county volunteer center distributes a calendar of service events to email subscribers. Here, we focus on intentional action brokering via an online platform, a community manager, and organizational support.

Action Brokering in an Era of Online Collective Action

Action brokering is an activity that occurs within the growing research areas of collective action, crowd-sourcing, participatory challenge platforms, and technology mediated social participation. Collective action is a concept used in sociology to help explain how individuals form around an idea to create action in a grassroots manner (Olson, 1965; Hardin, 1982; Tarrow, 1994). With the increase in social media tools that enable distributed collective action through online contexts, groups can form around any number of social issues in real time to achieve goals quickly and effectively. For example, Facebook and Twitter have been credited by some as key components to the Arab Spring uprisings of 2011 for their use in helping to organize protests against the local regimes and spread awareness about them (Khondker, 2011; Starbird & Palen, 2012). In a second example, during the aftermath of the Haiti Earthquake in 2010, relief workers both on-the-ground and far away used crowd-sourcing through social media and an open-source crisis mapping tool called Ushahidi[2] to map critical information about trapped persons, medical emergencies, and availability of supplies (Heinzelman & Waters, 2010).

With the unique role that new social technologies have been playing in such highly visible events, there is an increasing emphasis on understanding novel platforms designed to support active participation at all levels of civic and philanthropic life (Janssen & Estevez, 2008; Johnston & Hansen, 2011; Lathrop & Ruma, 2010; Liu, 2012). Researchers have given great attention to the technical aspects of platforms that support collective action through research on participatory platforms and in the area of technology-mediated social participation (Desouza, 2012; Kraut et al., 2010; Preece & Shneiderman, 2009). Action brokering as a theoretical construct advances our understanding of that broad literature by taking into account the facilitated nature of the collective actions that occur on participatory sites. Although technology is a mediating factor and we discuss its implications, here we shed light on the behind-the-scenes social practices of community managers, policy makers, and local citizens who use the technology to mediate social participation and increase collective action.

Action Brokering Environment

Brokering is typically mediated by a combination of technologies (e.g., Web-based platform and/or social media), people (e.g., community manager), social practices, and policies. A single individual or organization may serve as an action seeker at one time and an action provider at another time, depending on the situation. For example, a resident may seek out volunteer opportunities at the local homeless shelter and later provide such opportunities to others when organizing a neighborhood cleanup.

Platforms that Broker Action

Fueled largely by advances in social media, several platforms that broker action have emerged in the last few years. They serve citizens and communities from local to global levels and support the exchange of actions, ideas, and challenges in an effort to tackle social problems and improve civic life.[3] Despite the growing number of action brokering platforms, few studies have evaluated their effectiveness or the specific technical and social aspects that lead to success. ACTion Alexandria, a city-level action brokering platform, illustrates the potential that such platforms have for solving community problems.

ACTion Alexandria Platform

ACTion Alexandria is a platform designed to broker actions and ideas in Alexandria, VA. The platform includes a website (www.actionalexandria.org), a Twitter account (@ACTionAlexVA), a paid community manager responsible for the day-to-day operations and outreach efforts, a steering committee, and policies and procedures that underlie its use.

The platform is an initiative of the city's community foundation called ACT for Alexandria, which serves as a catalyst for increasing charitable donations in the community. ACT and the city of Alexandria helped raise funding for the development of the ACTion Alexandria platform from local nonprofit organizations and a Community Information Challenge grant.

ACTion Alexandria seeks to "empower citizens to take collective action on behalf of themselves and local organizations" (ACTion Alexandria, 2011). Its three main goals are to:

- Create a vibrant online platform that inspires offline action, where challenges are posted, solutions are debated, successes and failures are archived, data is both disseminated and captured, stories are shared, and essential civic relationships are developed.
- Improve the quality of life for its most vulnerable residents in a cost-efficient manner through a platform that provides everyone with a voice and the opportunity to identify problems and offer solutions.
- Engage residents and business people in problem solving to strengthen community ties and increase each individual's stake in creating positive outcomes for specific community problems (ACTion Alexandria, 2011).

ACTion Alexandria attempts to achieve these goals through a variety of mechanisms that help residents connect with local nonprofit organizations and government agencies. First, community members can seek out and complete *actions*—small donations of items, funds, or volunteer efforts that are posted and needed by a local nonprofit or government agency and often promoted as a Featured Action. Second, residents and organizations can help brainstorm and vote on *ideas* to Community Challenges identified by ACTion Alexandria and the greater Alexandria philanthropic community. *Actions* and *ideas*, as well as local events, such as in-person training for nonprofits, are promoted via the ACTion Alexandria website, Facebook page, and Twitter account, and emailed to a list of registered users. Finally, residents and organizations can communicate in less structured ways by blogging and commenting on the ACTion Alexandria website, posting on the Facebook wall, mentioning the Twitter account, or talking in person at local meetings and events. The majority of these activities are managed on a day-to-day basis by the community manager with input from the ACTion Alexandria steering committee and other volunteers (e.g., bloggers).

Findings

In this section, we present the major themes that emerged from our analysis of interviews, surveys, and web analytics related to the platform. More detail on these methods can be found in Hansen, et al. (2014). Here, we present themes related to the social practices and technical features of two critical aspects of ACTion Alexandria: Featured Action and Community Challenges. We provide evidence for the impact that these tools had on the success of the platform in its first year and discuss lessons learned along the way.

Featured Actions

Actions, a designated term for small acts of service such as donating goods and money or volunteering one's time, are some of the core elements brokered by the ACTion Alexandria platform. Local nonprofits or government agencies (i.e., action providers) sponsor actions, and community residents (i.e., action seekers) complete them. Actions require small contributions from a large number of action seekers over a short period of time (usually one week) to meet an immediate need. Actions are often in the form of small monetary donations, specific items such as diapers or Pedialyte, or one-time service opportunities. Any organization using the ACTion Alexandria system can post actions, which can be searched by community residents by topic on the website.

Actions that are facilitated and promoted by the community manager are referred to as Featured Actions. Featured Actions are week-long campaigns that seek to meet an urgent community need identified by a local nonprofit in conjunction with the community manager. Often the community manager helps identify a corporate sponsor that matches funds or items raised by the community. The community manager promotes Featured Actions via the ACTion Alexandria homepage, email list, Facebook page, and Twitter account, as well as by the organization sponsoring the specific Featured Action. Through fourteen Featured Actions promoted during the first year of the project, community members contributed $115,680 in online donations and 3,720 items (valued at $4,338) to City of Alexandria nonprofit organizations, making the overall donation total in excess of $120,000. Donated items included medicine, food, diapers, children's books, and toys.

Based on the number of items collected, money raised, and services provided, there is no question that the Featured Actions initiative has been successful. All but one of the Featured Actions promoted in the platform's first year met or exceeded their goal. In fact, Featured Actions through ACTion Alexandria enabled some nonprofits to accomplish more than they would have on their own. Several of the Featured Action nonprofits mentioned that they would not have launched a campaign like the one they launched for the Featured Action had it not been for ACTion Alexandria. This was particularly the case because many nonprofits did not have the technological sophistication and experience to perform online action-based initiatives. Smaller and newer organizations benefitted from gaining visibility in the community. Larger organizations used the Featured Action to bolster a campaign that they would have run on their own. But the promotion helped achieve greater efficiency and effectiveness, find new donors, and reach the goal quicker than they would have on their own.

The following sections discuss some of the key themes that emerged from the implementation of the Featured Actions on ACTion Alexandria, as well as potential problems that this approach raised.

Community Manager Time and Expertise

The role of the community manager was critical to the success of Featured Actions. Because this form of crowd-sourced charity through a Web-based platform was new to most organizations, it was not always clear to them how to craft a Featured Action that would appeal to residents (see Box 17.1). The community manager was able to use her experience to help organizations scope out and frame strong campaigns in this new online platform. In addition, resource-poor organizations benefited from the community manager's help in developing promotional materials, getting matching funds from corporations, and coordinating the launch of the action. This all resulted in a low cost to organizations that wished to participate, despite the "extra work" needed to coordinate with the community manager, which was not part of their regular workflow. It was clear from the interviews and comments that without the community manager's heavy involvement, many of the Featured Actions would not have occurred and would otherwise probably not have been as successful as they were.

Leveraging ACTion Alexandria's Social Network for Partner Organizations

ACTion Alexandria helped organizations extend their reach to a larger network of residents via the Featured Action emails sent to all ACTion Alexandria users,

---

BOX 17.1

Properties of a Successful Featured Action

Based on an interview with the ACTion Alexandria community manager, a successful Featured Action should:

- last for a short duration (e.g., one week);
- focus on a single, well-defined, and measurable goal (e.g., X number of diapers, Y dollar amount to go toward Z);
- allow contributors to make micro- and macro-contributions (i.e., no contribution is too small);
- clearly demonstrate how it will benefit the organization and the individuals it serves;
- relate to a popular need that resonates with the public (e.g., oriented around children);
- be a joint effort between the organization sponsoring the action and the action brokering platform (i.e., ACTion Alexandria); and
- have matching funds (e.g., a corporate sponsor) to increase incentives for organizations and to encourage citizens to participate.

and promotion on Facebook and Twitter. This helped the organizations reach their goals quicker, while also making more people aware of the organization. Thus, the messages from ACTion Alexandria complemented the work the organizations were already performing via their own outreach methods, which ranged from email lists to social media to on-the-ground outreach via church congregations and events. Organizations that were smaller or less well known for other reasons (e.g., they were new) had the most to gain from being featured because their existing network was limited.

Growing ACTion Alexandria's Social Network

ACTion Alexandria also extended its own network as a result of the Featured Action mechanism because the sponsoring organizations helped provide a continual source of new members to the site. Nonprofit organizations promoted their own actions through whatever existing channels they had, including email lists, newsletters, social media accounts, and contacts at religious organizations. Inevitably, this reached many people who were not yet part of ACTion Alexandria. Those who learned about the action were taken to ACTion Alexandria's website where there was a detailed description of the action and a plea to register and click "Take Action," directing the resident to an Amazon wish list, donation button, or other appropriate link. Website analytics and registration dates make it clear that Featured Actions account for a large proportion of the steady, if small, growth in membership throughout the first year (see Table 17.1). Unfortunately, this model can be difficult to completely enforce, e.g., when nonprofit organizations post a different link to their donation page or when the nature of the action makes tracking it online difficult (e.g., delivering books in person). A balance must be established between the desire to make actions as simple as possible to complete and the desire to extend the resident network associated with ACTion Alexandria.

TABLE 17.1 User activity and number of unique actors on ACTion Alexandria website (02/07/2011–02/07/2012)

	Activity (total number of actions)	Unique Actors
Website Visits	24,023	16,702
Actions Taken	374	282
Blog Posts	242	50
Blog Comments	73	64
Ideas Posted	187	36
Votes Cast	5,440	1,120

Competition in the Nonprofit Sector

When pressed to identify concerns about the platform, nonprofits mentioned competition as a potential issue. For example, some organizations expressed reservations about the idea of Featured Actions because they feared that it might unfairly privilege some organizations over others. Nonprofits often compete for the same scarce resources and, as a broker, ACTion Alexandria has the potential to drive attention and resources to one organization over another. An interviewee expressed concerns of "donor burnout" and wondered if all the nonprofits could get what they needed out of "a diminishing or exhausted pot [of funding and resources]." Although several organizations mentioned concerns about this before the site was launched, they still felt like the opportunities would outweigh the risks, particularly if ACTion Alexandria was fair in the way they decided what to feature.

In follow-up interviews, nonprofits indicated that competition had not arisen due to the use of Featured Actions, although we only interviewed featured organizations, so it is possible other non-featured organizations felt otherwise. Members of the ACTion Alexandria steering committee have not seen this concern explicitly expressed by nonprofits, but noted that some organizations were hesitant to drive their constituents to the ACTion Alexandria website because of the extra work involved or due to fear of losing their own website traffic (i.e., site visit analytics).

Sustainability of Featured Actions Moving Forward

One downside of using the Featured Action mechanism was that it drew so much attention toward the featured action that other actions posted on the site by organizations went unnoticed. Also, organizations used to receiving assistance from the community manager in crafting and promoting their Featured Action seemed less interested in creating their own non-Featured Actions on the site and several organizations did not know that actions and Featured Actions were different. Although additional training may alleviate this issue, the prominent use of Featured Actions may obscure the use of regular actions or set up the false expectation that the community manager needs to be heavily involved with all actions, leading to sustainability issues down the road.

Community Challenges

Another core activity of ACTion Alexandria is brokering *ideas* that address community-wide concerns or opportunities. The community manager, in conjunction with the local nonprofits and government agencies, may identify a problem or issue that could benefit from community input and posts a Community

Challenge[4] to the website. Outreach (online and offline) is conducted, and residents submit their *ideas* for solutions and then a voting round occurs to help identify the best ideas. The community manager follows up and appropriate steps are taken based on the top ideas. Overall, the process is similar to other idea generation sites used by some government agencies to gain citizen input and promote transparency and openness (Bertot, Jaeger, & Grimes, 2010, 2012). Community Challenges examples include proposals to make Alexandria a more sustainable community, resident feedback on a new set of community quality-of-life indicators, and collaborative identification of playgrounds that should be improved.

Role of the Action Brokering Management Team

The successes of the Community Challenges are largely a result of the ACTion Alexandria management team's connections, insights, and efforts. Since 2004, ACT for Alexandria, the primary sponsor of ACTion Alexandria, has been promoting philanthropic activity through training and helping organize the nonprofit and government sector in Alexandria. For example, they sponsor an annual Spring2ACTion event where they bring together nonprofit and local government organizations to network, share best practices, and teach new skills such as how to effectively use social media. Those skills are then put into action during a competitive fundraising drive that is community wide.

ACT steering committee members and the community manager sit on many community councils, participate in town hall meetings, and interface with local government. As a result of these endeavors, ACT for Alexandria, and by extension ACTion Alexandria, enjoys a great deal of social capital among the philanthropic community as evidenced by the very positive comments from interviewees.

ACT for Alexandria's pre-existing position as a network hub among the Alexandria nonprofit community places its steering committee, including the ACTion Alexandria community manager, in an ideal position to help broker and aggregate actions among residents and organizations. Their panoramic view of the entire community-giving network in Alexandria allows them to see problems that nonprofits share, as well as opportunities for collaboration. ACTion Alexandria Community Challenges act as a platform to draw attention to some larger social issues in the community, while simultaneously discovering fresh ideas from residents who can directly contribute their feedback. The Project Play Community Challenge is an example where ACTion Alexandria's community manager helped broker a partnership between the Childhood Obesity Action Network, Get Healthy Alexandria, and Smart Beginnings Alexandria/Arlington. The goal of this partnership was to create a taskforce on play spaces in Alexandria, submit a grant together, and solicit ideas from residents on which playgrounds should be prioritized for renovation. The project helped bring together groups working on similar issues that were previously unaware of each other.

A Competent Community Manager

As with Featured Actions, the community manager is critical in framing Community Challenges in a way that will resonate with residents and generate enthusiasm and ideas (see Box 17.2). The importance of having a competent community manager that nonprofits trust cannot be overstated. A large part of the success of the ACTion Alexandria project results from the foresight to hire an experienced, enthusiastic, full-time community manager (see Box 17.3).

One challenge with having a community manager who is so central to the successful operation of action brokering is that it introduces a single point of failure. In the case of ACTion Alexandria, this has not been a problem; but if a community manager without the right skill set is hired, the credibility of the entire action brokering system can be jeopardized. Additionally, relying on a single community manager to take on too many responsibilities may lead to burnout or a lack of scalability.

Increasing Capacity through Social Media

Another key factor leading to the success of the Community Challenges, similar to Featured Actions, was the effective use of social media. Indeed, social media is

---

BOX 17.2

Properties of a Successful Community Challenge

ACTion Alexandria provides the technical platform necessary to facilitate voting and idea generation for Community Challenges, which has the direct effect of increasing new registered users. However, not all Community Challenges are created equal.

- Popular Community Challenges tend to be ones where there is a "winner" associated with the ideas that are generated through a process of community voting.
- Community Challenges work well when they are sponsored by a project that spans multiple organizations and focuses on a social issue rather than a specific organization (e.g., childhood obesity, teen pregnancy, or affordable housing).
- It must be something that affects many people in the community or that many people in the community care about.
- Challenges last longer on the site (one to two months) and culminate in a week-long voting period by community members.

BOX 17.3

Role of the Community Manager in Action Brokering

The community manager is a key component of an action brokering platform. The community manager:

- Works with human service and civic organizations to identify and help shape opportunities for community actions and ideas.
- Posts actions, ideas, challenges, blog entries, Tweets, Facebook wall entries, events, and other community information.
- Identifies actions that fulfill an urgent human service need in the community and works with a nonprofit partner to run a week-long Featured Action campaign to meet a specific goal (100 books, 640 diapers, $500 for a room renovation at a shelter, etc.).
- Identifies community problems ripe for citizen-sourced solutions and posts them as challenges to the community on an ideation platform.
- Continually conducts community outreach through a variety of methods, both online and offline.
- Manages most aspects of the website and public relations (website administration, community organizing, email marketing, editorial content, sponsorships, organization partners, marketing, etc.).

---

the kind of tool that several interviewees said they wished they used more, but just did not have the time or capacity to do well. Since Community Challenges require large numbers of participants in order to be successful, strategic campaigning through social media greatly increases the word-of-mouth communication about the challenge and drives people to the website. In this way, ACTion Alexandria fills a critical gap for its community of nonprofit and government agencies.

Raising Awareness

The benefit of a Community Challenge in particular is that it allows organizations to obtain input from residents early in a project and build an audience around an issue that might not otherwise exist. Not only are Community Challenges useful for getting the word out about ACTion Alexandria and increasing its membership, but they are useful for raising awareness about social issues in the community.

Another challenge called the Quality of Life Indicators challenge raised awareness about social issues in the community, including employment opportunities, city and environmental cleanliness, and addressing the basic needs of Alexandria citizens. In this challenge, the ACTion Alexandria platform was used as one of three venues for discussion about quality of life in Alexandria (Partnership for a Healthier Alexandria and Virginia Tech University represented the other two venues). The results of the voting and participation revealed the issues that rose to the top on ACTion Alexandria: (1) employment opportunities, (2) clean city and environment, and (3) sense of community, helping one another, and basic needs, which were different than those that emerged through the other two venues (e.g., healthcare). These results, along with associated Web analytics, suggest that the ACTion Alexandria platform was able to reach a unique audience compared to the other two platforms, in addition to raising awareness about a different set of issues (Viselli, 2011).

Discussion

Nonprofits and government agencies at all levels in many nations are increasing their use of social technologies as a way to reach members of the public in new locations, extend government services, promote democratic participation and engagement, crowd-source solutions and innovations, and co-produce valuable community resources (Bertot, Jaeger, Munson, & Glaisyer, 2010; Jaeger, Paquette, & Simmons, 2010; Pirolli, Preece, & Shneiderman, 2010). Despite the well-articulated potential benefits of civically oriented social technologies, few studies have empirically demonstrated success. Some even argue that the Internet has had an overall negative effect on civic engagement (Besser, 2009).

Action brokering platforms provide a compelling model for helping citizens, nonprofits, and local government agencies collaboratively tackle important social challenges. They may be able to succeed where other approaches have failed. Unfortunately, they are still rare, and it is not clear how they will fit into the wide range of local civic environments. ACTion Alexandria succeeded in part due to the fertile conditions it was planted in: the willingness of the local government to help fund the project, the leadership of an existing nonprofit brokering organization (ACT), and the exceeding generosity of the people of Alexandria. Local communities without such support may prove too barren an environment for action brokering through Web-based platforms to thrive, even if such communities are the ones with the greatest need for it. Future studies of action brokering through other platforms and launched in different contexts will shed light on their potential benefits, limitations, and scalability.

One key policy question related to platforms for action brokering is how they relate to local government initiatives. ACTion Alexandria is an independent nonprofit organization, although they receive some funding from the local government. Another possible model would be to have the local government house the Web-based platform and serve the primary function of action brokering. Such an approach could be risky for several reasons. First, government agencies are restricted in their ability to use social media due to the complex, ambiguous, and increasingly outdated legal framework they must operate under

(Bertot, Jaeger, & Hansen, 2012). This has not stopped government agencies from using social media, which the public expects them to do (Jaeger & Bertot, 2010a, 2010b); however, it has likely contributed to the feeling shared by many government agencies and officials that they have been unsuccessful thus far with facilitating participation, engagement, and collaboration through social media (Bonson, Torres, Royo, & Flores, 2012; Ganapati & Reddick, 2012; Nam, 2012; Sandoval-Almazan & Gil-Garcia, 2012).

Second, to be an effective broker, one must be seen as independent of the decisions being brokered. If contributors thought the action brokering organization had an agenda of its own, such as a political agenda imposed by the incumbent administration, they may be less willing to participate. Fortunately, governments do not need to manage these engagement platforms directly to benefit from them, provided they have active participants, which can provide new avenues to promote transparency, openness, new ideas, and potential partners (Bertot, Jaeger, & Grimes, 2010; Dawes, 2010).

The increased success of action brokering initiatives that serve at the city level may increase their importance as long-term social structures that serve as a hub between governments, residents, and civic-oriented organizations. As this occurs, policy issues related to their implementation will need to be considered. One consideration is how nonprofit organizations should be chosen in Featured Actions. Not addressing this question could lead to increased competition among nonprofits, rather than increased collaboration. Also, how local government recognizes voting on such sites should be considered. The nature of online platform use may inadvertently exclude those without regular access, leaving populations underrepresented in the idea-sharing process. Further, processes to deal with deviant behavior, which often occurs once sites with user-generated content become successful, would need to be put into place.

As action brokering initiatives become more common, they can potentially offer different avenues of scholarly exploration and practical solutions that might begin to address the policy questions posed in this section. The growing field of policy informatics in particular is well positioned to examine these initiatives alongside other forms of technology-mediated social participation, exploring the impacts of information systems and information-based governance platforms for solving policy problems (Desouza, 2011; Johnston & Kim, 2011).

Conclusion

In this chapter, we defined the new concept of action brokering and illustrated it through the social and technical aspects of the ACTion Alexandria platform to identify emerging themes related to the successes and challenges of action brokering in a technologically mediated environment. Our primary goal was to better understand the role of action brokering as a mechanism for collective action in local social issues and to identify best practices and lessons learned for using such an approach, particularly during the early stages of a project's development.

Although our focus has been on one case, we believe many of our findings and recommendations should be considered by other sites engaged in action brokering. Some of the key takeaway messages include:

- Hire a competent community manager to serve as an action broker to act as a social network hub both online and offline, outreach expert, community organizer, and idea generator (see Box 17.3).
- Leverage organizations with existing social capital when launching new Web-based platforms for action brokering.
- Provide initial support for nonprofits that need help to craft compelling and reasonable campaigns for *actions* and *ideas* that will work in an online environment.
- Create win-win network-building opportunities where organizations and action brokering networks promote actions and ideas and, in so doing, drive their own networks toward others in a virtuous cycle. Beginning with popular actions and ideas and well-known sponsor organizations may be especially important in the early ramp-up stage.
- Broker actions and ideas between different nonprofits as well as between nonprofits and residents.

These takeaways demonstrate one successful model of action brokering. How the findings translate to other platforms, such as those not at a local city level or those that do not function with heavy influence from social aspects like a community manager, remains to be seen.

As a manifestation of the possibilities and opportunities provided by new information and communication technologies for governments to engage and empower residents, ACTion Alexandria and other action brokering initiatives provide compelling arenas for the study of policy informatics. In the introduction to this book, Johnston suggest viewing policy informatics as a way that new information and communication technologies can help individuals understand and address public policy issues and to foster innovative governmental processes. Through its methods of increasing awareness of community needs, soliciting resident feedback, and promoting volunteerism and contributions to community groups, brokering initiatives can serve as test-beds for better understanding the intersections of new technologies, governmental processes, and communities. In particular, these projects shed light on the role of these platforms for resident empowerment, empathy, and mobilization to address community needs and engage civic processes.

Based on the early data from the ACTion Alexandria project, action brokering is especially promising as a tool for community engagement. The results discussed in this chapter are important in light of the less-than-promising results thus far from other more commonly employed approaches to promote community engagement through new technologies (Bertot, Jaeger, & Grimes, 2012; Jaeger, Bertot, & Shilton, 2012). Such efforts focus on the existing structure and infrastructure of government, which requires specific literacy and competency to use effectively (Jaeger & Thompson, 2003, 2004; Jaeger, Bertot, Thompson, Katz, & DeCoster, 2012). In contrast, action brokering engages community members on terms that are more familiar to many residents and do not require a knowledge of governmental processes or terminology with which to participate, advocate, and engage.

It should be noted that this chapter only reviews the first year of the initiative, so these results, no matter how promising, are still early results. The future successes and failures of ACTion Alexandria are unknown. Subsequent years may reveal a new set of strategies necessary to successfully broker actions in a way that is scalable and sustainable over time. Future research will need to elucidate the changes in approach made by ACTion Alexandria, as well as other models of success from different platforms. Although our current understanding of action brokering is in its infancy, we hope the experience of ACTion Alexandria and the new theoretical focus will help inspire additional platforms designed to meet the needs of residents and service providers alike.

Acknowledgments

We would like to acknowledge the nonprofits, community members, government officials, and others who participated in interviews and surveys as part of this research effort. This work is sponsored in part by the Bruhn Morris Family Foundation. We would also like to thank members of the ACTion Alexandria team who reviewed early drafts of this work and provided useful insights.

Notes

1   Web: http://www.actionalexandria.com; Twitter: @ACTionAlexVA; Facebook http://www.facebook.com/ACTionAlexandria.
2   http://ushahidi.com
3   See: http://www.changemakers.com/; http://www.tigweb.org; http://www.giveaminute.org/; http://changeby.us; http://forums.e-democracy.org/twincities?beneighbors.
4   During the first year of the project, ACTion Alexandria used the term Community Challenges. The steering committee and materials now refer to Community Challenges more simply as "Ideas."

References

Action (n.d.). In *Oxford English Dictionary* (3rd ed.). Retrieved from http://dictionary.oed.com
ACTion Alexandria. (2011). *About ACTion Alexandria*. Retrieved from http://www.actionalexandria.org/about
Ansell, C., & Gash, A. (2007). Collaborative governance in theory and practice. *Journal of Public Administration Research and Theory, 18*, 543–571.
Bertot, J.C., Jaeger, P.T., & Grimes, J.M. (2010). Using ICTs to create a culture of transparency? E-government and social media as openness and anti-corruption tools for societies. *Government Information Quarterly, 27*, 264–271.

Bertot, J.C., Jaeger, P.T., & Grimes, J.M. (2012). Promoting transparency and accountably through ICTs, social media, and collaborative e-government. *Transforming Government: People, Process and Policy, 6,* 78–91.

Bertot, J.C., Jaeger, P.T., & Hansen, D.L. (2012). The impact of polices on government social media usage: Issues, challenges, and recommendations. *Government Information Quarterly, 29,* 30–40.

Bertot, J.C., Jaeger, P.T., Munson, S., & Glaisyer, T. (2010). Engaging the public in open government: The policy and government application of social media technology for government transparency. *IEEE Computer, 43*(11), 53–59.

Besser, T.L. (2009). Changes in small town social capital and social engagement. *Journal of Rural Studies, 25,* 185–193.

Bonson, E., Torres, L., Royo, S., & Flores, F. (2012). Local e-government 2.0: Social media and corporate transparency in municipalities. *Government Information Quarterly, 29*(2), 123–132.

Broker (n.d.). In *Oxford English Dictionary* (3rd ed.). Retrieved from http://dictionary.oed.com

Callahan, R. (2007). Governance: The collision of politics and cooperation. *Public Administration Review, 67,* 290–301.

Dawes, S.S. (2010). Information policy meta-principles: Stewardship and usefulness. In *Proceedings from the 43rd Hawaii International Conference on System Sciences.* Kauai, Hawaii: Computer Society Press.

Desouza, K.C. (2011, August 27). *Defining policy informatics.* Retrieved from http://kevindesouza.net/2011/08/defining-policy-informatics/

Desouza, K.C. (2012). Leveraging the wisdom of crowds through participatory platforms: Designing and planning smart cities. *Planetizen: Planning, Design & Development.* Retrieved from http://www.planetizen.com/node/55051

Ganapati, S., & Reddick, C.G. (2012). Open e-government in U.S. state governments: Survey evidence from Chief Information Officers. *Government Information Quarterly, 29*(2), 115–122.

Hansen, D. H., Koepfler, J. A., Jaeger, P. T., Bertot, J. C., & Viselli, T. (2014). Action brokering platforms: Facilitating local engagement with ACTion Alexandria. In *Proceedings of 17th ACM Conference on Computer Supported Cooperative Work and Social Computing* (Baltimore, MD, Feb. 15-19, 2014).

Hardin, R. (1982). *Collective action.* Baltimore: Johns Hopkins University Press.

Heinzelman, J., & Waters, C. (2010). *Crowd-sourcing crisis information in disaster-affected Haiti.* Washington, DC: United States Institute of Peace.

Jaeger, P.T., & Bertot, J.C. (2010a). Designing, implementing, and evaluating user-centered and citizen-centered e-government. *International Journal of Electronic Government Research, 6*(2), 1–17.

Jaeger, P.T., & Bertot, J.C. (2010b). Transparency and technological change: Ensuring equal and sustained public access to government information. *Government Information Quarterly, 27,* 371–376.

Jaeger, P.T., Bertot, J.C., & Shilton, K. (2012). Information policy and social media: Framing government-citizen Web 2.0 interactions. In C.G. Reddick & S.K. Aikins (Eds.), *Web 2.0 technologies and democratic governance: Political, policy and management implications* (pp. 11–25). London: Springer.

Jaeger, P.T., Bertot, J.C., Thompson, K.M., Katz, S.M., & DeCoster, E.J. (2012). The intersection of public policy and public access: Digital divides, digital literacy, digital inclusion, and public libraries. *Public Library Quarterly, 31*(1), 1–20.

Jaeger, P.T., Paquette, S., & Simmons, S.N. (2010). Information policy in national political campaigns: A comparison of the 2008 campaigns for President of the United States and Prime Minister of Canada. *Journal of Information Technology & Politics, 7*, 1–16.

Jaeger, P.T., & Thompson, K.M. (2003). E-government around the world: Lessons, challenges, and new directions. *Government Information Quarterly, 20*(4), 389–394.

Jaeger, P.T., & Thompson, K.M. (2004). Social information behavior and the democratic process: Information poverty, normative behavior, and electronic government in the United States. *Library & Information Science Research, 26*(1), 94–107.

Janssen, M., & Estevez, E. (2008). Lean government and platform-based government: Doing more with less. *Government Information Quarterly, 25*(3), 400–428.

Johnston, E., & Hansen, D.L. (2011). Design lessons for smart governance infrastructures. In A.P. Balutis, T.F. Buss, & D. Ink (Eds.), *Transforming American governance: Rebooting the public square*. Washington, DC: National Academy of Public Administration.

Johnston, E., & Kim, Y. (2011). Introduction to the special issue on policy informatics. *The Innovation Journal, 16*(1), 1.

Khondker, H.H. (2011). Role of new media in the Arab Spring. *Globalizations, 8*(5), 675–679.

Kraut, R., Maher, M.L., Olson, J., Malone, T.W., Pirolli, P., & Thomas, J.C. (2010). Scientific foundations: A case for technology-mediated social-participation theory. *IEEE Computer*, November, 22–27.

Lathrop, D., & Ruma, L. (Eds.). (2010). *Open government: Collaboration, transparency, and participation in action*. Sebastopol, CA: O'Reilly Media.

Liu, H.K. (2012). Open source, crowd-sourcing, and public engagement. In E. Downey & M.A. Jones (Eds.), *Public service, governance, and Web 2.0 technologies: Future trends in social media* (pp. 181–199). Hershey, PA: IGI Global.

Nam, T. (2012). Suggesting frameworks of citizen-sourcing via government 2.0. *Government Information Quarterly, 29*(1), 12–20.

O'Leary, R., Gerard, C., & Bingham, L.B. (2006). Introduction to the symposium on collaborative public management. *Public Administration Review, 66*(s1), 6–9.

Olson, M. (1965). *The logic of collective action: Public goods and the theory of groups*. Cambridge, MA: Harvard University Press.

Page, S. (2010). Integrative leadership for collaborative governance: Civic engagement in Seattle. *The Leadership Quarterly, 21*(2), 246–263.

Pirolli, P., Preece, J., & Shneiderman, B. (2010). Cyberinfrastructure for social action on national priorities. *Computer, 43*(11), 20–21.

Preece, J., & Shneiderman, B. (2009). The reader-to-leader framework: Motivating technology-mediated social participation. *Transactions on Human-Computer Interaction, 1*(1), 13–32.

Sandoval-Almazan, R., & Gil-Garcia, J.R. (2012). Are government internet portals evolving towards more interaction, participation, and collaboration? Revisiting rhetoric of e-government among municipalities. *Government Information Quarterly, 29*, S72–S81.

Starbird, K., & Palen, L. (2012). (How) will the revolution be retweeted? Information diffusion and the 2011 Egyptian uprising. In *Proceedings from the ACM 2012 Conference on Computer Supported Cooperative Work* (pp. 7–16). New York: ACM Press.

Tarrow, S. (1994). *Power in movement: Social movements, collective action and politics*. New York: Cambridge University Press.

Viselli, T. (2011, April 29). The results are in! Alexandria Quality of Life Indicators voting. [Blog post.] Retrieved from http://www.actionalexandria.org/blog/results-are-alexandria-quality-life-indicators-voting

18

BREAKING THE SILOS OF SILENCE

The Importance of New Forms of Knowledge Incubation for Policy Informatics

Ines Mergel

Introduction

Policy informatics is a new field of inquiry defined by Johnston in this volume as "the study of how computation and communication technology is leveraged to understand and address complex public policy and administration problems and realize innovations in governance processes and institutions" (p. 3 of this volume). Policy informatics focuses on the need of scholars and public organizations alike to understand how knowledge is governed in public sector organizations, how these processes can be analyzed, and how their impact on the efficiency and effectiveness of government organizations can be measured, interpreted, and revised based on the insights gained from social computational data analysis methods.

Many new Web technologies entered the public sector during the last 5–10 years. This has brought an influx of innovative knowledge through new forms of citizen interactions that must be processed by government agencies. The most prominent example are social media technologies that are hosted on third party websites and used by government (Mergel, 2012b). These social media applications have become acceptable practice in the public sector and the changes connected with these new practices of interaction and communication are part of the emergence of the policy informatics field. Public sector organizations are experimenting with the use of Facebook pages to represent their agency, Twitter for short updates, wikis for collaborative co-production of content, YouTube and Flickr to share and incorporate videos and photos from citizens, or online contest and challenge platforms to access innovative ideas and solutions from government's diverse audiences. The new technologies provide government organizations with additional channels for information creation, dissemination, and sourcing to fulfill the mandate of the Open Government Initiative to make government more transparent, participatory, and collaborative (Mergel, 2010; The White House, 2009a).

However, the main strategy of current observable use of social media applications is targeted toward broadcasting and educating government's audiences—pushing information out, instead of actively integrating innovative knowledge extracted out of interactions on social media channels (Dunn, 2010). Managing the influx of innovative knowledge that flows into government and between government agencies through social media channels has indeed posed unsolvable challenges for government. It is unclear how knowledge provided by the public is incorporated into new policies, how the interactions with government might change perceptions of citizens about the degree of responsiveness, accountability, or transparency of government, or otherwise impact the effectiveness and efficiency of standard government operating procedures.

Knowledge is only codifiable to a certain extent—not everything that an organization knows is searchable in databases, handbooks, manuals, or standard operating procedures, or obtainable from experts within the focal agency (Anand, Glick, & Manz, 2002; Grant, 1996). The transfer of knowledge, i.e., information that is relevant, actionable, and in part based on experience, is difficult and flows primarily through informal processes such as socialization and internalization (Morrison, 2002; Nonaka & Takeuchi, 1995, 1996).

Two barriers prevent the incorporation of innovative knowledge into government. First, the current information paradigm focuses on one-directional release of highly vetted information to the public. Innovative knowledge is usually acquired only from contractors through pre-defined acquisition processes that prevent free-flowing and emergent forms of innovative knowledge creation. Second, organizational and governance processes must be designed to support bi-directional knowledge flows and the incorporation of newly acquired knowledge into the existing organizational knowledge base. The opportunities to use social media technologies to access innovative knowledge both from inside government and from the public can be defined as a new form of knowledge incubation, i.e., the active solicitation of innovative knowledge from unlikely sources. These include horizontal sources from across the public sector, including other government agencies, and vertical sources, including government employees, and other types of knowledge flowing into government from external sources such as stakeholders.

This chapter provides insights into emergent forms of knowledge incubation from selected pockets within government. It highlights measurement mechanisms for effective knowledge incubation of each of the social media channels. Finally, challenges are analyzed and open research questions are provided to highlight opportunities for future inquiry in the field of policy informatics.

Understanding the Shift in Knowledge Access in the Public Sector

Driven by the mandate of the Transparency and Open Government memo, United States federal government departments and agencies are confronted with the task of identifying ways to increase their organization's transparency, participation, and collaboration efforts (The White House, 2009b). The memo highlights that information is seen as a national asset that needs to be widely shared with citizens, within government, and across governmental boundaries. Unlike previous open government movements, the United States administration recognizes that not all knowledge necessary to run government operations in an efficient and effective manner is available from within government; innovative ideas are widely spread throughout society. The recent Open Government Initiative in the United States has also inspired a worldwide movement. Currently, over 60 countries have come together under the umbrella of the Open Government Partnership to design similar open government processes with a current focus on open data.[1] The European Union has created an open government project to develop open democratic political processes to support efficient governance processes and open transparency and participation processes.[2]

What all of these projects have in common is that they are employing new technologies to increase transparency, participation, and collaboration within and across government and its diverse stakeholders. Using mostly free and open-source social networking platforms, many public sector organizations have begun to explore the usefulness of Web 2.0 tools (Mergel, 2011b; O'Reilly, 2007). The main difference between current open government initiatives and previous waves of e-government is that the new generation of web applications allows government and stakeholders to reach each other in non-traditional ways—interacting with citizens where *they* prefer to receive their information: through the newsfeeds of social networking platforms (Bretschneider & Mergel, 2010; Gladwell & Shirky, 2011). Much of the effort must still be attributed to early experimentation. Outcomes are therefore unclear at this point. In fact, even the most innovative social media directors within United States federal government agencies state they primarily collect anecdotes of success stories to support the resources and capacities used for the implementation of social media applications (Mergel, 2012a).

At the same time, the innovative format of accessing opinions and solutions from sources that are otherwise not part of the decision-making process in the public sector has the potential for knowledge creation and sharing. Knowledge must be used in meaningful ways in government, absorbed into the existing organizational knowledge base, and made available for broader use within the organization. This poses major challenges for government organizations. Innovative knowledge is traditionally not freely shared or widely distributed across government organizations or even within the same department. Knowledge is stuck in so-called "silos of silence"; it is only shared through a predefined email distribution list, hidden in hard drives with limited accessibility, or via restricted distribution in accordance with written rules and regulations (Mergel, 2011b). The notion until now was that innovative knowledge is "sticky"—it is costly to acquire and remains where it was created (von Hippel, 1994). This restricts knowledge transfer and use for problem solving in different locations within the system. The same phenomenon is observable in the public sector, where innovative solutions are outsourced and must be acquired from external contractors. With new social media applications and other technologies, the public sector for the first time has the opportunity to invite the knowledge from the public into the problem solving process. For example, Facebook pages are used to ask citizens for their feedback and open innovation platforms are used to pose problems for which government does not have a solution or cannot provide the answers (Chesbrough, 2003). Moreover, as government embraces social media applications, the public becomes aware of the knowledge needs of the public sector and can volunteer solutions.

The change in attitude toward the use of social media applications as an accepted tool to interact with stakeholders has been primarily market-driven and citizen-centric (Mergel & Bretschneider, 2013). Until recently, the success of social media applications has been attributed to the popularity among teenagers and young adults. However, the use among teens has increased slowly (from 75 percent to 93 percent). The greatest adoption increase has occurred in the older age groups who represent the majority of voters and taxpayers (Carr, 2011). Recent polls show that the use of social media by public agencies has increased each year since the 2008 elections. In 2010, only 14 percent of government workers were willing to use social media to support the mission of their agency, but by 2011, this number had increased to 79 percent (Anderson, 2009). This trend can also be seen among members of Congress. The Congressional Management Foundation reported in 2011 that more than 80 percent of congressional websites now link to external social media sites, e.g., Twitter or Facebook (Hoffman & Fodor, 2010). By comparison, in 2009, the number was only 20 percent. In addition, 50 percent of all congressional committees link to a committee Twitter account, up by 30 percent since 2009, and 40 percent of the committees also provide a link to a committee Facebook page (Hoffman & Fodor, 2010). Although most activities on these accounts focus on traditional broadcasting and sharing of information, a small percentage of members and committees allow interaction and engagement to a certain degree and have opened their Facebook pages to receive comments from the public.

The increased acceptance of Internet use and ubiquitous access to social media platforms have also changed the way that citizens search and access online government information. Government officials recognize that portions of their audience no longer access official government information through federal websites. Other media consumption patterns support the move to social media as a major source of information. The mere popularity of Facebook's newsfeed has reduced direct website hits and subscriptions to paper versions of newspapers. In fact, the rise of social media has resulted in major income loss for the news industry. Many people have stopped subscribing to newspapers. Indeed, 41 percent of respondents to a Pew Research Center survey said they received their news on the Web (Edmonds, 2008). Specifically, 70 percent of the respondents receive news from their family and friends on Facebook and 36 percent receive information via their Twitter newsfeed (Nolan, 1973).

The shift toward social validation of information of news through the connections citizens make on social networking platforms is paired with an increased distrust in government operations. A recent Gallup poll shows that about 80 percent of United States citizens distrust that government will do the right thing (King & Kraemer, 1984; Naaman, Boase, & Lai, 2010). Citizens turn to their trusted contacts on social networking sites to vet the information they read in the news or receive through official government channels before they act on the information. The extreme degree of distrust in government operations indicates a true shift in the information paradigm among citizens. This change in citizen attitude and preference for accessing information opens new opportunities for the public sector to reach citizens where they are and how they prefer to receive their information. New opportunities to rebuild trust also exist through the provision of additional easy-to-use communication channels that take into account the changing behavior of citizens and make it easy for them to transport their opinions, solutions, and suggestions collaboratively into government.

Knowledge Creation, Dissemination, and Sourcing in the Public Sector

Knowledge sharing in the public sector is regulated through rules and regulations. There is a clear sense of hierarchy with predefined reporting structures, standard operating procedures, and laws that channel the flow of knowledge through a set of acceptable technologies and hierarchical structures. This limits or even prohibits free flow of knowledge across organizational boundaries (Desouza, 2009; Trkman & Desouza, 2012). The result is that information produced in one agency might not be available to entities in other corners of the system (Binz-Scharf, Lazer, & Mergel, 2012). The limitations of knowledge sharing impact the spread of best practices or innovative ideas across government. Innovation is flowing into government through costly outsourcing processes only: Government contractors are paid to serve as conduits for innovation creation (Brown, Potoski, & Van Slyke, 2010). Oftentimes, contractors uncover innovation in one part of the system and implement it slightly customized in other parts of the system.[3] Government is spending valuable human and financial resources to reinvent the wheel on a daily basis based on the fact that innovative knowledge is detained within knowledge silos (Noveck, 2009). The Open Government Initiative asserts that government must break these silos of knowledge hoarding and use new technologies creatively to collaborate across organizational boundaries and with the public to make knowledge about innovations accessible.

The Traditional Information Creation and Dissemination Paradigm in the Public Sector

The traditional information paradigm in the public sector has a long history and can be justified based on arguments that usually include accountability or the need to reduce ambiguity in the decision-making processes (Byars, 2012). Government organizations are operating within traditional bureaucratic interactions bound to predefined federal, state, and local levels, departments, or even team boundaries, with very clearly divided tasks and reporting structures. The nature of the tasks to be accomplished is usually so complex that they must be divided into fine grained independent components that can be treated separately while still integrating with the overall objective of the task: service delivery to the public (Blau, 1970; Simon, 1976, p. 69). This bureaucratization has led to elaborate rules and regulations that every member of a government organization must follow.

The traditional information paradigm in the public sector focuses on broadcasting and educational or informational activities. Information is internally vetted through many iterations of an initial draft, signed off by top management, and finally distributed to the public. An elaborate internal process leads to the creation of a final product—such as an announcement of a new policy—that is uploaded in the form of a formal press release to a website or emailed to journalists and only then can be used by the public. Unfortunately, the process of content creation, including background research, decision-making, and the vetting processes are oftentimes not transparent and it is unclear who participated in creating the final document. Similarly, the use and usefulness of the output provided to the public is a black box as well. Very few government agencies attempt to measure who their audiences are and how they access and process the information that is provided. In addition, government rarely provides a transparent or participatory channel to interact with its public about such content. It is often unclear who receives the official statements, how they are reused and processed, and how the public feels about the information they receive.

Social media tools are challenging this traditional "need to know" information-sharing paradigm in which all authority over content and knowledge to be shared resides with government (Dawes, Cresswell, & Pardo, 2009; Desouza & Awazu, 2004). Moreover, the use of social media applications in government has the potential to increase the degree of participation of all stakeholders in the process of creating, maintaining, sourcing, and sharing knowledge. The resulting—partially informal—emerging interactions between the public and government itself are creating opportunities for increased transparency, accountability, participation, and collaboration (OMB, 2009).

Incubating knowledge from these interactions that are facilitated through social media is disruptive and may have transformative effects on knowledge sharing that have not been fully covered in public administration literature (Mergel, Schweik, & Fountain, 2009). Every knowledge sharing process that goes beyond structured information broadcasting that is not systematically covered by the existing information paradigm disrupts regulated information flows and must be absorbed by other information sharing mechanisms, such as informal networks, market mechanisms, and even ad hoc bazaars, without any rules at all (Demil & Lecocq, 2006).

Prior research on knowledge sharing and advice seeking has shown that seeking knowledge from others has clear informational benefits, such as access to solutions, meta-knowledge, problem reformulation, validation, and legitimation (Cross, Borgatti, & Parker, 2001). That is, in part, the reason members of an organization often rely on knowledge from external third parties and must reach out to their informal network contacts (Anand et al., 2002). Informal as well as formal networks of professionals help access knowledge to conduct the tasks within a professional environment that is not accessible in codified form due to its highly intangible and tacit nature (Cross, Rice, & Parker, 2001; Hays & Glick, 1997; Kram & Isabella, 1985; Morrison, 2002). Creative interactions using social media channels can lead to the incubation of innovative knowledge in the public sector when organizations realize the value of permeable boundaries and allow their stakeholders and employees to reach across organizational boundaries to access innovative information that is not available in their own agency. These knowledge networks supported by innovative technologies can help break down knowledge silos by supporting horizontal and vertical knowledge incubation needs.

Social media technology is used in innovative ways in the public sector to include stakeholders in the policy formation and improvement process, in the preparation of decisions, and in the solution sourcing of policy innovations. In addition to the formal traditional information broadcasting model, new forms of social interactions with government stakeholders that are supported by online social networking platforms help large-scale n-2-n interactions. Information is no longer merely pushed out to the public to inform them. Instead, additional channels include a participatory feedback loop to promote co-creation of policies, input of solutions to existing government problems, and an interactive reuse of government-provided information. As an example, many government agencies use social media applications to actively seek out commentary on policy drafts from their audience. The most prominent example is the U.S. Army's call to rewrite its manual using a wiki platform (Cohen, 2009).

The new knowledge incubation paradigm in the public sector does not abandon the traditional information broadcasting model, but instead extends the interaction possibilities and provides conduits for new forms of knowledge exchanges in the public sector. The participative feedback loop shows the new channel for knowledge flowing back and forth between citizens and government.

The following examples portray both intra-organizational as well as inter-organizational knowledge incubation processes with the use of wikis.

Intra-organizational Knowledge Incubation

A prominent example for an innovative intra-organizational form of knowledge incubation in the public sector is Intellipedia, often called the CIA's Wikipedia (Andrus, 2004; Lawlor, 2008). It was initiated by the Office of the Director of National Intelligence and is designed to collaboratively capture all knowledge available by content area across all 18 intelligence agencies in the United States. Users have access through three different security levels, and knowledge areas are connected to knowledge specialists using a design component that is internally compared to leaving bread crumbs: Instead of revealing potentially top secret sources, the authors of Intellipedia leave only enough information to be identified as knowledge experts and their contact information as a way to vet information directly to the knowledge seeker. The goal of Intellipedia is to provide access to a wider knowledge pool more efficiently, and ultimately reach a more informed level of decision accuracy. The goal is to integrate knowledge created on the vertical as well as the horizontal level across different agencies within government to break up knowledge silos and build a basis for improved decision-making using a wiki application. The information-sharing environment is supported by a host of Web 2.0 applications, such as iVideo, blogs, shared documents, collaboration spaces, and photo galleries.

A knowledge requestor must follow the hierarchical request structure, requesting knowledge access from superiors in other parts of the system. On the top level, information is requested and knowledge experts deeper in the hierarchy are identified, a process that can take days or even weeks.

Within the new knowledge incubation paradigm, a wiki serves as an accelerator for knowledge acquisition. Mostly independent of the hierarchical position or geographic location of a knowledge expert in government, participants collaboratively create knowledge and are using the wiki to make it widely available to everyone in government who is working on a related problem (Mergel, 2011a, 2011c).

Inter-organizational Knowledge Incubation

Similarly, prominent wikis include Diplopedia in the State Department (Cohen, 2008), or Techipedia in the Department of Defense. United States soldiers in the battlefield have implemented wiki technology to speed up peer-to-peer information about battlefield conditions. Initial concern by higher-level officers about the break in chain-of-command information-flow and loss of control over the message were assuaged with the success of the wiki. These examples of increased collaboration in unlikely environments such as the highly regulated and compartmentalized command-and-control culture of the intelligence community can serve as a model for other inter-organizational collaboration and may move government from a need-to-know to a need-to-share-and-cocreate knowledge paradigm.

While government organizations are experimenting with the use social media applications, such as the microblogging service, Twitter, or Facebook, it has become clear that unsolicited comments and reuse of government information leads to unmanageable information overload. Agencies have very little means to streamline the inflow of comments via social media applications and legitimacy of the government's use of social media is therefore suffering.

Alternative means of joint knowledge incubation have tackled this problem. Driven by a White House initiative to offer prizes and challenges to promote open government, the General Services Administration (GSA) set up a new platform called Challenge.gov. As part of the Strategy for American Innovation, the White House promotes the use of online participation platforms and the policy instrument with prizes and challenges to harness innovation from the public (The White House, 2010).

Following the logic of innovation incubation, online participation platforms provide a controlled approach for accessing and promoting innovative knowledge creation from unlikely places that until now were inaccessible by government. Agencies are using Challenge.gov as a platform to seek solutions to government problems, ranging from technology solutions to image and branding campaigns. Comparable platforms even allow the public to vote on solutions—so that government can choose among the most popular solutions. Depending on the governance model, government can go back to the solution creator and ask them to collaborate on improving the solution from an early idea to the final implementation stage (Schweik, Mergel, Sanford, & Zhao, 2011).

Other forms of fast access to information and knowledge across hierarchical boundaries are emerging in government; forms that can be described as innovation accelerators. The privately run social networking site, GovLoop.com, can be described as an informal knowledge incubator. Members are government professionals who organize themselves around issues such as acquisition, leadership, careers, and communications, and freely share advice and insights from their own work environments. They discuss and share problem solutions that have been successfully implemented in other parts of the vast government system. But practices have not spread or have not been approved as a new standard operating procedure in the public sector. One example is the Acquisition 2.0 discussion group on GovLoop where government officials, industry, and journalists discussed the redesign of the federal acquisition process, which was ultimately implemented at the General Services Administration.[4]

These examples show that new technological solutions provide innovative forms of social interactions in the public sector and create opportunities to enhance transparency, communication, and collaboration in government, and may promote deeper levels of civic engagement.

Challenges for Knowledge Incubation in the Public Sector

Changing an existing information paradigm involves challenges and barriers on many levels, including organization culture, governance, and measurement challenges.

Challenges concerning the existing organizational culture focus on the incentives that must be in place for innovative knowledge incubation and acquisition. On one hand, government actors must be willing to step outside existing knowledge-sharing routines and tap into innovative sources of novel information. On the other hand, they need to design user rewards that support scalable contributions from the public. As an example, gamification elements, such as voting mechanisms and platform rewards, can help create an inner competitiveness within the platforms that motivate people to vote on each other's ideas or provide additional help to improve solutions (Oxley, 2011). Other platforms work with social recognition leadership elements to provide updates on how knowledge is weighed in comparison to others. Accumulated points can be traded in for benefits, such as lunch with government executives or financial incentives. Nevertheless, there will always be the potential for selection bias: Those participants willing to take the time to contribute their knowledge may be primarily digital natives or newcomers who believe that this form of knowledge incubation is the existing standard. Other participants may try to game the system by collecting votes on ideas that are off-topic and would need to be removed from the knowledge incubation process. Even if ideas are not implemented, continuous feedback loops and increased transparency in the decision-making process can motivate participants to check back.

Governance challenges for the integration of innovative knowledge into the existing organizational knowledge base may require changes to existing government routines. These challenges include the not-invented-here syndrome and myopic thinking, where government is still seen as the sole source of innovative information and outside knowledge is not accepted (Katz & Allen, 1982). To make the boundaries of public sector organizations more permeable, government organizations will need to allow non-experts and outliers to infuse innovative ideas into the policy making process. The most difficult process will be to integrate innovative external knowledge into the existing organizational knowledge base and into existing governing routines, to facilitate knowledge sharing across organizational boundaries and all levels of government.

One of the most important and least tackled challenges is the data measurement challenge. The policy informatics field must develop, evaluate, and test new metrics that help gain insights into *big data* and transfer the insights to the actionable micro-level of government. Big data discussions have been ongoing for some time and focus on the analysis of "social phenomena" regarding computing and advanced information processing technologies (Rasmussen, 2009). Large-scale analyses of the use of social networking sites, such as Twitter, Facebook, and citizen interactions on participatory open innovation platforms, will help researchers understand the emerging processes of online interactions. Beyond actual interactions, it will also be a challenge to identify quality content and decide whether innovative knowledge extracted from online interactions is of sufficiently high quality that should be integrated into governing processes. The main platforms, Facebook and Twitter, do not allow automatic data downloads to facilitate analysis. This hurdle must be approached directly with the third-party platform providers similar to how government-compatible Terms of Service agreements were negotiated.

Similar to analyses run on Wikipedia documents and crowd-sourced and large-scale contributions, the challenge will be to extract the golden nuggets, the truly innovative knowledge that is worth extracting. Current measurement techniques focus on quantitative contributions, but ignore the socialness of knowledge, who contributed it, who supported the contributions, and how the contributions were finally accepted into government. Finally, it will be a challenge to interpret the results of the massive amounts of data now available to be analyzed, to employ the right people for its interpretation, and to convince top management to act on the interpretation of the results (Burke, 2008).

Future Research Opportunities in Policy Informatics

The field of policy informatics opens many fresh venues for future research. Among them are insights into the challenges listed in the previous section, including the need to gain a better understanding of government innovation and knowledge incubation processes. Governance mechanisms should be designed for public managers to easily identify and open pathways of innovation that might question existing principles of knowledge and information vetting in government and potentially lead to changes in existing knowledge-sharing procedures of the public sector. Designing processes that streamline innovative knowledge incubation in the public sector must become accepted parts of organizational routine and must be integrated into the information-sharing hierarchy that is still the standard in most bureaucracies.

Finally, the main challenge for any knowledge incubation process under the policy informatics research angle is the need for more research on the impact of innovation on the public sector. Researchers and public managers alike need to understand whether innovative knowledge accepted into the organization is truly an improvement and whether it will help make government more effective and efficient. Exploring, developing, and testing appropriate metrics will be the necessary basis for formulating interventions into policy informatics processes. In addition, evaluating and predicting behavioral changes based on the insights derived from the data collected will help to design online emergent social process interventions (Calabresi, 2009; Mergel, 2012a).

Notes

1 See http://www.opengovpartnership.org for more information about the Open Government Partnership and participating countries.
2 See http://eu-opengovernment.eu/opengovernment/ for a description of the project goals and the Open Government Data Initiative: http://www.govdata.eu/.
3 See for example, the design of websites by contractors among Members of Congress in the United States, *Gladwell and Shirky (2011).*
4 See http://www.govloop.com/group/acquisition20.

References

Anand, V., Glick, W.H., & Manz, C.C. (2002). Thriving on the knowledge of outsiders: Tapping organizational social capital. *Academy of Management Executive, 16*(1), 87–101.

Anderson, C. (2009). *The long tail.* Retrieved from http://www.longtail.com/the_long_tail/

Andrus, D.C. (2004). The wiki and the blog: Toward a complex adaptive intelligence community. *SSRN—Social Sciences Research Network.* Retrieved from http://ssrn.com/abstract=755904

Binz-Scharf, M., Lazer, D., & Mergel, I. (2012). Searching for answers: Networks of practice among public administrators. *American Review of Public Administration, 42*(2), 202–225.

Blau, P.M. (1970). A formal theory of differentiation in organizations. *American Sociological Review, 35*(2), 201–218.

Bretschneider, S.I., & Mergel, I. (2010). Technology and public management information systems: Where have we been and where are we going. In D. C. Menzel & H.J. White (Eds.), *The state of public administration: Issues, problems and challenges* (pp. 187–203). New York: M.E. Sharpe, Inc.

Brown, T.L., Potoski, M., & Van Slyke, D.M. (2010). Contracting for complex products. *Journal of Public Administration Research and Theory*, 20(Suppl. 1), i41–58.

Burke, D. (2008). CIA presenting Intelipedia—Social media for govt conference (video). Retrieved from http://www.youtube.com/watch?v=wuSrbZpXu40

Byars, N. (2012). Government websites. In M. Lee, G. Neeley, & K. Stewart (Eds.), *The practice of government public relations.* Boca Raton, FL: CRC Press.

Calabresi, M. (2009, April 8). Wikipedia for spies: The CIA discovers Web 2.0. *Time Magazine.*

Carr, D.F. (2011, June 21). Social software needs metrics that matter. *Information Week.* Retrieved from http://www.informationweek.com/thebrainyard/news/community_management_development/231000075

Chesbrough, H.W. (2003). *Open innovation: The new imperative for creating and profiting from technology.* Cambridge, MA: Harvard Business Press.

Cohen, N. (2008, August 3). On Web, U.S. diplomats learn to share information. *International Herald Tribune.* Retrieved from http://www.iht.com/articles/2008/08/03/technology/link04.php

Cohen, N. (2009, August 13). Care to rewrite Army doctrine? With ID, log on. *New York Times.* Retrieved from http://www.nytimes.com/2009/08/14/business/14army.html

Cross, R., Borgatti, S.P., & Parker, A. (2001). Beyond answers: Dimensions of the advice network. *Social Networks, 23*(3), 215–235.

Cross, R., Rice, R.E., & Parker, A. (2001). Information seeking in social context: Structural influences and receipt of information benefits. *IEEE Transactions on Systems, Man and Cybernetics, Part C-Applications and Reviews, 31*(4), 438–448.

Dawes, S.S., Cresswell, A. M., & Pardo, T.A. (2009). From "Need to Know" to "Need to Share": Tangled problems, information boundaries, and the building of public sector knowledge networks. *Public Administration Review, 69*(3), 392–402.

Demil, B., & Lecocq, X. (2006). Neither market nor hierarchy nor network: The emergence of bazaar governance. *Organization Studies, 27*(10), 1447–1466.

Desouza, K. (2009). Information and knowledge management in public sector networks: The case of the US intelligence community. *International Journal of Public Administration, 32*(4: Special Issue: Security Issues for Public Administration), 1219–1267.

Desouza, K., & Awazu, Y. (2004). "Need to Know"—Organizational knowledge and management perspective. *Journal Information-Knowledge-Systems Management, 4*(1), 1–14.

Dunn, B. (2010). Best Buy's CEO on learning to love social media. *Harvard Business Review*, December, 43–48.

Edmonds, R. (2008). *Is the newspaper industry dying?* Retrieved from http://stateofthemedia.org/2007/newspapers-intro.

Gladwell, M., & Shirky, C. (2011). From innovation to revolution: Do social media make protests possible? *Foreign Affairs*, March/April. Retrieved from http://www.foreignaffairs.com/articles/67325/malcolm-gladwell-and-clay-shirky/from-innovation-to-revolution

Grant, R.M. (1996). Toward a knowledge-based theory of the firm. *Strategic Management Journal, 17*, 109–122.

Hays, S.P., & Glick, H.R. (1997). The role of agenda setting in policy innovation: An event history analysis of living-will laws. *American Politics Research, 25*, 497–516.

Hoffman, D., & Fodor, M. (2010). Can you measure the ROI of your social media marketing? *MIT Sloan Management Review*, Fall, 41–49. Retrieved from http://sloanreview.mit.edu/the-magazine/2010-fall/52105/can-you-measure-the-roi-of-your-social-media-marketing/

Katz, R., & Allen, T.J. (1982). Investigating the not invented here (NIH) syndrome: A look at the performance, tenure, and communication patterns of 50 R & D project groups. *R & D Management, 12*(1), 7–19.

King, J.L., & Kraemer, K. (1984). Evolution and organizational information systems: An assessment of the Nolan stage model. *Communications of the ACM, 27*(5), 466–475.

Kram, K.E., & Isabella, L.A. (1985). Mentoring alternatives: The role of peer relationships in career development. *Academy of Management Journal, 28*(1), 110–132.

Lawlor, M. (2008). Web 2.0 intelligence. *Signal, 62*(8), 63.

Mergel, I. (2010). Government 2.0 revisited: Social media strategies in the public sector. *PA Times, 33*(3), 7–10.

Mergel, I. (2011a). Crowdsourced ideas make participating in government cool again. *PA Times, American Society for Public Administration, 34*(4), 4–6.

Mergel, I. (2011b). The use of social media to dissolve knowledge silos in government. In R. O'Leary, S. Kim, & D. VanSlyke (Eds.), *The future of public administration, public management, and public service around the world* (pp. 177–187). Washington, DC: Georgetown University Press.

Mergel, I. (2011c). Using wikis in government: A guide for public managers. In IBM Center for The Business of Government (Ed.), *E-Government*. Washington, DC: IBM.

Mergel, I. (2012a). Measuring the effectiveness of social media tools in the public sector. In E. Downey & M. Jones (Eds.), *Public service, governance and Web 2.0 technologies* (pp. 48–64). Hershey, PA: IGI Global.

Mergel, I. (2012b). *Social media in the public sector: Participation, collaboration, and transparency in the networked world*. San Francisco, CA: Jossey-Bass/Wiley.

Mergel, I., & Bretschneider, S.I. (2013). A three-stage adoption process for social media use in government. *Public Administration Review, 73*(3), 390–400.

Mergel, I., Schweik, C.M., & Fountain, J. (2009). The transformational effect of Web 2.0 technologies on government. *SSRN—Social Science Research Network*. Retrieved from http://ssrn.com/abstract=1412796

Morrison, E.W. (2002). Newcomers' relationships: The role of social network ties during socialization. *Academy of Management Journal, 45*(6), 1149–1160.

Naaman, M., Boase, J., & Lai, C.-H. (2010). Is it really about me? Message content in social awareness streams. In *Proceedings from CSCW '10: ACM Conference on Computer Supported Cooperative Work*. Retrieved from http://research.microsoft.com/en-us/um/redmond/groups/connect/CSCW_10/docs/p189.pdf

Nolan, R.L. (1973). Managing the computer resource: A stage hypothesis. *Communications of the ACM, 16*(7), 399–405.

Nonaka, I., & Takeuchi, H. (1995). *The knowledge-creating company*. New York: Oxford University Press.

Nonaka, I., & Takeuchi, H. (1996). The knowledge-creating company: How Japanese companies create the dynamics of innovation. *Long Range Planning, 29*(4), 592.

Noveck, B. (2009). *Open government laboratories of democracy*. Retrieved from http://www.whitehouse.gov/blog/2009/11/19/open-government-laboratories-democracy

O'Reilly, T. (2007). What is Web 2.0: Design patterns and business models for the next generation of software. *Communications & Strategies, 65*(1), 17–37.

Office of Management and Budget (OMB). (2009, December 8). *Open government directive*. Washington, DC: Executive Office of the President of the United States. Retrieved from http://www.whitehouse.gov/open/documents/open-government-directive

Oxley, A. (2011). *A best practices guide for mitigating risk in the use of social media*. Retrieved from http://www.businessofgovernment.org/report/best-practices-guide-mitigating-risk-use-social-media

Rasmussen, C. (2009). *Living intelligence*. Retrieved from http://www.youtube.com/watch?v=nbgQ1V2BLEs

Schweik, C., Mergel, I., Sanford, J., & Zhao, J. (2011). Toward open public administration scholarship. *Journal of Public Administration Research & Theory, 21*(Supplement 1 [Minnowbrook III: Special Issue]), i175–i198.

Simon, H. (1976). From substantive to procedural rationality. In S.J. Latsis (Ed.), *Method and appraisal in economics*. Cambridge: Cambridge University Press.

Trkman, P., & Desouza, K. (2012). Knowledge risks in organizational networks: An exploratory framework. *The Journal of Strategic Information Systems, 21*(1), 1–17.

von Hippel, E. (1994). "Sticky information" and the locus of problem solving: Implications for innovation. *Management Science, 40*(4), 429–439.

The White House. (2009a). *The open government initiative*. Retrieved from http://www.whitehouse.gov/open

The White House. (2009b). *President's memorandum on transparency and open government - interagency collaboration*. Retrieved from http://www.whitehouse.gov/sites/default/files/omb/assets/memoranda_fy2009/m09-12.pdf

The White House. (2010). *A strategy for American innovation: Securing our economic growth and prosperity*. Retrieved from http://www.whitehouse.gov/innovation/strategy, http://www.whitehouse.gov/sites/default/files/omb/assets/memoranda_2010/m10-11.pdf

PART VI
Conclusion

19

THE FUTURE OF POLICY INFORMATICS

Justin Longo, Dara M. Wald, and David M. Hondula

The preceding pages in this volume represent the first attempt to bring together in book form a collection of scholars, their thoughts, and evidence, with the objective of illustrating various essential elements of what policy informatics is and what it offers society. This edited collection set out some of the themes in this emerging field, demonstrated some of the specific methodologies and approaches under the policy informatics banner, and provided specific examples set in context for appreciating the contribution policy informatics can make in addressing complex public policy challenges now and in the future.

At the outset, policy informatics was defined as *the study of how computation and communication technology is leveraged to understand and address complex public policy and administration problems and realize innovations in governance processes and institutions.* Beyond the 'how' and 'why' of policy informatics, however, lies the 'what,' 'where from,' and 'where to' of policy informatics as a field of study. In this concluding chapter, at this point in the early development of the field, it is useful to consider where policy informatics comes from, where it appears to be heading, and what the field can hope to offer the future. We start by considering the multidisciplinary origins of policy informatics: how the fields of information science, mass communication, and policy analysis—three key foundations of policy informatics— have each developed independently, and how they together influence the character of policy informatics and frame future development of the field. In transitioning between this assessment of its past and a consideration of its future, we pause to evaluate the place of policy informatics in the spectrum of disciplines, interdisciplinary fields, and research areas. We then look to the future, noting some emerging complex policy challenges before considering how the field is positioned to respond in a way that takes advantage of both its foundations and emerging exogenous forces, specifically accelerating technology development and momentum toward opening governance.

The Foundations of Policy Informatics

In its most basic sense, policy informatics is an informatics approach to the study of public policy. Since the term 'informatik' was first coined by Karl Steinbuch in 1957 to describe the then-field of computer science, 'informatics' has expanded beyond its narrow interest in the evaluation of scientific information to include a number of subfields, including bioinformatics, health informatics, and now policy informatics. We expand on the informatics background ahead, but those unfamiliar with the term policy informatics need not feel embarrassed as the field is certainly new by the standards of many fields. And owing to its newness, the depth of the field is relatively thin—for now.

For an indication of the relative newness of policy informatics as a distinct field, we need only glance at the arbiter of relevance and salience in the digital era: Google. Whether through the *Ngram Viewer*,[1] which searches for the presence of particular phrases in the over five million books digitized by Google (Greenfield, 2013), or in *Trends*,[2] which measures how often a particular search term is entered relative to all searches and thus provides a window into what issues appear to matter across society (Choi & Varian, 2012), meaning, identity, and even existence are often determined by the data amassed by the search engine giant. Based on those measures, policy informatics barely registers. We certainly find the concept useful, but must admit that the term 'policy informatics' can hardly be said to be in common usage.

Policy informatics is also an understandably thin field. The authors collected in this volume are among the core of policy informatics with very few of the field's leaders absent from this book. The field's early innovators can also be found in a small number of other institutions such as the Virginia Bioinformatics Institute (Advanced Computing and Informatics Lab[3]), articulations of the policy informatics approach (e.g., Dawes & Janssen, 2013; Helbig, Nakashima, & Dawes, 2012), and collections such as a special issue on policy informatics in the *Public Sector Innovation Journal* (Johnston & Kim, 2011), and a recent special issue on policy informatics in the *Journal of Policy Analysis and Management* (Desai & Kim, 2015). The overlap among these initiatives and authors confirms that the policy informatics community is both relatively small and cohesive, although the sustained and increasing level of activity in this space points toward the field's rapid growth. Within that community, the conceptualization of policy informatics is also relatively cohesive. Some emphasize the data and computational aspects of informatics as applied to policy problems (Barrett et al., 2011), whereas others point toward the shifting notions of governance as being key to the field (e.g., Johnston & Hansen, 2011), but we do not anticipate that many would object to this volume's definition repeated at the top of this chapter.

We also believe that the way we have segmented the broad disciplines and fields that provide policy informatics with its foundations would not cause many in the community to register a strenuous objection. Although our characterization here is obviously influenced by our own disciplinary perspectives, and we do not suggest that these cover all of the traditions that inform the policy informatics movement, the following builds on some of the keywords in the foregoing definition—specifically "computation," "communication," and "public policy"—to consider three disciplinary foundations for policy informatics: information science, mass communications, and policy analysis approaches. Although these are by no means comprehensive across the entire field of policy informatics, nor do they represent the entire breadth of the community, we argue that they address much of what is implied under the heading policy informatics.

The first foundational leg of the emerging field of policy informatics might be best described as *information science*, a body of scholarship that has itself faced identity challenges throughout its growth that persist into modern academia. The concept of information management (and the study of that process) has existed since the advent of human society. Although important lessons from its past (e.g., the advent of the scroll) offer valuable lessons for our future (e.g., Rayward 1996), our interest is in the evolution of information science in recent decades. In the post–World War II period, the advances in disciplines connected to information science that policy informatics draws on are significant: ever-expanding computational capabilities with respect to the speed and volume at which information can be stored, processed, analyzed, and visualized; the spread of mobile technologies that permit access to information from a seemingly infinite number of locations and situations; technological improvements that allow for smaller and more affordable sensors with a wide array of capabilities; and methods by which such information and analysis can be archived, organized, accessed, preserved, and communicated. These advances all provide opportunities for improving societal capacity to address its most pressing problems, with information science aiding the transition of data to knowledge and ultimately to societal wisdom (see chapter 3, in this volume).

Common terminology that we consider proximate to our definition of information science includes "computer science," "data science," "big data," "general systems theory," "information theory," and "informatics" (e.g., Hjørland, 2014; Rayward, 1996). We stress that policy informatics, with its shared roots in policy analysis and communication, offers a more humanist perspective than the bleak outlook painted by Anderson (2008) for big data, in which data collection and analysis replace the theoretical foundations of nearly all disciplines. But is it possible to isolate information science as its own, independent entity, from which policy informatics may build? And where does it sit with respect to other disciplines with a firmer theoretical foundation?

In some aspects, information science provides no sense of identity beyond science itself, given Rayward's (1996, p. 4) suggestion that "almost everything could be argued to be information." Machlup and Mansfield (1983, p. 22) articulate an only slightly refined vision, arguing that information science is "a rather shapeless assemblage of chunks picked from a variety of disciplines that happen to talk about information in one of its many meanings." A widely applicable definition of information science is, they contend, impractical, based on the different interpretations and meanings of information within different disciplinary settings (Machlup & Mansfield, 1983). Indeed the concept of interdisciplinary is a central tenet of many of the other definitions for information science brought forward in the 20th century, including the notion of information science as an "interdiscipline" (Rayward, 1996). We identify the aspects of information science most relevant for policy informatics to be those that are between the subfields of computer-and-information science and library-and-information science identified by Machlup and Mansfield (1983), the former of which focuses on the design and use of computers, and the latter on the improvement of systems by which records and documents are acquired, stored, retrieved, and displayed (Rayward, 1996). Alternatively, Machlup and Mansfield offer a tightly focused view that sits at the intersection of the two, deemed "narrow" information science, covering key elements of our vision of policy informatics (allowing for a broad consideration of the term 'information system'), including novel methods of information exchange, control of access to information, modeling and computer simulation of information systems and networks, and studies of the character and behavior of users of information systems and services.

Connections to the other foundations of policy informatics—mass communication and policy analysis—are obvious, but the contributions from information science rest in the technological and theoretical infrastructure within which the questions of the modern policy informatician are addressed. Surrounding the elements presented earlier, Rayward's (1996, p. 11) definition of information science becomes attractive: "[modern] attempts to study in a formal and rigorous way processes, techniques, conditions, and effects that are entailed in improving the efficacy of information, variously defined and understood, as deployed and used for a range of purposes related to individual, social and organizational needs."

From a more utilitarian perspective, it is the roots in information science that place the word "informatics" in policy informatics. Definitions for informatics date at least as far back as the middle of the 20th century, including Mikhailov, Chernyi, and Giljarevskij's (1967, p. 238) suggestion that "informatics is the discipline of science which investigates the structure and properties of scientific information, as well as the regularities of scientific information activity, its theory, history, methodology, and organization." Hjørland (2014) recognizes that many sources consider informatics and information science to be synonyms, but ultimately concludes that the term informatics in and of itself, which has a connotation closer to computer science than library science, has little value except for notable exceptions: the use of so-called 'compound terms,' like medical informatics, social informatics, and, of course, policy informatics.

Like its academic 'compound term' relatives, policy informatics faces challenges within academic institutions. Among these are concepts of identity and terminology:

> social informatics studies are scattered in the journals of several different fields, including computer science, information systems, information science, and some social sciences. Each of these fields uses somewhat different nomenclature. This diversity . . . makes it hard for many nonspecialists . . . to locate important studies.
>
> *(Kling 2007, p. 205)*

More substantially, this lack of identity and the appropriate institutional framework for emerging areas of scholarship presents problems in the identification of merit, as reported by Greenes and Shortliffe (1990, p. 1119) in the field of medical informatics: "The unique nature of the medical informatics field was exemplified when the thesis committee unanimously acknowledged that the work was original and fully worthy of a doctorate, although none felt that the scope, content, and emphasis would have matched precisely with the advanced degree requirements of their own departments." Our own experience suggests that researchers in policy informatics could face similar stresses in the years ahead as most academic institutions are limited in their capacity to rapidly adapt their framework to appropriately recognize such 'interdisciplines.' It is likely that the emerging policy informatics community at large faces analogous challenges.

Just as information science struggled to define itself as an 'interdiscipline,' the second foundational leg of policy informatics—communication research—has also faced serious questions about the range of fields, research areas, and disciplines it can claim. The aforementioned definition of policy informatics that frames the discussion in this book centers on the importance of communication in the development of innovative governance arrangements. An understanding of how information is processed and communicated is a prerequisite to fostering effective and informed decision-making. In this respect, policy informatics also derives much of its intellectual infrastructure from the field of communication research. Mass communication research as a field of inquiry also owes much to Harold Lasswell's work in policy sciences (see ahead). After all, it was Lasswell who first described social science as "for the intelligence needs of an age" (Peters, 1986, p. 535). It was from this tradition that leading interdisciplinary scholars at the University of Illinois and University of Chicago—including Wilbur Schramm (considered by many to be the founder of communication studies), the Hutchins Commission on Freedom of the Press, and Douglas Waples (a friend and collaborator of Lasswell's)—drew the boundaries of the field of communications studies (Rogers & Chaffee, 1994; Wahl-Jorgensen, 2004). During World War II, both Schramm and Waples worked in the Office of War Information, the domestic bureau in charge of wartime propaganda (Peters, 1986; Wahl-Jorgensen, 2004). Much of the early research in the field of communication stemmed from political concerns about persuasive communication and the proper role of the media in a democracy. In this respect, communication research originated as an exemplar of the policy sciences (Peters, 1986). Initially, the field attempted to conform to existing disciplinary boundaries, developing boundary organizations like the Institute of Communications Research[4] at the University of Illinois (established by Wilbur Schramm in 1947) and the Committee on Communication at the University of Chicago (1947–1960) to mediate the differences between the disciplines (Herbst, 2008; Wahl-Jorgensen, 2004). These efforts ultimately led to the founding of the International Communication Association in 1950 and its flagship *Journal of Communication* in 1951 (Herbst, 2008). Such organizations, journals, and identified subfields such as political communications (Chafee & Hochheimer, 1985) were developed in an attempt to define the boundaries of what was to become the field of mass communication (Herbst, 2008).

But tension between interdisciplinary research driven by shared interests and the institutionalization of the field along traditional disciplinary boundaries resulted in clashes among scholars. These conflicts culminated in a public debate published in *Public Opinion Quarterly* in 1959. The debate featured Wilbur Schramm and colleagues writing in response to Bernard Berelson, a professor of behavioral sciences recognized as a preeminent scholar of public opinion and communications and one of the founders and leaders of the Committee on Communication at the University of Chicago (Sils, 1980). In what has come to be described as his "obituary of communication study" (Wahl-Jorgensen, 2004, p. 561) Berelson famously asserted "as for communication research, the state is withering away" (1959, p. 1), to which Schramm responds that in death the communication field is in "a somewhat livelier condition than I had anticipated" (Schramm, Riesman, & Bauer, 1959, p. 6) and suggests that, although not without its problems, communication research "is an extraordinarily vital field at the moment, with a competent and intellectually eager group of young researchers facing a challenging set of problems" (Schramm et al., 1959, p. 9). Despite criticism of Schramm's response as a "self-celebration he constructed himself" (Wahl-Jorgensen, 2004, p. 561), his role as an advocate for a fledgling field helped propel the field of mass communication forward to establish it as a legitimate discipline recognized and supported at many of the top institutions across North America (Herbst, 2008; Wahl-Jorgensen, 2004).

Due to the "historically permeable borders and openness of Communication as a discipline" (Herbst, 2008, p. 607), it developed with a "determined eclecticism" (Menand, 2001). Some have argued that the broad origins and disciplinary nature of communication theory led to a richness of ideas without a guiding domain, set of theories, or disciplinary goals (Craig, 1999). However, others have argued that the conscious embrace of "the epistemological proposition of determined eclecticism" (Herbst, 2008, p. 608)—instead of a determined effort to ground the field within the boxes outlined by other disciplines—contributed to an expeditious recognition of communication as a legitimate discipline organized around a diverse set of methods, questions, and theories. As a new field struggling to create a position for itself within the academy, policy informatics could learn much from the disciplinary struggles fought by fields like mass communication. Just as communication scholars battled over how to develop a field grounded in a unified set of theories (see Craig, 1999, 2001; Myers, 2001), the development of policy informatics is likely to create a similar theoretical skirmish. In this era of information and post-disciplinarity, where "organizing structures of disciplines themselves will not hold" (Case, 2001, p. 150), does it make sense for policy informatics to try to justify the boundaries of our discipline or instead embrace the pursuit of novel questions and disciplinary eclecticism? Before we address this question directly, we explore a third foundational leg of policy informatics.

Turning to the part of this volume's definition of policy informatics focused on complex public policy and administration problems, we see how policy informatics owes much to the interdisciplinary field of *the policy sciences*, or what has ultimately come to be called *policy analysis*.[5] Interest in increasing the relevance of the social and natural sciences for informing government decision-making preceded World War II (Hall, 1989), but it was through the publication of Lerner and Lasswell's edited volume *The Policy Sciences* (1951) that an integrated, multidisciplinary approach to the study of public problems first took shape. Lasswell drew on what he saw as the best elements of the social sciences—principally, the disciplines that emphasized quantitative methods in their inquiry—and, adapting the American pragmatism of John Dewey and others, hoped for a scientific approach to studying "the fundamental problems of man in society, rather than upon the topical issues of the moment" (Lasswell, 1951, p. 8). Lasswell displays a particular respect for the advances made in economics and psychology during the first half of the 20th century, a perspective that lies at the root of the rational approach to policy analysis and the belief that human behavior can be objectively observed, quantitatively analyzed, and accurately predicted—a perspective that continues to influence the field (Morçöl, 2001).

It seems a reflection of the particular point in history at which Lasswell was writing—immediately following the formative experiences of the Great Depression and World War II, immersed in a "crisis of national security" and "the urgency of national defense" (1951, p. 3) as motivators for a more rational and scientific approach to governing and benefiting from the advanced state of social science method—that led him to highlight the policy sciences as the great hope for the advancement of the human condition at about the same time that Vannevar Bush was promoting the potential contribution of computer technology to the same end (Bush, 1945) and Wilbur Schramm was establishing the field of communications research (see earlier).

Standing as a bookend to Lasswell's seminal conceptualization of the policy sciences in 1951 was his introductory article in the inaugural issue of the journal

Policy Sciences (Lasswell, 1970), and the subsequent expansion of those ideas in book length (Lasswell, 1971), where he characterized the policy approach as problem-oriented, multidisciplinary, set within a wider social context, and explicitly normative. Edward Quade's introductory editorial to the first issue of *Policy Sciences*, although seeking to advance the quantitative revolution that motivated the policy science approach, goes to some lengths to downplay the expectations that can be placed upon the management and decision sciences as they are further deployed in public policy areas. He calls the policy sciences an effort "simply to augment, by scientific decision methods and the behavioral sciences, the process that humans use in making judgments and taking decisions" (Quade, 1970, p. 1). Although the new journal sought to publish "hard" papers that "keep the analytical component up," Quade stressed that the new discipline of the policy sciences must also recognize "extrarational and even irrational processes as sources of knowledge" (1970, pp. 1–2).

This distinction proved to be prescient for the future of the policy analysis movement. Following the dominance of analytical methods throughout the 1970s (Yang, 2007), profound shifts away from traditional analytical activities undertaken by policy analysts, and toward public management functions (Howlett, 2011) and the providing of support for the political agendas of ruling parties (Forester, 1995), began to take hold—the very status Lasswell sought to rescue political economy from in the 1950s. As much as policy analysis is usually considered distinct from politics, the post-positivist policy perspective highlights the normative basis of policy analysis and the crucial role that politics plays in the process (Fischer, 2003; Mayer, van Daalen, & Bots, 2004; Meltsner, 1976; Mouffe, 2000; Stone, 1997). Policy analysis continues to struggle with the alternative views from within the field as to how it should evolve, between a return to its quantitative roots and a further embrace of post-positivist efforts to democratize policy analysis (Morçöl, 2001).

These existential struggles define the boundaries and illuminate the various perspectives within policy analysis, but are also useful for helping us understand the emerging field of policy informatics. With the world of policy analysis marked by "ambiguity, relativism and self-doubt" (Lawlor, 1996, p. 120), policy informatics re-engages these debates by pursuing the contribution that both information sciences and communications research can make in resolving emerging complex policy problems. To be clear, policy informatics should not be thought of as a technological solution to the problems of society, but rather as the appreciation of the role that technology can play as part of a toolset for helping society find solutions to complex problems. That policy informatics searches for those solutions equally in databases, algorithms, formal models, participatory platforms, and civic deliberation confirms policy informatics as a direct descendant of policy analysis.

One of the questions that often gets asked of new academic subject areas is whether it represents an emerging discipline, a status that many wear as a badge of honor. There is no shortage of fields discussing whether theirs is "an emerging academic discipline"; a sample of the many examples would include medical informatics (Greenes & Shortliffe, 1990), knowledge management (Grossman, 2007), supply-chain management (Cousins, Lawson, & Squire, 2006), and nano-toxicology (Oberdörster, Oberdörster, & Oberdörster, 2005). Among the many long-established disciplines in academia to which these emerging fields aspire are the standards of most university campuses, such as economics, psychology, biology, physics, philosophy, and history. Disciplines matter so much because academic research and university teaching have traditionally been organized according to disciplines, and resource allocations are often made based on those categories (Becher & Trowler, 2001).

It is not always clear what constitutes a discipline, or distinguishes it from a field, subject area, or sub-discipline. Take, for example, the three foundation fields we surveyed earlier. Information science continues to experience contentious disagreement over the placement of the term "science" (Rayward, 1996), exactly what constitutes "information," and thus what precisely is being studied (Hjørland, 2014). Despite describing communication as both a new discipline and a field, Herbst (2008) suggests it still struggles to justify itself, has failed to develop a unified theoretical framework, and continually struggles to avoid seclusion. And even after more than 60 years of activity, Radin (2013, p. 6) refers to policy analysis as "not an exact science but rather an art." Some criteria have been proposed for determining when a subject becomes a discipline: a specific topic or object of research (which may be shared with other disciplines); a body of accumulated specialist knowledge (usually unique to the discipline); theories and concepts to organize the discipline's knowledge; a specific technical language; specific research methods aligned with the discipline's research requirements; and, crucially, an institutional presence, such as courses taught at universities, academic departments, professional associations, and dedicated academic journals (Krishnan, 2009, p. 9).

Based on those criteria, does policy informatics represent an emerging discipline? The answer, in part, lies in our field's interest in governance processes and outcomes over institutions of government (Johnston, 2010). One central issue that distinguishes policy informatics from, for example, e-government, is its focus on governance over government, as this volume's definition of policy informatics draws attention to with its reference to *innovations in governance processes*. To make explicit the distinction, a government is an institution with formal authority in a geopolitical jurisdiction run by a combination of public servants and political leaders who have the power to enforce their decisions, whereas governance describes how a range of institutions, actors, rules, and norms, often operating across geopolitical boundaries, come together to influence, negotiate, and arrive at shared decisions (Rhodes, 1996). Four features of new governance configurations are that they operate through partnerships rather than enforced arrangements, are multi-jurisdictional, have a plurality of stakeholders, and are network-based (Bevir, 2012).

This interest in process and outcome over institutions is one reason why the status of discipline should not be a primary concern of policy informatics (Kersbergen & Waarden, 2004). As new approaches to organizing academic inquiry come to focus less on the name of the discipline or department (a perspective that would align with a focus on government), instead becoming more oriented toward agile reconfigurations of interest and inquiry, policy informatics comes to exemplify its own focus on governance as a way of understanding how the field is conceptualized. The question then evolves to not whether policy informatics is a discipline, but rather whether that status is an appropriate goal. In the world of Mode 2 science—characterized by Gibbons et al. (1994) as knowledge production that happens outside disciplinary and academic contexts, and is problem-oriented—and in settings where innovation in research and knowledge production is promoted (Crow, 2010; Stehr & Weingart, 2000), disciplines no longer have the relevance and authority they once did. Just as governments are challenged to explain their relevance in a world of governance (Peters & Pierre, 1998), we propose that policy informatics can just as usefully explore its future as a field and contribute to the pursuit of science without the distraction of trying to claim for itself the status of emerging discipline.

The Future of Policy Informatics

These foundations of the policy informatics movement discussed earlier—information science, mass communication, and policy analysis—provide a sketch of some of the origins of the field. In this final section, we turn to the question of where the field might go in the future. The development of policy informatics will draw on its disciplinary traditions, but its future will be forged in response to several exogenous forces, including the appreciation of complex policy challenges, continuing advances in technology, and changing expectations of governance. Policy informatics is coming of age in this environment and—perhaps more acutely with policy informatics than other academic communities—external forces will strongly influence the field's development.

Because of the uncertainty of the future and the immediacy of the present, policy informatics is, as much as any other field, largely a servant of current problems. But if policy informatics is to remain relevant, it must prove its effectiveness in the face of emerging, important, complex policy challenges that will confront us in the future. Complex policy challenges are systems-level problems that exhibit features such as profound uncertainty, rapid emergence, multiple issue interconnectedness, and a diversity of stakeholder interests (Geyer & Rihani, 2010). Specific conditions of complex policy problems include partial order (Kim, 2012), profound uncertainty (Dryzek, 1983), and often-rapid emergence that challenges our mental models and predictive capacity (Howlett & Ramesh, 1995). They are thermodynamically open and non-linear (Homer-Dixon, 2010), have whole-system implications (Kendall, 2000), and have probabilistic rather than deterministic outcomes that are subject to interpretation (Fischer, 2003). Owing largely to the systems dynamics strengths of policy informatics (see especially chapters 5–8 in this volume), the field is particularly well positioned to address complex policy challenges.

Because we cannot know this future, we instead point toward a brief sample of current complex public policy challenges that can serve as examples of the types of critical issues that will face future policy makers and that policy informatics, among other fields, will be challenged to respond to. Climate change, more accurately anthropogenic global warming, is one such issue where long times-cale increases in the average temperature of the Earth's atmosphere are attributed largely to increasing concentrations of greenhouse gases that result from human activities (Stocker et al., 2013). The effects of increasing global temperatures include rising sea levels, changing precipitation patterns, and more frequent extreme weather events, including heat waves, floods, and droughts. The climate challenge is representative of an emerging, important, complex policy challenge because of the lack of current impacts that can be attributed with certainty, the longtime lags between action and effect, the possible devastating risks to human and natural systems, the uncertainty as to the impact of actions taken today on future outcomes and the uncertainty as to future impacts, the role of natural cycles in the context of anthropogenic forcing, and the role that future technology may play in mitigation and adaptation. One emerging response to the climate change problem is global-scale geoengineering or climate engineering, which involves interventions that seek to modify the atmosphere and climate system at the global scale to counteract anthropogenic global warming. Two broad types of climate engineering efforts involve removing carbon dioxide from the atmosphere and reducing the amount of sunlight reaching the earth. The challenge of climate engineering represents a complex policy challenge because of the untested effects of these interventions, the possible destructive impacts from miscalculation and unintended consequences, the unknowns with respect to how countries might respond to unilateral actions by other countries, and the absence of a global governance regime to regulate action (Keith, Parson, & Morgan, 2010). Examples of other issues that fall into this category of emerging, important, complex policy challenges include large landscape and cross-boundary resource management, terrorism and armed conflict (see chapter 9 in this volume), advanced robotics and artificial intelligence, pandemic disease (see chapter 15 in this volume), inequality and inequity, and sexual violence.

In addition to complex policy challenges, society will continue to witness accelerating technology development in coming years. New developments will be built upon previous technologies, and the pace of change will accelerate. Assuming the continued general thrust of Moore's law—that the number of transistors on an integrated circuit doubles approximately every two years—we can anticipate further reductions in the cost, size, and power consumption of computer devices. From these advances in the basics of technology hardware, the capacity, power, and reach of computer technology will continue to develop. The technology outputs from basic and applied research will, in turn, be adopted, deployed, and reconfigured by inventors and entrepreneurs seeking new functions and business opportunities. Services, functions, and applications unknown to us today will someday soon become commonplace, whereas some technologies currently occupying our focus will fail to materialize. If we were writing this 15 years ago, following the failure of the Y2K bug to wreak havoc as predicted (Backus, Schwein, Johnson, & Walker, 2001), we may or may not have focused on the coming dominance of social media, the rapid decline of traditional media and the resultant freeing up of cognitive surplus directed toward content production and collaboration (Shirky, 2010), the business requirement that much Internet content and services be offered free of charge (Anderson, 2009), the phenomenon of "commons-based peer production" (Benkler & Nissenbaum, 2006), the ubiquity of powerful mobile devices, or the accumulation of massive datasets and enthusiasm for predictive analytics. As we write this today, the 'Internet of Things'—the idea that devices in our homes, workplaces, and public spaces, not traditionally thought of as computer devices, will have their own IP address and be connected to and controlled by other devices on the Internet—is predicted to be the next great advance in technology (Atzori, Iera, & Morabito, 2010). This may indeed turn out to be true. Or not. What is likely true is that some of the most significant changes to occur in coming years are unknown to anyone today.

However, these changes are likely to become additional tools in the policy informatics arsenal for addressing emerging, important, and complex policy challenges that may confront us in the future, while simultaneously giving rise to some of those future policy challenges. Whatever the future of technology development holds, policy informatics is well situated to take advantage of significant changes because the central premise of the field is built upon the application of new computation and communication technologies. This does not mean that the field should blindly adopt and promote every new technology. Indeed, part of the field's origins in policy analysis and communication research demonstrates a willingness to question the negative social implications of some technology and media developments, and the inequities that some new technologies create and exacerbate. But to the extent that new computation and communication technologies can be employed to help understand and address complex public policy problems, policy informatics is seeking to use those tools in the service of devising governance innovations.

Although changing technology will shape society, the history of technology use and social development indicates that society is also capable of harnessing technology in support of its own preferences. We are witnessing two seemingly contradictory although simultaneous trends with respect to power and control: consolidation and decentralization. The use of new technologies in support of greater freedom, transparency, democracy, and openness illustrates this point. From Arab Spring revolutions in the Middle East to political reconfigurations in

America, social media has had a profound effect on once stable governing regimes, although governments have proven adept at using technology to reassert their authority. Once insurmountable corporate hegemonies have been undermined by newly available technologies, replaced in some cases by consumer power but also by new corporate giants. Citizen voice has been amplified through new social media channels, weakening the power of centralized government institutions and strengthening demands for wider involvement in decision-making, although the dismissal of these diverse voices as noise has weakened their contribution to policy-making. Transparency has become an expectation, with the onus on public administrators to argue why public information should not be regularly and routinely released (McIvor, McHugh, & Cadden, 2002), although there is skepticism over whether transparency alone contributes to democratic legitimacy (Lindstedt & Naurin, 2010). The open data movement continues to promote the regular release and availability of government-held digital data repositories for unrestricted reuse, with at least three objectives: to encourage third-party developed citizen services, to expand policy networks for knowledge creation, and to increase government transparency and accountability (Longo, 2011). Platforms for engaging the capacity of citizens to be active participants in knowledge discovery, innovation, and decision-making have the potential to strengthen our societies and our democracies (Noveck, 2009). Participation in research and knowledge creation can also be facilitated by extending mechanisms such as citizen science beyond their use to date as ways to engage volunteer labor inputs in research activities, including citizen scientists in the design and conduct of research and the interpretation of results (Shum et al., 2012). That these movements have gained significant momentum in a short period is due to the combination of advances in technology and fuelled by the access expectations that Web users have. Policy informatics can continue to promote this movement toward openness not as advocates for a normative position but by testing what works and when, and identifying platforms that demonstrate success.

Conclusion

The preceding has been an attempt to set the policy informatics movement in context and to describe some of the disciplinary traditions that inform its past and highlight what the field can hope to offer the future. We described a sample of the multidisciplinary origins of policy informatics, sketching the view from the fields of information science, mass communication, and policy analysis. Each of these fields continues to influence the character of policy informatics and will continue to do so in the future. In looking toward that future, we considered two particular strengths of policy informatics: its willingness and ability to adopt and shape technology development, and its support for the principles of opening governance.

These forces are not unequivocal benefits on the side of policy informatics, however. Both technology development and open governance are essentially value-neutral forces. When ubiquitous video technology and social networks are used to generate a viral campaign raising millions of dollars in donations through the "Ice Bucket Challenge," most would laud the power of networked technology to raise awareness and generate charitable giving (Townsend, 2014). But when technology allows anyone with a 3D printer to acquire instructions over the Internet for constructing their own handgun, bypassing whatever regulations may exist, we may question the supposed benefits that knowledge-sharing over the Internet enables (Jensen-Haxel, 2011). And as much as open governance enables policy informatics to address complex public policy problems and develop innovative governance solutions, it also raises policy problems of its own and may give rise to governance challenges. The Internet allows citizens direct access to their political leaders and civic debate. But when Internet trolls are able to violate the rights of others, acting behind a mask of anonymity to threaten violence and promote hatred, we might ask whether online discussion is the broken part of the Internet (Buckels, Trapnell, & Paulhus, 2014). Advances in technology are thus neither good nor bad; rather, they are simply inevitable. Society will benefit greatly from having publics and scholars that can fluently understand and guide these future consequences from the ambiguous to the intentional.

As we noted previously, no one knows what will happen in the future, including us. Our only prediction is that policy informatics will have no shortage of public policy challenges to address. But we will enjoy the company of a vibrant community of scholars to investigate them with. This book is just the start of a long conversation about what policy informatics is and what it can offer in the future. We are excited to be part of that future, and to help build it together with colleagues, governance leaders, practitioners, and citizens, to explore ways to leverage technology to help understand and address complex public policy and administration problems, and promote innovations in governance.

Notes

1 https://books.google.com/ngrams/graph?content=policy+informatics.
2 http://www.google.com/trends/explore#q=policy%20informatics.
3 http://www.vbi.vt.edu/about/division/Advanced-Computing-and-Informatics-Laboratories.
4 http://media.illinois.edu/icr/history.
5 Harold Lasswell's preference for the term *policy sciences*, grounded in his noticeable admiration for the advancement of social science methods in the first half of the 20th century, has not adhered widely in the policy literature. The term *policy analysis* is the more common term for the field that Lasswell thought of as the policy sciences (Parsons, 1995).

References

Anderson, C. (2008). The end of theory. *Wired Magazine, 16*(7), 16–07.
Anderson, C. (2009). *Free: The future of a radical price*. New York: Random House.

Atzori, L., Iera, A., & Morabito, G. (2010). The Internet of things: A survey. *Computer Networks, 54*(15), 2787–2805.

Backus, G., Schwein, M.T., Johnson, S.T., & Walker, R.J. (2001). Comparing expectations to actual events: The post mortem of a Y2K analysis. *System Dynamics Review, 17*(3), 217–235.

Barrett, C.L., Eubank, S., Marathe, A., Marathe, M.V., Pan, Z., & Swarup, S. (2011). Information integration to support model-based policy informatics. *Innovation Journal: The Public Sector Innovation Journal, 16*(1), 5–23.

Becher T., & Trowler, P. R. (2001). *Academic tribes and territories.* Buckingham, UK: Society for Research into Higher Education / Open University Press.

Benkler, Y., & Nissenbaum, H. (2006). Commons based peer production and virtue. *Journal of Political Philosophy, 14*(4), 394–419.

Berelson, B. (1959). The State of Communication Research. *The Public Opinion Quarterly,* 23(1), 1–6

Bevir, M. (2012). *Governance: A very short introduction.* Oxford, UK: Oxford University Press.

Buckels, E.E., Trapnell, P.D., & Paulhus, D.L. (2014). Trolls just want to have fun. *Personality and Individual Differences, 67,* 97–102.

Bush, V. (1945). As we may think. *Atlantic Monthly, 176*(1), 101–108.

Case, S. (2001). Feminism and performance: A post disciplinary couple. *Theatre Research International, 26,* 145–152

Chaffee, S. H., & Hochheimer, L. H. (1985). The beginnings of political communication research in the United States: Origins of the "limited effects" model. In E.M. Rogers & F. Balle (Eds.), *The media revolution in America and Western Europe* (pp. 267–298). Newbury Park, CA: SAGE.

Choi, H., & Varian, H. (2012). Predicting the present with google trends. *Economic Record, 88*(s1), 2–9.

Cousins, P.D., Lawson, B., & Squire, B. (2006). Supply chain management: Theory and practice–the emergence of an academic discipline? *International Journal of Operations & Production Management, 26*(7), 697–702.

Craig, R.T. (1999). Communication theory as a field. *Communication Theory, 9*(2), 119–161.

Craig, R.T. (2001). Minding my metamodel, mending Myers. *Communication Theory, 11*(2), 231–240.

Crow, M.M. (2010). Differentiating America's colleges and universities: A case study in institutional innovation in Arizona. *Change: The Magazine of Higher Learning, 42*(5), 36–41.

Dawes, S.S., & Janssen, M. (2013, June). Policy informatics: Addressing complex problems with rich data, computational tools, and stakeholder engagement. In *Proceedings of the 14th Annual International Conference on Digital Government Research* (pp. 251–253). New York, NY: ACM.

Desai, A., & Kim, Y. (2015). Symposium on policy informatics. *Journal of Policy Analysis and Management.* doi: 10.1002/pam.21823

Dryzek, J.S. (1983). Don't toss coins in garbage cans: A prologue to policy design. *Journal of Public Policy, 3,* 345–367.

Fischer, F. (2003). *Reframing public policy: Discursive politics and deliberative practices.* New York: Oxford University Press.

Forester, J. (1995). Response: Toward a critical sociology of policy analysis. *Policy Sciences, 28,* 385–396.

Geyer, R., & Rihani, S. (2010). *Complexity and public policy: A new approach to twenty-first century politics, policy and society.* New York: Routledge.

Gibbons, M., Limoges, C., Nowotny, H., Schwartzman, S., Scott, P., & Trow, M. (1994). *The new production of knowledge: The dynamics of science and research in contemporary societies.* London: SAGE.

Greenes, R.A., & Shortliffe, E.H. (1990). Medical informatics: An emerging academic discipline and institutional priority. *JAMA, 263*(8), 1114–1120.

Greenfield, P. M. (2013). The changing psychology of culture from 1800 through 2000. *Psychological Science, 24*(9), 1722–1731.

Grossman, M. (2007). The emerging academic discipline of knowledge management. *Journal of Information Systems Education, 18*(1), 31–38.

Hall, P.A. (1989). *The political power of economic ideas: Keynesianism across nations.* Princeton, NJ: Princeton University Press.

Helbig, N., Nakashima, M., & Dawes, S.S. (2012, June). Understanding the value and limits of government information in policy informatics: A preliminary exploration. In *Proceedings of the 13th Annual International Conference on Digital Government Research* (pp. 291–293). New York, NY: ACM.

Herbst, S. (2008). Disciplines, intersections, and the future of communication research. *Journal of Communication, 58*, 603–614.

Hjørland, B. (2014). Information science and its core concepts: Levels of disagreement. In F. Ibekwe-SanJuan & T. M. Dousa (Eds.), *Theories of Information, Communication and Knowledge* (pp. 205–235). Dordrecht: Springer Netherlands.

Homer-Dixon, T. (2010). *Complexity science and public policy.* New Directions Series. Toronto: Institute of Public Administration of Canada.

Howlett, M. (2011). Public managers as the missing variable in policy studies: An empirical investigation using Canadian data. *Review of Policy Research, 28*(3), 247–263.

Howlett, M., & Ramesh, M. (1995). *Studying public policy: Policy cycles and policy subsystems.* Toronto: Oxford University Press.

Jensen-Haxel, P. (2011). 3D printers, obsolete firearm supply controls, and the right to build self-defense weapons under Heller. *Golden Gate University Law Review, 42*, 447.

Johnston, E. (2010). Governance infrastructures in 2020. *Public Administration Review, 70*(s1), s122–s128.

Johnston, E.W., & Hansen, D.L. (2011). Design lessons for smart governance infrastructures. *American Governance, 3*, 197–212.

Johnston, E., & Kim, Y. (2011). Introduction. Policy informatics. Special issue of *Innovation Journal: The Public Sector Innovation Journal, 16*(1), 1–4.

Keith, D.W., Parson, E., & Morgan, M.G. (2010). Research on global sun block needed now. *Nature, 463*(7280), 426–427.

Kendall, J. (2000). The mainstreaming of the third sector into public policy in England in the late 1990s: Whys and wherefores. *Policy & Politics, 28*(4), 541–562.

Kersbergen, K.V., & Waarden, F.V. (2004), "Governance" as a bridge between disciplines: Cross-disciplinary inspiration regarding shifts in governance and problems of governability, accountability and legitimacy. *European Journal of Political Research, 43*, 143–171.

Kim, H.-S. (2012). Climate change, science and community. *Public Understanding of Science, 21*(3), 268–285.

Kling, R. (2007). What is social informatics and why does it matter? *Information Society, 23*(4), 205–220.

Krishnan, A. (2009). *What are academic disciplines? Some observations on the disciplinarity vs. interdisciplinarity debate.* Southampton, UK: University of Southampton, ESRC National Centre for Research Methods. Retrieved from http://eprints.ncrm.ac.uk/783/1/what_are_academic_disciplines.pdf

Lasswell, H.D. (1951). The policy orientation. In D. Lerner & H. D. Lasswell (Eds.), *The policy sciences* (pp. 3–15). Stanford, CA: Stanford University Press.

Lasswell, H.D. (1970). The emerging conception of the policy sciences. *Policy Sciences, 1,* 3–14.

Lasswell, H.D. (1971). *A pre-view of the policy sciences.* New York: American Elsevier.

Lawlor, E.F. (1996). Reviews of the books *The argumentative turn in policy analysis and planning, Narrative policy analysis: Theory and practice,* and *Policy change and learning: An advocacy coalition approach. Journal of Policy Analysis and Management, 15*(1), 110–121.

Lerner, D., & Lasswell, H. D. (Eds.). (1951). *The policy sciences.* Stanford, CA: Stanford University Press.

Lindstedt, C., & Naurin, D. (2010). Transparency is not enough: Making transparency effective in reducing corruption. *International Political Science Review, 31*(3), 301–322.

Longo, J. (2011). #OpenData: Digital-era governance thoroughbred or new public management Trojan horse? *Public Policy and Governance Review, 2*(2), 38–51.

Machlup, F., & Mansfield, U. (Eds.). (1983). *The study of information: Interdisciplinary messages.* New York: Wiley.

Mayer, I.S., van Daalen, C.E., & Bots, P.W. (2004). Perspectives on policy analyses: A framework for understanding and design. *International Journal of Technology, Policy and Management, 4*(2), 169–191.

McIvor, R., McHugh, M., & Cadden, C. (2002). Internet technologies: Supporting transparency in the public sector. *International Journal of Public Sector Management, 15*(3), 170–187.

Meltsner, A.J. (1976). *Policy analysts in the bureaucracy.* Berkeley: University of California Press.

Menand, L. (2001). The marketplace of ideas. American Council of Learned Societies Occasional Paper No. 49, pp. 1–23. Retrieved November 2014 from http://archives.acls.org/op/49_Marketplace_of_Ideas.htm

Mikhailov, A.I., Chernyi, A.I., & Giljarevskij, R.S. (1967). Informatics new name for theory of scientific information. *FID News Bulletin, 17*(7), 70–4.

Morçöl, G. (2001). Positivist beliefs among policy professionals: An empirical investigation. *Policy Sciences, 34*(3–4), 381–401.

Mouffe, C. (2000). *The democratic paradox.* London: Verso.

Myers, D. (2001). A pox on all compromises: A reply to Craig (1999). *Communication Theory, 11*(2), 218–230.

Noveck, B. S. (2009). *Wiki government: How technology can make government better, democracy stronger, and citizens more powerful.* Washington, DC: Brookings Institution Press.

Oberdörster, G., Oberdörster, E., & Oberdörster, J. (2005). Nanotoxicology: An emerging discipline evolving from studies of ultrafine particles. *Environmental Health Perspectives, 113*(7), 823–839.

Parsons, W. (1995). *Public policy: An introduction to the theory and practice of policy analysis.* Cheltenham: Edward Elgar.

Peters, B.G., & Pierre, J. (1998). Governance without government? Rethinking public administration. *Journal of Public Administration Research and Theory, 8*(2), 223–243.

Peters, J.D. (1986). Institutional sources of intellectual poverty in communication research. *Communication Research, 13*(4), 527–559.

Quade, E. S. (1970). Why policy sciences? *Policy Sciences, 1,* 1–2.

Radin, B.A. (2013). *Beyond Machiavelli: Policy analysis reaches midlife.* Washington, DC: Georgetown University Press.

Rayward, W.B. (1996). The history and historiography of information science: Some reflections. *Information Processing & Management, 32*(1), 3–17.

Rhodes, R.A.W. (1996). The new governance: Governing without government. *Political Studies, 44*(4), 652–667.

Rogers, E.N., & Chaffee, S.H. (1994). Communication and journalism from "Daddy" Bleyer to Wilbur Schramm: A palimpsest. *Journalism Monographs*, 148, 1–50.

Schramm, W., Riesman, D., & Bauer, R. A. (1959). The state of communication research: Comment. *Public Opinion Quarterly, 23*(1), 6–17.

Shirky, C. (2010). *Cognitive surplus: How technology makes consumers into collaborators.* New York: Penguin.

Shum, S.B., Aberer, K., Schmidt, A., Bishop, S., Lukowicz, P., Anderson, S., Charalabidis, Y., Domingue, J., De Freitas, S., Dunwell, I., Edmonds, B., Grey, F., Haklay, M., Jelasity, M., Karpistsenko, A., Kohlhammer, J., Lewis, J., Pitt, J., Sumner, R., & Helbing, D. (2012). Towards a global participatory platform: Democratising open data, complexity science and collective intelligence. *European Physical Journal-Special Topics, 214*(1), 109–152.

Sils, D. L. (1980). In memoriam: Bernard Berelson, 1912–1979. *Public Opinion Quarterly, 44*(2), 274–275.

Stehr, N., & Weingart, P. (Eds.). (2000). *Practising interdisciplinarity.* Toronto: University of Toronto Press.

Stocker, T.F., Qin, D., Plattner, G.K., Tignor, M., Allen, S.K., Boschung, J., Nauels, A., Xia, Y., Bex, V., & Midgley, P. M. (2013). Climate change 2013: The physical science basis. *Intergovernmental Panel on Climate Change, Working Group I Contribution to the IPCC Fifth Assessment Report (AR5).* New York: Cambridge University Press.

Stone, D. (1997). *Policy paradox: The art of political decision making.* New York: W.W. Norton.

Townsend, L. (2014). How much has the ice bucket challenge achieved? *BBC News Magazine, 2.*

Wahl-Jorgensen, K. (2004). How not to found a field: New evidence on the origins of mass communication research. *Journal of Communication, 54*(3), 547–564.

Yang, K. (2007). Quantitative methods for policy analysis. In F. Fischer, G. J. Miller, & M. S. Sidney (Eds.), *Handbook of public policy analysis: Theory, politics, and methods* (pp. 349–368). Boca Raton, FL: CRC Press.

CONTRIBUTORS

Petra Ahrweiler, Ph.D., is Professor of Technology Assessment at the Johannes-Gutenberg University of Mainz, Germany and Director of the European Academy of Technology and Innovation Assessment. She studied law, sociology, journalism, and political science at the University of Hamburg, finishing with her Ph.D. in the area of science and technology studies at the Free University Berlin. Since her habilitation thesis at the University of Bielefeld on social simulation of innovation processes, she worked as a Heisenberg Fellow of the Deutsche Forschungsgemeinschaft DFG and as a professor of economic sociology at the University of Hamburg. From 2007–2012, she was Professor of Technology and Innovation Management at the Michael Smurfit School of Business at University College Dublin in Ireland, where she led the UCD Innovation Research Unit (IRU). Her main research interests are innovation networks, science and technology studies, policy modeling, complex social systems, and agent-based simulation.

Christopher Barrett, Ph.D., is Director of the Network Dynamics and Simulation Science Laboratory at the Virginia Bioinformatics Institute, the Scientific Director of the Virginia Bioinformatics Institute, and Professor in the Department of Computer Science at Virginia Tech. For more than 25 years, he has led advanced research and development programs about computationally augmented decision-making in complex social and socio-technical systems, including detailed epidemiological simulation methods that use high-performance computing..

John C. Bertot, Ph.D., is Professor and Co-director of the Information Policy and Access Center (iPAC) in the University of Maryland College Park iSchool. His research spans information and telecommunications policy, e-government, government agency technology planning and evaluation, and library planning and evaluation. He is President of the Digital Government Society of North America (DGSNA), Editor of *Government Information Quarterly,* and Co-editor of *The Library Quarterly.* Dr. Bertot's research has been funded by the National Science Foundation, the Bill & Melinda Gates Foundation, the Government Accountability Office, the American Library Association, and the U.S. Institute of Museum and Library Services.

Christopher Bronk, Ph.D., is the Baker Institute Fellow in information technology policy and an adjunct associate professor of computer science in Rice's George R. Brown School of Engineering. He previously served as a career diplomat with the U.S. Department of State on assignments both overseas and in Washington, D.C. His last assignment was in the Office of eDiplomacy, the department's internal think tank on information technology, knowledge management, and interagency collaboration. Since arriving at Rice, Dr. Bronk has divided his attentions among a number of areas, including cybersecurity, technology for border management, broadband policy, online collaboration, and the militarization of cyberspace. He is also a senior fellow with the Canada Centre for Global Security Studies at the University of Toronto. Holding a Ph.D. from the Maxwell School of Syracuse University, Dr. Bronk also studied international relations at Oxford University and received a bachelor's degree from the University of Wisconsin–Madison.

Britte H. Cheng is Senior Education Researcher at SRI International's Center for Technology in Learning. Her research focuses on design of learning technologies, instruction, and assessment in K-12 science, but also on systemic issues of educational practice and policy. She investigates processes and designs that cross learning contexts and settings. She is PI of the Modeling Social Complexity in Education project which integrates agent-based and systems dynamics modeling to understand STEM education as a complex, adaptive system. She was also co-PI of the NSF project Developing and Testing Theories of Implementation: A Workshop on Design Research with Educational Systems, which posits a systemic approach to design and research of educational interventions that account for local implementation.

Ivan Damnjanovic, Ph.D., is an associate professor in the Zachry Department of Civil Engineering at Texas A&M University. He received his Ph.D. in civil engineering in 2006 from the University of Texas at Austin in the area of infrastructure management. Dr. Damnjanovic is an expert in the areas of project financing, public-private partnerships, performance-based specifications, system risk analysis, and innovative project delivery. He is also a member of ASCE where he serves on the Infrastructure Systems Committee, TRB where he serves on the Construction Management Committee, and Construction Industry Institute (CII) where he serves on the Academic Committee. Dr. Damnjanovic is a member of the Institute for Operations Research and the Management Sciences (INFORMS) and involved in two professional sections of this institute: Transportation and Logistics, and Decision Analysis.

Sharon S. Dawes, Ph.D., is senior fellow at the Center for Technology in Government (CTG) and Professor Emerita of Public Administration and Policy and Informatics at the University at Albany, State University of New York. Her research interests are cross-boundary information sharing and collaboration, international digital government research, and government information strategy and management. As the founding director of CTG, from 1993–2007, she led the Center to international prominence in the field of digital government research. She is a fellow of the US National Academy of Public Administration.

Anand Desai, Ph.D., is a professor at the John Glenn School of Public Affairs at The Ohio State University where he teaches basic statistics, policy modeling and logic of inquiry. His interests include research methods, the use of formal models to support public policy decisions and action, pattern analysis in data, and archery. He was educated in India, the UK, and the US and obtained his doctorate at the University of Pennsylvania.

William H. Dutton is the Quello Professor of Media and Information Policy in the College of Communication Arts and Sciences at Michigan State University, where he is Director of the Quello Center. Dutton was the first professor of Internet studies at the University of Oxford where he was founding director of the Oxford Internet Institute, and a fellow of Balliol College. He has recently edited the *Oxford Handbook of Internet Studies* (OUP 2013), four volumes on politics and the Internet (Routledge 2014), and a reader titled *Society and the Internet*, with Mark Graham (OUP 2014).

Stephen Eubank, Ph.D., is a Deputy Director of the Network Dynamics and Simulation Science Laboratory at the Virginia Bioinformatics Institute at Virginia Tech. His research interests are in building and analyzing mathematical and computational models of complex socio-technical systems. For the past decade, he has focused on providing epidemiological modeling support to policymakers as PI of an NIGMS-funded Modeling Infectious Disease Agents Study (MIDAS) research group.

David N. Ford, Ph.D., PE, is an associate professor in the Construction Engineering and Management Program in the Zachry Department of Civil Engineering at Texas A&M University and a Research Associate Professor of Acquisition in the Graduate School of Business and Public Policy at the US Naval Postgraduate School. He researches development project dynamics, managerial real options, public–private partnerships, and sustainability. He teaches strategic engineering management; engineering project control; and project planning, estimating, and control. Dr. Ford was previously on the Department of Information Science faculty at the University of Bergen, Norway. For more than 14 years he designed and managed the development of constructed facilities in industry and government. He received his Ph.D. from the Massachusetts Institute of Technology and his master's and bachelor's degrees from Tulane University.

Navid Ghaffarzadegan, Ph.D., is an assistant professor at Virginia Tech's Department of Industrial and Systems Engineering. His research interests include studies of complex socio-technical systems, and applications of system dynamics modeling in decision-making, management, and public policy contexts. Ghaffarzadegan is the recipient of the Lupina Young Researcher Award and the Dana Meadows Award for health system dynamics modeling. His research has been supported by various organizations including NIH and NSF. Prior to arriving at Virginia Tech, he was a postdoctoral researcher at MIT, Engineering Systems Division, where he conducted research on modeling research enterprises, research policy, and higher education systems. Navid received his Ph.D. in Public Administration and Policy from the State University of New York (SUNY) at Albany in 2011.

Nigel Gilbert, Ph.D., is Professor of Sociology and Director of the Centre for Research in Social Simulation at the University of Surrey, United Kingdom. He has a first degree in engineering, and a doctorate in sociology. He is Editor of the *Journal of Artificial Societies and Social Simulation* and has written and edited 21 books and about 160 academic papers on topics ranging from the sociology of science to the computerization of social security benefits and human-computer interaction. He is an author of the standard textbook on social simulation, *Simulation for the Social Scientist* (Open University Press, 2nd edition, 2005, with Klaus G. Troitzsch), and more recently, *Agent-Based Models* (Sage, 2007), and is editor of *Computational Social Science* (four volumes, Sage, 2010).

Derek L. Hansen, Ph.D., is Associate Professor of Information Technology at Brigham Young University. Hansen completed his Ph.D. in the University of Michigan's School of Information, where he was an NSF-funded interdisciplinary STIET Fellow focused on understanding and designing effective online socio-technical systems. His research focuses on technology-mediated social participation, social network analysis, and human–computer interaction. He has published one book on social media analytics and more than 30 journal articles and conference proceedings. He also serves on the board of directors for the Social Media Research Foundation.

Kristin Harlow is a doctoral student at the John Glenn School of Public Affairs at The Ohio State University. Her research focuses on health care policy, nonprofit organizations, and administrative law. She also holds advanced degrees in law and social work.

Natalie Helbig, Ph.D., is Assistant Research Director for the Center for Technology in Government (CTG) and is an affiliate faculty in Informatics, and teaches courses in public administration and government information and strategy at the University at Albany. Her primary research interests examine how public agencies design, manage, and use data and information in support of better governance and management. She has expertise in cross-disciplinary collaboration, evaluation, and research-practice partnerships.

David M. Hondula, Ph.D., is Assistant Research Professor at Arizona State University with appointments in the Center for Policy Informatics and the School of Geographical Sciences and Urban Planning. Hondula's research explores the impact of weather and climate on human health and societal outcomes, with a particular focus on the health impacts of extreme heat events. He is pursuing technology-based solutions to reduce adverse health outcomes associated with extreme heat through novel applications of personal sensors and visualization/modeling techniques. Hondula received his Ph.D. in 2013 from the Department of Environmental Sciences at the University of Virginia and welcomes engagement on Twitter @ASUHondula.

Paul T. Jaeger, Ph.D., J.D., is Associate Professor and Diversity Officer of the College of Information Studies and Co-director of the Information Policy and Access Center at the University of Maryland. Dr. Jaeger's research focuses on the ways in which law and public policy shape information behavior, particularly for underserved populations. He is the author of more than 125 journal articles and book chapters, along with seven books. His research has been funded by the Institute of Museum and Library Services, the National Science Foundation, the American Library Association, the Smithsonian Institution, and the Bill & Melinda Gates Foundation, among others. Dr. Jaeger is Co-editor of *Library Quarterly*, Co-editor of the Information Policy Book Series from MIT Press, and Associate Editor of *Government Information Quarterly*.

Scott T. Johnson is the founder and president of SYSDYNX, LLC, a consulting company specializing in the use of system dynamics modeling and simulation for research, projects, and education. He is also a principal advisor at the Institute for Energy Studies (IES) and instructor of petroleum engineering, University of North Dakota (UND). Prior to his current activities, Scott completed a 30-year oil and gas industry career with BP/Amoco where he recently focused on developing assurance technology for large engineering projects (LEP) to demonstrate the robustness of plans and schedules to perform against expected risks and uncertainties. He also developed and delivered curriculum to support LEP executive education objectives. Scott has completed all course work for a master's degree in system dynamics philosophy (ABT) from the University of Bergen, Norway, and he earned a bachelor's of science in mechanical engineering from South Dakota School of Mines and Technology in 1980.

Erik W. Johnston, Ph.D., is Associate Professor in the School of Public Affairs and Co-founder and Director of the Center for Policy Informatics, Arizona State University. His research centers on three areas: open governance, participatory modeling, and smarter governance infrastructures. As a dedicated action researcher, Dr. Johnston has received external funding from the National Science Foundation, the MacArthur Foundation, and the Virginia G. Piper Charitable Trust. He helps to lead a research team at ASU that is studying how people come together to collaborate using 10,000 Solutions, a university-wide challenge platform to propose answers to problems from education to human rights. He was also an organizer behind the ASU Policy Challenge, an ideation contest for contributing policy suggestions to the White House. With undergraduate degrees in computer science and psychology as well as a Master of Business Administration and a Master of Science in information technology from the University of Denver, Dr. Johnston holds a Ph.D. in information from the University of Michigan with a certificate in complex systems. He is a two-time NSF IGERT Fellow, and in 2014 he received the ASU Faculty Achievement Award for Best Professional Application for his work on opening governance.

Jes A. Koepfler, Ph.D., is Managing Director at Intuitive Company (www.intu itivecompany.com), a user-centered research, design, and development firm in Philadelphia, PA. She received her doctorate in Information Studies from the University of Maryland. Her dissertation research looked at the intersection of human values, online communication, and online identity related to homelessness. Her applied research efforts explore ways in which digital experiences can support meaningful human to human interactions across verticals from education and healthcare to the financial services industry.

Christopher Koliba, Ph.D., is an associate professor in the Community Development and Applied Economics Department at the University of Vermont (UVM) and Director of the Master of Public Administration (MPA) Program. He possesses a Ph.D. and an MPA from Syracuse University's Maxwell School of Citizenship and Public Affairs. His research interests include governance networks and complex adaptive systems, organizational learning and development, action research methods, and performance and accountability indicators. His current research program focuses on the development of complex adaptive systems models of regional planning, watershed governance, food systems, transportation planning, and smart grid energy networks. He is the lead author of *Governance Network in Public Administration and Public Policy* published in the fall of 2010 by Taylor

& Francis. He is chair of the Complexity and Network Studies section of the American Society of Public Administration, and he teaches courses pertaining to public policy and public affairs, public administration, systems analysis, governance networks, collaborative management, and the intersection of science and society.

Gerard P. Learmonth, Sr., Ph.D., is Research Associate Professor in the Department of Systems and Information Engineering at the University of Virginia, where he is a member of the Computational Statistics and Simulation group. He holds a secondary appointment in the Department of Public Health Sciences in the School of Medicine. Learmonth's principal research area is complex system modeling. Current activity is focused on the design, development, and use of participatory simulation models. Among these are the UVa Bay Game® and its successor, Global Water Games; a participatory simulation model of the U.S. Healthcare system based on the Affordable Care Act; and a model of the provisioning of clean water to improve the long-term health and growth of children in sub-Saharan Africa. Learmonth holds a B.S. and an M.B.A. from New York University; an M.S. from the Naval Postgraduate School; and a Ph.D. from the University of Michigan–Ann Arbor.

Evert Lindquist is Professor at the School of Public Administration at the University of Victoria in British Columbia, Canada, serving as its Director since 1998. He received his Ph.D. in Public Policy from UC Berkeley and was a faculty member of the University of Toronto's Department of Political Science from 1988–1998. During 2010–11 he held the ANZSOG-ANU Chair in Applied Public Management Research at the Australian National University. He is an adjunct faculty member with ANU's Crawford School of Economics and Government. He has published widely on topics relating to public sector reform, governance and decision-making, central agencies and initiatives, policy capability, think tanks and consultation processes, horizontal management, and recently on policy visualization. His work on policy visualization has been undertaken with the support of the HC Coombs Policy Forum at Australian National University and the Australia and New Zealand School of Government. He is Editor of *Canadian Public Administration*, the Institute of Public Administration of Canada's flagship journal.

Justin Longo, Ph.D., is the postdoctoral fellow in open governance at the Center for Policy Informatics at Arizona State University. He has a Ph.D. from the University of Victoria in public policy and public administration (2013) where he researched the use of enterprise social collaboration platforms inside government policy analysis settings. In his postdoctoral work, he focuses on developing research related to opening governance (as processes of social decision-making) to more diverse sources of knowledge, more avenues of interaction, and enhanced social understanding; and opening government (or, more broadly, formal public institutions) as knowledge organizations, promoting the conditions where knowledge is shared and used, collaboration encouraged, and capacity throughout the policy cycle enhanced. His academic blog is at jlphd.wordpress.com and he can be found on Twitter @whitehallpolicy.

John Lyneis, Ph.D., is a management consulting specializing in the use of system dynamics to solve complex business and public policy problems. He received his Ph.D. in 2012 from the System Dynamics Group at the MIT Sloan School of Management.

Achla Marathe, Ph.D., is the lead economist and social scientist at the Network Dynamics and Simulation Science Laboratory at the Virginia Bioinformatics Institute, and Associate Professor in the Agriculture and Applied Economics Department at Virginia Tech. Her research interests are in network science, socioeconomic systems, behavioral economics, and the economics of epidemiology.

Madhav Marathe, Ph.D., is the Deputy Director at the Network Dynamics and Simulation Science Laboratory at the Virginia Bioinformatics Institute and Professor in the Department of Computer Science at Virginia Tech. His research interests are in network science, modeling and simulations of large societal systems, high performance computing, theoretical computer science, and policy and decision informatics.

Ines Mergel, Ph.D., is Associate Professor of Public Administration and International Affairs at Syracuse University's Maxwell School of Citizenship and Public Affairs. Her research focuses on the use of innovative technologies in the public sector, especially the strategic, managerial, and administrative decisions that lead to the adoption of new technologies. Her work has been published in the *Journal of Public Administration Research and Theory, Public Administration Review, International Public Management Journal, Public Management Review, Information Polity and Government Information Quarterly,* and others. She is the author of *Social Media in the Public Sector* published by Jossey-Bass/Wiley in 2012. Her thoughts on new media applications can be read on her blog: inesmergel.wordpress.com and on Twitter @inesmergel.

William R. Penuel, Ph.D., is Professor of Educational Psychology and Learning Sciences in the School of Education at the University of Colorado Boulder. His research focuses on the design, implementation, and evaluation of educational innovations in schools and other settings. A strand of his research focuses on conditions that facilitate organizational transformation. He employs social network analysis and agent-based modeling of educational systems to gain insight on key leverage points for improving teaching and learning. Dr. Penuel has been a leading advocate for developing collaborative approaches to intervention research in educational systems, including an approach he and his co-authors (along with Barry Fishman of the University of Michigan) call design-based implementation research. Prior to joining the faculty of the School of Education, Dr. Penuel served as Director of Evaluation Research at the Center for Technology in Learning at SRI International.

Jeffrey Plank, Ph.D., recently retired as Associate Vice President for Research at the University of Virginia where he coordinated large-scale, multi-sector initiatives, including the pan-University sustainability institute. As leader of the Global Water Games project, Dr. Plank pioneered development of an innovative participatory simulation that demonstrates the potential of digital and social media to motivate personal and collective behavior change for environmental conservation. His scholarly publications include a three-part book series on architectural photography, building restoration, and architectural history, using the American modern architect Louis Sullivan and long-neglected photograph archives of his work as his case study. Prior to joining UVA in 1993, Dr. Plank held academic and administrative positions at the University of Southern California, Georgia Tech, and the University of Chicago. Dr. Plank holds a BA (high honors), MA, and Ph.D. from UVA.

Andreas Pyka, Ph.D., graduated in economics and management at the University of Augsburg in 1998 and spent two years as a postdoc in Grenoble, France, participating in a European research project on innovation networks. Following the postdoc, he worked as an assistant professor at the chair of Professor Horst Hanusch at the University of Augsburg. Dr. Pyka's fields of research are neo-Schumpeterian economics and evolutionary economics with a special emphasis on numerical techniques of analyzing dynamic processes of qualitative change and structural development. From October 2006 to March 2009, he worked at the University of Bremen as a professor in economic theory. Since April 2009, Dr. Pyka holds the chair for innovation economics at the University of Hohenheim, Stuttgart.

George P. Richardson, Ph.D., is Professor Emeritus of Public Administration, Public Policy, and Informatics at the University at Albany. He is the author of *Introduction to System Dynamics Modeling with DYNAMO* (1981) and *Feedback Thought in Social Science and Systems Theory* (1991), both of which were honored with the System Dynamics Society's Jay W. Forrester Award, and the edited two-volume collection *Modeling for Management: Simulation in Support of Systems Thinking* (1996). He founded the *System Dynamics Review* and later served for seven years as its executive editor. He has been honored with awards for Excellence in Teaching (2003) and Academic Service (2010) by the chancellor and trustees of the State University of New York.

Derek Ruths, Ph.D., is an assistant professor of Computer Science at McGill University. He joined the faculty in 2009 after completing his Ph.D. in computer science at Rice University. A major research direction in his work considers the problem of characterizing and predicting the large-scale dynamics of human behavior from unstructured online data. His ongoing work in this area includes quantitatively modeling how communities change over time, measuring and predicting group demographics from unstructured user-generated content, and computational methods for assessing discussion topics within a collection of users. His work has been published in top-tier journals and conferences including *Science, ICWSM,* and *PLoS Computational Biology.* His research is funded by a wide array of government agencies, tech companies, and private organizations—underscoring the broad, interdisciplinary nature of his work.

Nora H. Sabelli holds a Ph.D. in theoretical organic chemistry from the University of Buenos Aires, Argentina. Prior to joining SRI International, Dr. Sabelli was Senior Program Director for the NSF's Directorate for Education and Human Resources, working on NSF-wide and cross-agency initiatives on research, education, technology, and science. At NSF, she focused on research on the use of current scientific and technological advances to provide quality STEM education for all students, and was instrumental in the development and support of several learning science and technology research programs. During that period, Dr. Sabelli participated in interagency education research issues at the U.S. Office of Science and Technology Policy (OSTP). Her recent interests include educational policy and the integration of education research and practice. Former positions include Senior Research Scientist and Assistant Director for Education at the National Center for Supercomputing Applications, University of Illinois at Urbana-Champaign, and Associate Professor of Chemistry, University of Illinois at Chicago.

Samarth Swarup, Ph.D., is a research scientist at the Network Dynamics and Simulation Science Laboratory at the Virginia Bioinformatics Institute at Virginia Tech. His research interests are in artificial intelligence, machine learning, cognitive science, policy and decision informatics, and computational social science.

Kimberly M. Thompson, Ph.D., serves as Professor of Preventive Medicine and Global Health, University of Central Florida, College of Medicine and President of Kid Risk, Inc., which is a nonprofit organization focused on improving children's lives by understanding, characterizing, and communicating about the real risks that children face around the world and on empowering policy makers, parents, kids, and others to use the best available information to manage the risks and improve children's lives (www.kidrisk.org). Dr. Thompson served on the faculty of the Harvard School of Public Health, where she created and directed the Kids Risk Project for nearly a decade before spinning it out into Kid Risk, Inc. She also served as a visiting faculty member at the Massachusetts Institute of Technology Sloan School of Management in the System Dynamics Group, and she is a past-president and fellow of the Society for Risk Analysis.

Tracy Viselli is the Director of Digital Strategy and Community Engagement for ACT for Alexandria and manages ACTion Alexandria, an online citizen engagement project based in Alexandria, Virginia. Viselli's career has been focused on increasing citizen engagement and advocacy through technology, and she has worked on several groundbreaking projects including the award-winning TwitterVoteReport and TweetProgress. Viselli has an MA in teaching writing and literature from George Mason University.

Dara M. Wald, Ph.D., is a postdoctoral fellow at Arizona State University with joint positions in the Center for Policy Informatics and the Decision Center for a Desert City. Her research explores the contextual and psychological drivers of socio-political conflict over environmental policy and management, the barriers to effective science and environmental communication, and how technology (e.g., interactive games and computer simulated models) can be used to build consensus, promote effective communication, and inspire collaborative behavior. She completed her Ph.D. in the Department of Wildlife Ecology and Conservation at the University of Florida (2012).

David Wheat, Ph.D., is an associate professor of system dynamics at the University of Bergen in Norway, where he teaches simulation-based policy design, system dynamics modeling, and macroeconomic dynamics. He is also a visiting professor of economics at institutions in Lithuania and the United States. His research interests include economic demography, macro and regional economic modeling, and policy implementation analysis in collaboration with traditionally trained public policy analysts. He is developing system dynamics modeling capacity among economists in Ukraine, and developing a macroeconomic forecasting model in collaboration with economists at the Bank of Lithuania. After a career as a policy consultant, Wheat moved to Norway and earned his Ph.D. in system dynamics at the University of Bergen. He received his public policy master's degree from Harvard University and his political science degree from Texas Tech University. During the 1970s, he served at the White House as staff assistant to the President of the United States.

Asim Zia, Ph.D., is currently serving as an assistant professor in the Department of Community Development and Applied Economics at the University of Vermont. He has a Ph.D. in public policy from the Georgia Institute of Technology. His research is focused on the development of computational and complex systems-based approaches for policy analysis, governance informatics, and adaptive management. His published research spans the substantive policy domains of transportation, air quality, water quality and land-use planning, climate policy, and international development and biodiversity conservation. He is the author of the book *Post-Kyoto Climate Governance: Confronting the Politics of Scale, Ideology and Knowledge*, published by Routledge Press in 2013. His professional career began as a civil servant in the Ministry of Finance and Economic Affairs of the Federal Government of Pakistan. Prior to joining the University of Vermont, he has also worked as a postdoctoral scientist at the National Center for Atmospheric Research and as an assistant professor at San Jose State University.

INDEX

Page numbers in *italics* indicate figures, tables, and boxes.

Milton Keynes UK
Ingram Content Group UK Ltd.
UKHW022057141024
449569UK00031B/1675

9 781138 832084